Endoglycosidases

Biochemistry, Biotechnology, Application

Endoglycosidases

Biochemistry, Biotechnology, Application

Edited by
M. Endo, S. Hase, K. Yamamoto and K. Takagaki

With 123 Figures and 40 Tables

Editors:
Professor Masahiko Endo
President, Hirosaki University
Aomori 036-8560, Japan
e-mail: endo-m@cc.hirosaki-u.ac.jp

Professor Sumihiro Hase
Graduate School of Science, Osaka University
Osaka 560-0043, Japan
e-mail: suhase@chem.sci.osaka-u.ac.jp

Professor Kenji Yamamoto
Graduate School of Biostudies, Kyoto University
Kyoto 606-8502, Japan
e-mail: yamamotk@kais.kyoto-u.ac.jp

Professor Keiichi Takagaki
Department of Biochemistry, Hirosaki University School of Medicine
Aomori 036-8562, Japan
e-mail: keiichi@cc.hirosaki-u.ac.jp

ISBN 4-06-212192-1 Kodansha Ltd., Tokyo

ISBN-10 3-540-34494-2 Springer Berlin Heidelberg New York
ISBN-13 13 978-3-540-34494-0 Springer Berlin Heidelberg New York

Library of Congress Control Number: 2006925938

Springer is a part of Springer Science+Business Media
springer.com

Printed in Japan

Coverdesign: *design & production*, Heidelberg, Germany

Printed on acid-free paper – 5 4 3 2 1 0

List of Contributors

Numbers in parentheses refer to chapters and/or sections.

Ajisaka, Katsumi (4) Niigata University of Pharmacy and Applied Life Sciences, Niigata 956-8603, Japan

Ashida, Hisashi (2.2.2) Graduate School of Biostudies, Kyoto University, Kyoto 606-8502, Japan

Bhavanandan, Veer P. (2.2.1) Center for Glycosciences and Technology, The Biodesign Institute, The Arizona State University, Arizona 85287-6001 U.S.A.

*Endo, Masahiko (1.1.3, 2.3, 5, 7.3) President, Hirosaki University, Aomori 036-8560, Japan

Fujita, Kiyotaka (3.2.2) Department of Biochemical Science and Technology, Faculty of Agriculture, Kagoshima University, Kagoshima 890-0065, Japan

Haneda, Katsuji (3.1.2) Department of Applied Chemistry, Kanagawa Institute of Technology, Kanagawa 243-0292, Japan

*Hase, Sumihiro (1.1.1, 1.1.2, 1.2, 2.1.2, 3.3, Appendix) Graduate School of Science, Osaka University, Osaka 560-0043, Japan

Higashi, Hideyoshi (7.4) Neuronal Circuit Mechanisms Research Group, Brain Science Institute RIKEN, and CREST, JST, Saitama 351-0198, Japan

Horibata, Yasuhiro (6) Neuronal Circuit Mechanisms Research Group, Brain Science Institute RIKEN, Saitama 351-0198, Japan

Ishii-Karakasa, Ikuko (7.2) Department of Biochemistry, School of Medicine, Kitasato University, Kanagawa 228-8555, Japan

Ishimizu, Takeshi (2.1.2, 3.3, Appendix) Graduate School of Science, Osaka University, Osaka 560-0043, Japan

Ito, Kazuo (7.1) Department of Biology, Graduate School of Science, Osaka City University, Osaka 558-8585, Japan

Ito, Makoto (1.1.4, 2.4.1, 2.4.2, 6, 7.4) Department of Bioscience and Biotechnology, Graduate School of Kyushu University, Fukuoka 812-8581, Japan

Li, Su-Chen (2.2.2) Department of Biochemistry, Tulane University School of Medicine, Louisiana 70112, U.S.A.

Li, Yu-Teh (2.2.2) Department of Biochemistry, Tulane University School of Medicine, Louisiana 70112, U.S.A.

Sakaguchi, Keishi (2.4.1) Division of Biological Sciences, Graduate School of Science, Hokkaido University, Hokkaido 060-0810, Japan

*Takagaki, Keiichi (1.1.3, 2.3, 5, 7.3) Department of Biochemistry, Hirosaki University School of Medicine, Aomori 036-8562, Japan

Takegawa, Kaoru (1.4, 2.1.1, 3.1.3, 3.2.1) Department of Life Sciences, Faculty of Agriculture, Kagawa University, Kagawa 761-0795, Japan

*Yamamoto, Kenji (1.3, 2.1.1, 3.1.1, 3.2.2) Graduate School of Biostudies, Kyoto University, Kyoto 606-8502, Japan

*Editor

Preface

This volume reviews the unlimited possibilities that appear to be opening up as a result of developments in glycotechnology and the physiological significance of endoglycosidases leading to new avenues in the field.

Endoglycosidases are one of the general types of glycosidases and hydrolyze internal glycosidic bonds of oligosaccharide and polysaccharide chains, resulting in the release not of monosaccharides, but oligosaccharides.

On the other hand, exoglycosidases (which act on nonreducing end sites of carbohydrate chains and as a result release a monosaccharide) corresponding to almost all of the glycosidic bonds of the known carbohydrate chains of glycoconjugates have been discovered.

Some endoglycosidases, for example, amylase, cellulase and hyaluronidase, were previously known. However, except for hyaluronidase, the endoglycosidases which act on a long carbohydrate chain of glycoconjugates (glycoproteins, glycolipids and proteoglycans) have not been investigated.

The original enzymes for the inborn error of metabolism on the carbohydrate chains of glycoconjugates are exoglycosidases and related enzymes on the degradation processes of carbohydrate chains, and the mechanism of glycoconjugate metabolism is explained by the exoglycosidases. Therefore, there was little or no awareness of the many kinds of endoglycosidases that exist.

In order to elucidate the degradation mechanism of glycoconjugates, the urinary carbohydrates were examined, resulting in the identification of many kinds of oligosaccharides and polysaccharides which are free from amino acids and peptides. Thus it was suggested that many kinds of endoglycosidases which act on internal sites of carbohydrate chains exist in the animal body.

The cause of the delay in endoglycosidase discoveries was most probably the difficulty in determining their enzyme activities. In general, the exoglycosidase activities are easily determined using artificial substrates which are glycosidically bound to *p*-nitrophenol, 4-methylumbelliferon and so on as aglycons.

Before the use of artificial substrates, exoglycosidase activities were determined by measuring the reducing power in the reaction solution. However, these methods are not applicable to the determination of endoglycosidase activities because the reaction of the reducing power determination is usually carried out in a boiling water bath under alkaline conditions. Therefore, the reaction conditions cause the peeling reaction of the carbohydrate chain, i.e., depolymerization. As a result, reducing power that is much higher than the actual reducing power is shown. Moreover, it is very difficult to determine a special carbohydrate residue which newly appears at the reducing end of the carbohydrate chain by the reaction of an endoglycosidase. Therefore, it should be understood that each determination method of endoglycosidase activity described in the following text has been devised with a great deal of effort.

The physiological functions of the endoglycosidases are as follows. In the early stage of the catabolic processes of glycoconjugates, endoglycosidases play an important role in releasing carbohydrate chains from the peptide and lipid moiety as a core of glycoconjugates and from the carbohydrate moiety of glycoconjugates. In the same way, the enzyme is responsible for the depolymerization of carbohydrate chains.

The endoglycosidases catalyze a transfer reaction of oligosaccharide residue through a process called transglycosylation. For example, the branching enzyme acting on glycogen and amylopectin catalyzes both hydrolysis and transglycosylation at the same time. Then the enzyme hydrolyzes a branch having an α1-6 bond in a glycan and directly transfers it to the nonreducing end, producing a straight carbohydrate chain.

One glycotechnological application using endoglycosidases is to cut off an intact carbohydrate chain having a free reducing end from glycoconjugates. With a carbohydrate chain having a free reducing end, it is possible to mark the reducing end with radioisotopes, fluorescence and other means. As a result, a detailed analysis of the chemical structure of the carbohydrate chain can be easily performed.

The endoglycosidases catalyze the transference of a carbohydrate chain into another substance, somewhat the reverse of the hydrolysis reaction. Using this reaction, the endoglycosidases are useful for the reconstruction of carbohydrate chains and for the transference of carbohydrates to peptides and lipids. The chemical synthesis of carbohydrate chains presently requires great cost and time, but it is possible to synthesize even a high molecular carbohydrate chain using endoglycosidases.

Recently, genetic engineering has made great progress. However, it is known that some recombinant proteins encounter problems in biological activity because of the incompleteness of their carbohydrate chains. Therefore, it is important to glycosylate these proteins artificially. Furthermore, it has been clarified that endoglycosidases act on the linkage region between a core protein and carbohydrate moiety, such as endo-β-xylosidase acting on the proteoglycan, endo-β-*N*-acetylglucosaminidase acting on *N*-glycan, and so on. Using these enzymes, it is possible to attach a carbohydrate chain to a protein core. This reaction is regarded as a specific reaction using the reverse reaction of hydrolysis. It is also possible that a protein may take on a new biological function by attaching a carbohydrate chain to which the protein bears no relation naturally.

The glycotechnological application of endoglycosidases is expected to greatly expand. Endoglycosidases have many functions useful in the field, such as cutting off long chain oligosaccharides from polysaccharides and glycoconjugates, the transglycosylation or attachment of oligosaccharides between carbohydrate chains and peptides and the reconstruction of carbohydrate chains. In gene manipulation the endoglycosidases have both enzyme restriction and ligation activities.

Although few endoglycosidases have been discovered to date, great progress is expected and there is little doubt that the endoglycosidases will open up new avenues in glycotechnology.

Most of the contributing authors were members of a project team led by M. Endo working under a grant from the Ministry of Education, Science, Culture, and Sports of Japan, on a study entitled, "Developments in New Glycotechnology Using the Specific Activity of Endoglycosidases," and speakers at the international symposium of the same name held in Kyoto, Japan, October 20, 2000.

April 2006

Masahiko Endo
Sumihiro Hase
Kenji Yamamoto
Keiichi Takagaki

Contents

1

Biochemistry of Glycoconjugates

This chapter is an introduction to carbohydrate chemistry necessary for the understanding of endoglycosidases.

1.1 Chemical Structures

1.1.1 Structures and Basic Properties of Carbohydrates

A. Carbohydrates[1-9)]

Monomer units of carbohydrates are monosaccharides. The definition of monosaccharides varies, but one example is as follows. Monosaccharides are noncyclic hydroxy-oxo-compounds capable of forming intramolecular hemiacetal linkages and have at least 4 oxygen atoms but not more than 2 oxo(carbonyl)groups.[10)] Oligosaccharides contain 2 - 6 monosaccharides. Some chemical characteristics are described below using glucose (Glc) as an example.

B. Monosaccharides

The structure of D-glucose ($C_6H_{12}O_6$) was determined by E. Fisher, and the structure is shown in Fig. 1.1 using the Fisher projection. Glucose contains four asymmetric carbon atoms (C2 ~ C5) and an aldehyde group (C1) which is written on the top. If the hydroxyl group at the highest numbered asymmetric carbon atom (C5, the configurational carbon atom) is written to the right, the monosaccharide belongs to the D-series; if it is written to the left, it belongs to the L-series.

An aldehyde group reacts with an alcohol group to form a hemiacetal and the further reaction of another alcohol group leads to a mixed acetal (Scheme 1.1).

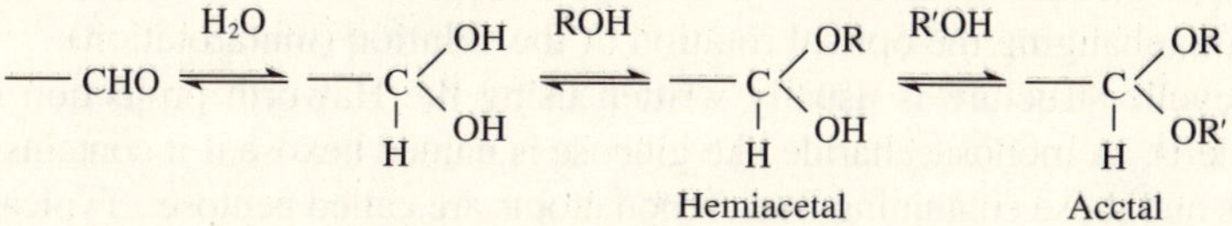

—CHO $\overset{H_2O}{\rightleftharpoons}$ —CH(OH)(OH) $\overset{ROH}{\rightleftharpoons}$ —CH(OR)(OH) $\overset{R'OH}{\rightleftharpoons}$ —CH(OR)(OR')

Hemiacetal Acctal

Scheme 1.1 Chemical reactivity of aldehyde

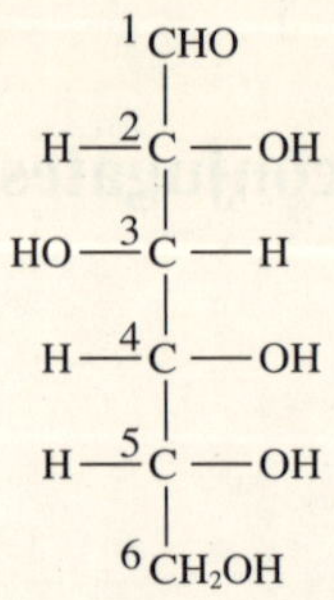

```
      1 CHO
        |
   H —2 C — OH
        |
  HO —3 C — H
        |
   H —4 C — OH
        |
   H —5 C — OH
        |
      6 CH2OH
```

Fig. 1.1 Fisher projection of D-glucose.

The aldehyde group of glucose reacts with one of its own alcohol groups to form the hemiacetal, a cyclic structure (Fig. 1.2). The six-membered cyclic structure is named pyranose (glucopyranose, Glc*p*) for pyran, and the five-membered cyclic structure is named furanose (glucofuranose, Glc*f*) for furan. As a result of forming a cyclic hemiacetal, a new asymmetric center appears at the C1 atom (Fig. 1.2). Thus glucopyranose has two isomers, named anomer, α and β in the case of carbohydrates. The α-anomer rotates polarized light more dextrorotatory than the β-anomer of the D-series. The hemiacetal hydroxyl group at C1 of the α-anomer is written on the same side as that of the configurational carbon atom (C5) (Fig. 1.2). The open-chain form and glucopyranose (α and β) are in equilibrium in water, but the ratio of the open-chain form is negligible.

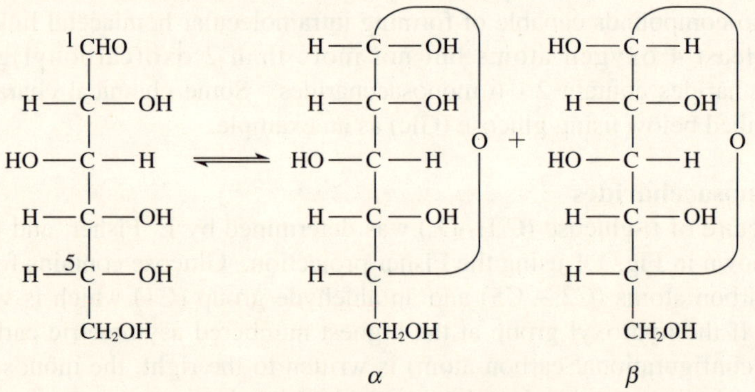

```
      1 CHO                  H — C — OH  ┐          HO — C — H   ┐
        |                        |       |               |       |
   H — C — OH                H — C — OH  |           H — C — OH  |
        |                        |       |               |       |
  HO — C — H      ⇌         HO — C — H   O   +      HO — C — H   O
        |                        |       |               |       |
   H — C — OH                H — C — OH  |           H — C — OH  |
        |                        |       |               |       |
   H —5 C — OH               H — C ──────┘           H — C ──────┘
        |                        |                       |
      CH2OH                    CH2OH                   CH2OH
                                 α                       β
```

Fig. 1.2 Ring structures of D-glucose.

When α-glucose is dissolved in water, it is converted to β-glucose to reach equilibrium, changing the optical rotation of the solution (mutarotation).

The cyclic structure is usually written using the Haworth projection formula (Fig. 1.3, left). A monosaccharide like glucose is named hexose if it contains six carbon atoms and those containing five carbon atoms are called pentose. Typical monosaccharides and their abbreviations are shown in Fig. 1.4. Sialic acid is a collective name covering the derivatives of neuraminic acid (Neu), such as *N*-acetylneuraminic

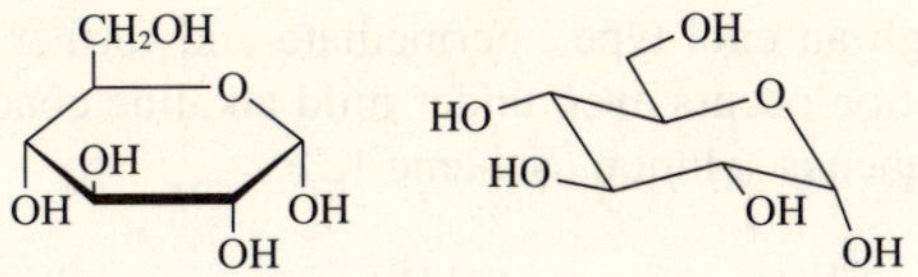

Fig. 1.3 Haworth projection of α-D-glucopyranose (left).

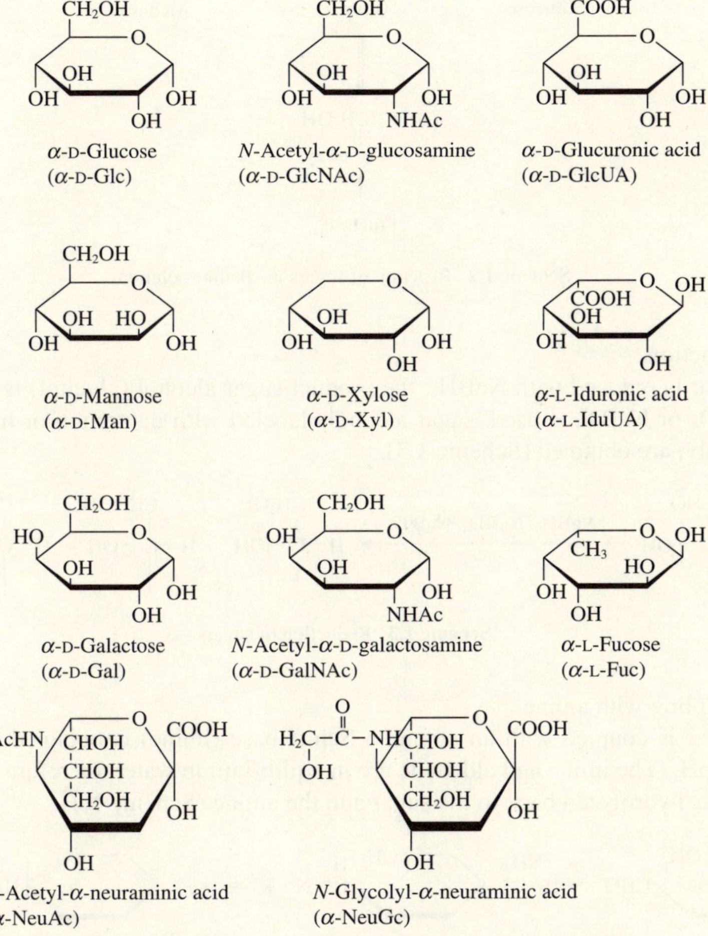

Fig. 1.4 Typical monosaccharides often found in glycoconjugates.

acid (NeuAc), *N*-glycolylneuraminic acid (NeuGc), and so on (Fig. 1.4).

C. Chemical Characteristics of Reducing Sugars

1) In alkaline aqueous solution

Reducing sugars and reducing end sugar residues of sugar chains are unstable in alkaline aqueous solution in contrast to the stable glycosidic linkages, leading to isomerization through an enol-type intermediate and further leading to degradation. The isomerization occurs even under mild alkaline conditions such as those found in $Ca(OH)_2$ aqueous solution (Scheme 1.2).

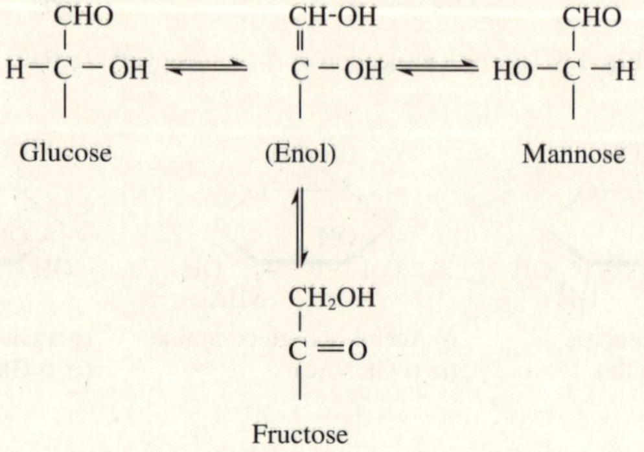

CHO | H−C−OH | ⇌ CH-OH ‖ C−OH | ⇌ CHO | HO−C−H |

Glucose (Enol) Mannose

⇅

CH_2OH | C=O |

Fructose

Scheme 1.2 Reaction of sugars in alkaline solution

2) Reduction

If a sugar is reduced with $NaBH_4$, the product sugar alcohol (glycitol) is obtained. If $NaBD_4$ or $NaBT_4$ is used, sugar alcohols labeled with deuterium or tritium, respectively, are obtained (Scheme 1.3).

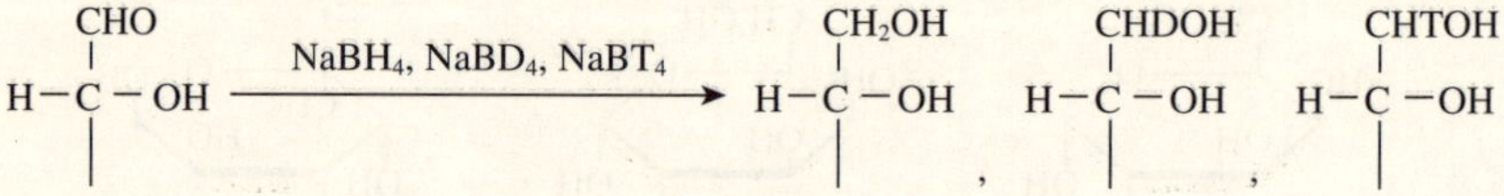

CHO | H−C−OH | —($NaBH_4$, $NaBD_4$, $NaBT_4$)→ CH_2OH | H−C−OH | , CHDOH | H−C−OH | , CHTOH | H−C−OH |

Scheme 1.3 Reduction of sugars

3) Coupling with amine

If a sugar is coupled with an amine, a Schiff base (imine) is obtained at around neutral pH. The imine and aldehyde are in eqilibrium in water, therefore the imine formed is hydrolyzed back to the sugar and the amine (Scheme 1.4).

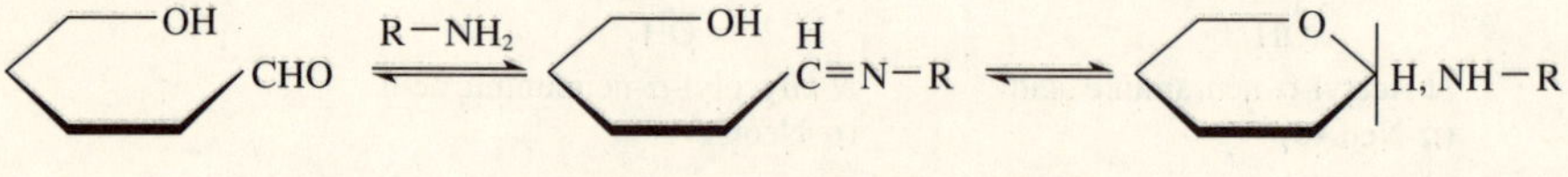

OH, CHO ⇌ (R−NH_2) OH, H, C=N−R ⇌ O, H, NH−R

Scheme 1.4 Coupling with amine

4) Maillard reaction

On further incubation, a Schiff base (imine) proceeds to a 2-keto derivative. The reaction is called Amadori rearrangement (Fig. 1.5). The reaction proceeds further to the decomposition of sugars (fumin, black precipitates).

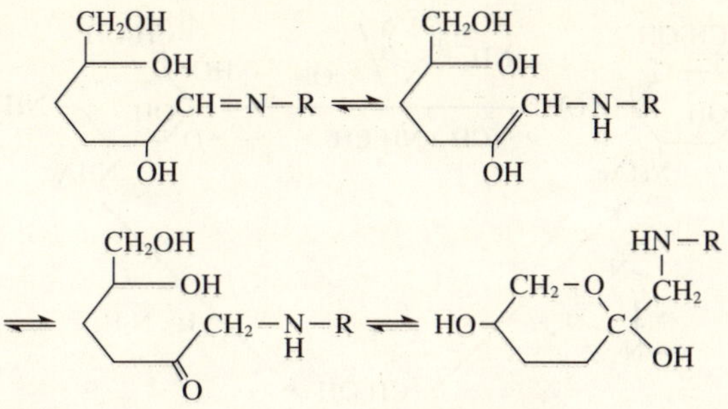

Fig. 1.5 Amadori rearrangement.

5) Reductive amination

A Schiff base is reduced to 1-amino-1-deoxy-derivatives (secondary amine) (Scheme 1.5). The reaction can be achieved with $NaBH_4$, $NaBH_3CN$ or dimethylamine-borane complex ($Me_2NH{\cdot}BH_3$). If a fluorescent reagent is used, a sugar chain is tagged with fluorescence, and the chemical reaction can be used for sensitive analysis of sugar chains as described below.

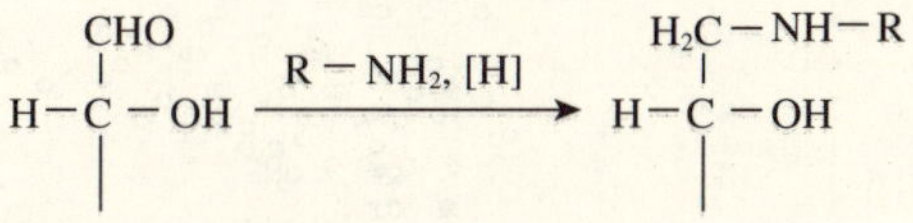

Scheme 1.5 Reductive amination

6) Fluorescence labeling method

The reductive amination of sugars is used for fluorescence tagging at the reducing end of sugar chains. The basic reaction using 2-aminopyridine is shown in Fig. 1.6. The chemical structures of pyridylamino derivatives of (PA-) sugar chains are analyzed by two-dimensional sugar mapping (Fig. 1.7) coupled with exoglycosidase digestion. PA-sugar chains are converted to 1-amino-1-deoxy-derivatives (Fig. 1.6), which can be used for coupling with other derivatives such as fluorescein, resins, lipid, protein and others. 1-Amino-1-deoxy-derivatives are further converted to reducing sugars (Fig. 1.6).

7) Boronic esters (Boronates)

In aqueous alkaline solution, sugars are coupled with borate or phenylboronate derivatives (Scheme 1.6). The stability of the complexes depends on the stereochemistry of a 1,2-diol (glycol) part of saccharides; *cis*-oriented glycol complexes are more stable than the *trans*-oriented ones. This reaction is used for purification of sugars by electrophoresis and by anion-exchange chromatography.

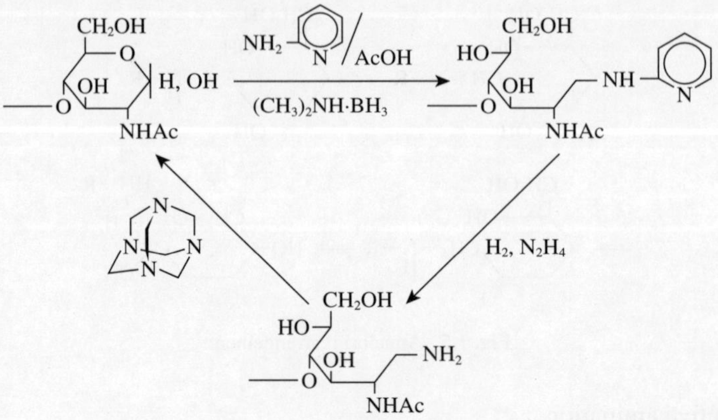

CH_2OH / O / OH / H, OH / O / NHAc

NH_2–(pyridine) / AcOH, $(CH_3)_2NH{\cdot}BH_3$ →

CH_2OH / HO / OH / NH–(pyridine) / O / NHAc

H_2, N_2H_4

CH_2OH / HO / OH / NH_2 / O / NHAc

Fig. 1.6 Three basic chemical reactions used for pyridylamino glycotechnology. [Reprinted with permission from Takahashi, C. *et al*. (2003) *J. Biochem*. **134**, 53]

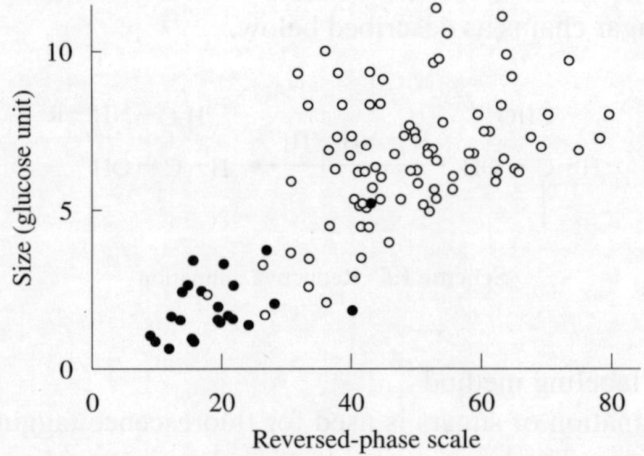

Fig. 1.7 Two-dimensional sugar map of PA-sugar chains. PA-*N*-linked (○) and *O*-linked (●) sugar chains are separated by two kinds of HPLC (size-fractionation HPLC and reversed-phase HPLC), and their elution positions are plotted.

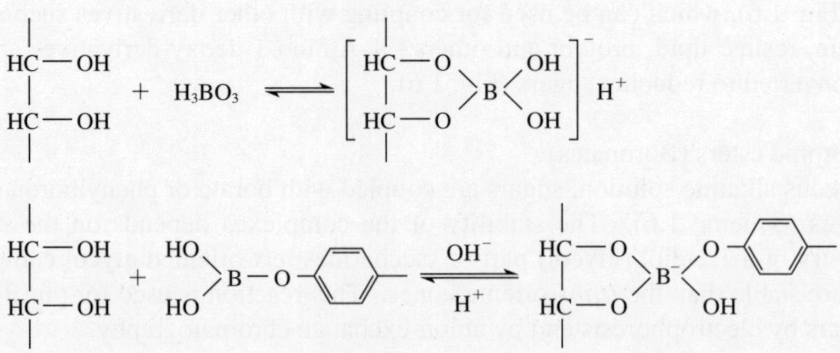

HC—OH, HC—OH + H_3BO_3 ⇌ [HC—O, HC—O >B< OH, OH]$^-$ H^+

HC—OH, HC—OH + HO, HO >B—O—(tolyl) ⇌ (OH^- / H^+) HC—O, HC—O >B^-< O—(tolyl), OH

Scheme 1.6 Boronic ester

8) Other esters

Phosphate ester and sulfate ester are hydrolyzed with acid, but stability depends largely on the neighboring chemical structures. Carboxylic ester is easily hydrolyzed with alkali and acid (Scheme 1.7).

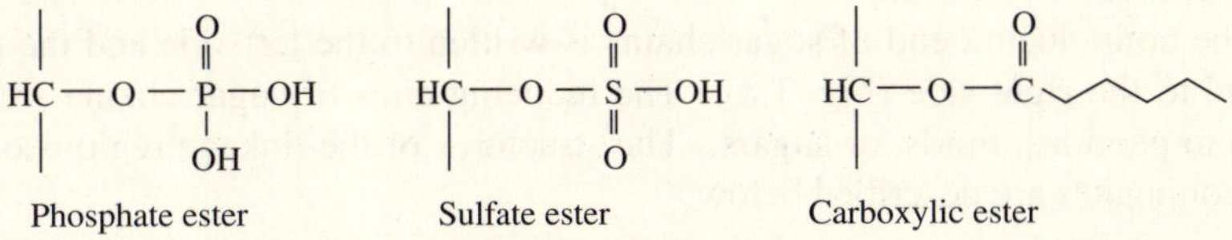

Scheme 1.7 Other esters

9) Galactose oxidase

D-Galactose is oxidized to an aldehyde with galactose oxidase (EC 1.1.3.9). Reduction of the product gives D-galactose (Scheme 1.8). The enzyme also oxidizes *N*-acetyl-D-galactosamine. The reaction is used to label the terminal galactose or *N*-acetylgalactosamine residues of sugar chains when reduced with $NaBD_4$ or $NaBT_4$.

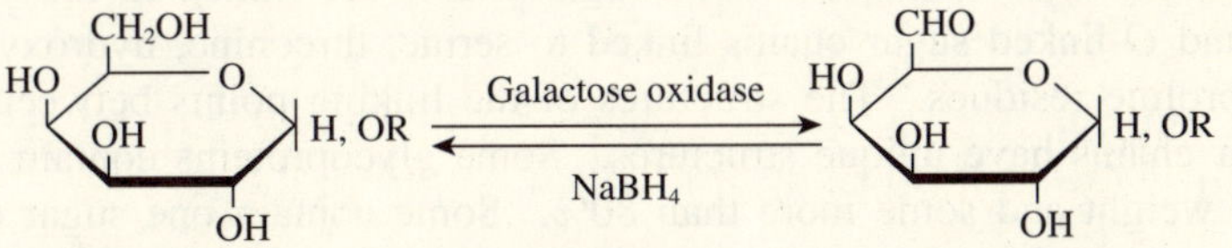

Scheme 1.8 Oxidation with galactose oxidase and reduction with $NaBH_4$

D. Sugar Chains

A hemiacetal hydroxyl group can react with an alcohol group of another monosaccharide, and a disaccharide is formed by the acetal linkage. This is usually called a glycosidic linkage. The sugar part of sugar chains is called the sugar residue. Monosaccharides are joined by glycosidic linkages to form oligosaccharides, polysaccharides and sugar chains (Fig. 1.8). The glycosidic linkages are not of one kind but several kinds are possible due to the anomers (α, β) and several available hydroxyl groups of a monosaccharide leading to branched chains, in contrast to the peptide bond of proteins and the phosphodiester linkage of nucleic acids.

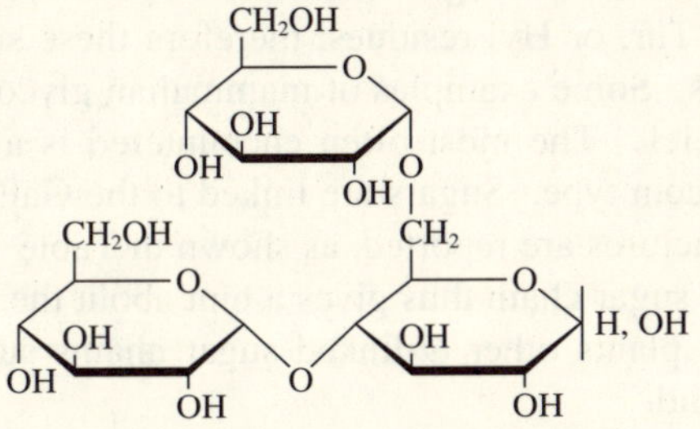

Fig. 1.8 An oligosaccharide; Glcα1-4(Glcα1-6)Glc.

Chemical characteristics of glycosidic linkage:
Inherent to the chemical characteristics of acetals, the glycosidic linkage is stable against alkali and hydrolyzed with acid. However, glycosidic linkages are more stable in acid than aliphatic acetals, probably due to cyclization that renders the molecule more stable.

The nonreducing end of sugar chains is written to the left side and the reducing end to the right side (Fig. 1.8). The reducing ends of sugar chains are often linked to peptides, lipids, or sugars. The structures of the linkage regions of such glycoconjugates are described below.

1.1.2 Structures of Glycoproteins

A. Introduction[5,11)]

Proteins covalently joined by a sugar chain or more than one sugar chain are called glycoproteins. Proteins with polysaccharides with repeating disaccharides are usually called proteoglycans (see section 1.1.3). Most proteins present in mammals are glycoproteins. Sugar chains of glycoproteins are classified mainly into *N*-linked sugar chains in which sugar chains are linked to the asparagine residue and *O*-linked sugar chains linked to serine, threonine, hydroxylysine, or hydroxyproline residues. The structures of the linking points between peptides and sugar chains have unique structures. Some glycoproteins contain 1% sugar chain by weight and some more than 80%. Some contain one sugar chain and some more than 800. A typical feature of sugar chains is that sugar chains have microheterogeneity and their structures are composed of several sugar chains with similar structures even though a glycoprotein has a single sugar chain binding site. One part of a sugar chain, especially the reducing end *N*-acetylglucosamine residue, is linked to proteins by hydrogen bonds to stabilize the three-dimensional structure of proteins.

B. Structures of Sugar Chains of Glycoproteins

In asparagine-linked (*N*-linked) sugar chains, the reducing end GlcNAc residue is linked to the amido group of an asparagine residue in the following sequence of proteins Asn-Xaa-Ser/Thr (Asn-Xaa-Cys in very rare cases). The structures of typical *N*-linked sugar chains are shown in Fig. 1.9 and Table 1.1.

The reducing ends of some sugar chains are bound to proteins through the hydroxyl group of Ser, Thr, or Hyl residues; therefore these sugar chains are called *O*-linked sugar chains. Some examples of mammalian glycoproteins are shown in Fig. 1.10 and Table 1.1. The most often encountered is a GalNAα1-*O*-Ser/Thr type referred to as mucin type. Sugars are linked to the GalNAc residue, and several types of core structures are reported, as shown in Table 1.1. Determination of the reducing end of a sugar chain thus gives a hint about the entire structure (type) of a sugar chain. In plants other *O*-linked sugar chains such as L-Araα1-*O*-hydroxyproline was found.[5)]

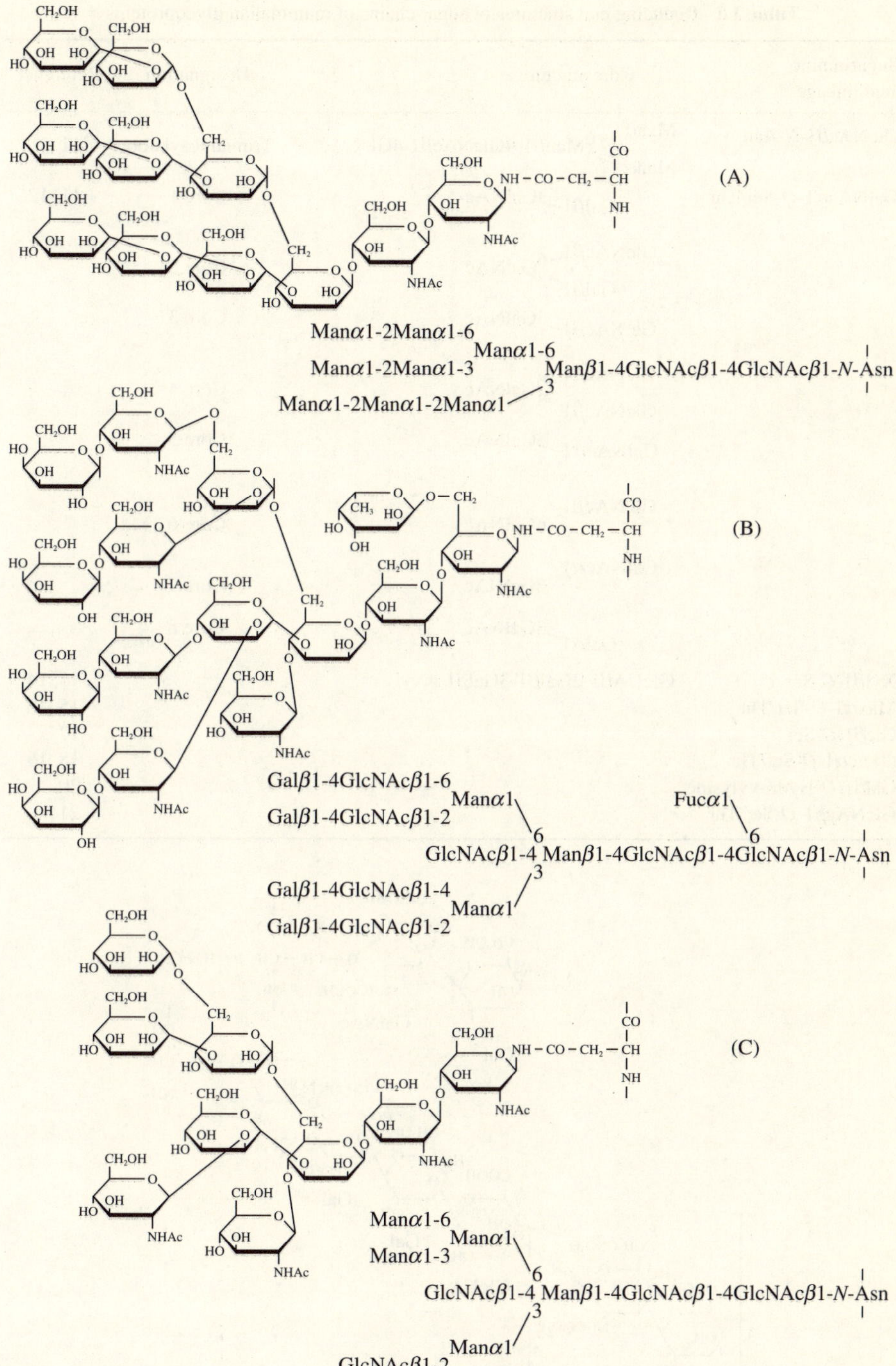

Fig. 1.9 Typical chemical structures of *N*-linked sugar chains. A, high-mannose-type sugar chains; B, complex type sugar chains; C, hybrid type sugar chains.

Table 1.1 Reducing end stuctures of sugar chains of mammalian glycoproteins

Sugar-amino acid linkage	Core stucture	Designation	Reference
GlcNAcβ1-*N*-Asn	Manα1–6 Manβ1-4GlcNAcβ1-4GlcNAc Manα1–3	Trimannosyl core	11
GalNAcα1-*O*-Ser/Thr	Galβ1–3GalNAc	Core 1	11, 12
	GlcNAcβ1–6 GalNAc Galβ1–3	Core 2	
	GlcNAcβ1–3GalNAc	Core 3	
	GlcNAcβ1–6 GalNAc GlcNAcβ1–3	Core 4	
	GalNAcα1–3GalNAc	Core 5	
	GlcNAcβ1–6GalNAc	Core 6	
	GalNAcα1–6GalNAc	Core 7	
	Galα1–3GalNAc	Core 8	
Xylβ1-*O*-Ser	GlcUAβ1-3Galβ1-3Galβ1-4Xyl		13, 14
Manα1-*O*-Ser/Thr			15, 16
Glcβ1-*O*-Ser			17
L-Fucα1-*O*-Ser/Thr			18, 19
Galβ1-*O*-Hydroxylysine			20
GlcNAcβ1-*O*-Ser/Thr			21

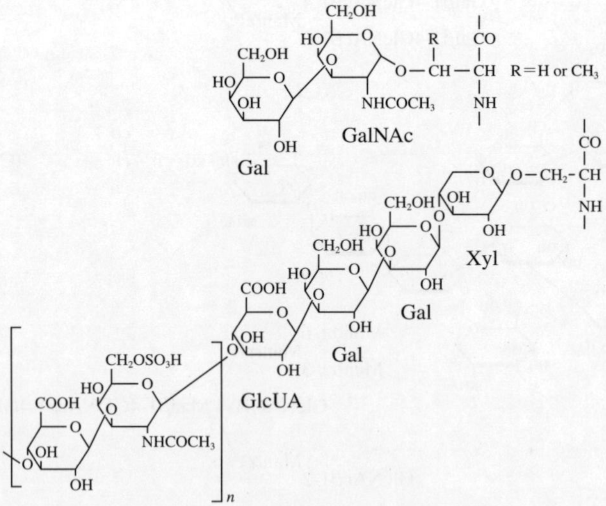

Fig. 1.10 Typical *O*-linked sugar chains.

C. Liberation of Sugar Chains from Glycoproteins with Enzymes and by Chemical Methods

N-Linked sugar chains are liberated by peptide *N*-glycanases, endo-β-*N*-acetylglucosaminidases, endo-β-mannosidases, and also by chemical means such as hydrazinolysis-*N*-acetylation. *N*-Linked sugar chains with *N*-acetylglucosaminitol at the reducing ends were obtained by treating glycoproteins with NaOH-$NaBH_4$ followed by *N*-acetylation using stronger reaction conditions than those required for the liberation of *O*-linked sugar chains.

Some *O*-linked sugar chains (especially Galβ1-3GalNAc) are liberated with an enzyme. Some *O*-linked sugar chains are also liberated by anhydrous hydrazine followed by *N*-acetylation, and in this case reducing sugar chains are obtained and can be used for fluorescence labeling. Some *O*-linked sugar chains are liberated by the β-elimination reaction using NaOH-$NaBH_4$, and sugar chains with glycitol at the reducing ends are obtained. For both chemical methods, during the liberation of *O*-linked sugar chains, further β-elimination reaction (peeling reaction) takes place more or less, especially when a reducing end residue is 3-*O*-substituted, *e.g.*, Galβ1-3GalNAcα1-*O*-Ser/Thr. However, sugar chains are not liberated when they are not linked to the β-position of amino acid residues such as Gal-Hyl.

1.1.3 Structures of Proteoglycans and Glycosaminoglycans

Among the glycoconjugates, proteoglycans have the largest variety of structures and functions. Their sugar chains are called glycosaminoglycans (GAGs), and mostly link to serine (Ser) residues on the proteoglycan core protein through *O*-glycosylation. Molecules having GAG chains, which are characteristic of proteoglycans, have not been found in glycolipids. Here, the structures of proteoglycans and their GAG chains are described.

A. Proteoglycans

Glycoconjugates with covalently linked GAG chains as a sugar moiety on their proteins are generically named proteoglycans.[22] Linkage regions between core proteins and GAG chains of proteoglycans have a common structure, GlcUAβ1-3Galβ1-3Galβ1-4Xylβ1-*O*-Ser, (GlcUA-Gal-Gal-Xyl-Ser) (Fig. 1.11).

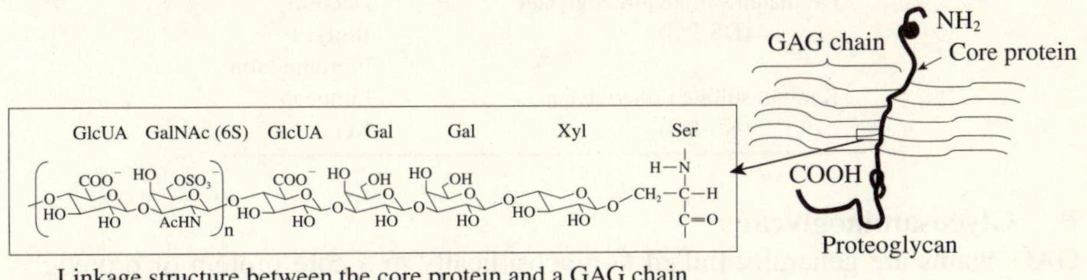

Fig. 1.11 Structure of a proteoglycan and its linkage region between a core protein and a GAG chain.

Chondroitin sulfate/dermatan sulfate can be synthesized if GalNAc is transferred to the linkage tetrasaccharide structure, while heparan sulfate/heparin can be synthesized if GlcNAc is transferred. The linkage structure shown in Fig. 1.11 is that of the chondroitin sulfate proteoglycan.

Proteoglycans have been identified: small proteoglycans with molecular weight of around 200 k, such as decorin, and large proteoglycans with molecular weight of around 3000 k, such as aggrecan. A proteoglycan is composed of a core protein with molecular weight of from 10 k to 400 k, with one to hundreds of GAG chains, each of which can have a molecular weight of up to thousands. Therefore, proteoglycans are huge and complex compared with molecules consisting of only a protein such as albumin, or only a sugar chain such as hyaluronan (HA). Actually, proteoglycan structures have great variety depending on the kind and number of GAGs, and differences in the core protein. There are also hybrid-type proteoglycans that have more than two kinds of GAG on a core protein. Table 1.2 shows a classification of proteoglycans based on their GAG chains: chondroitin sulfate proteoglycan (ChS-PG), dermatan sulfate proteoglycan (DS-PG), heparan sulfate proteoglycan (HS-PG) and keratan sulfate proteoglycan (KS-PG). Simplified diagrams of proteoglycans, whose chemical structures have been well studied, are shown in Fig. 1.12. The primary structures of the core proteins of a number of proteoglycans have been determined by cDNA cloning. Some potential glycosylation sites have also been proposed.

Table 1.2 Classification of well-known proteoglycans based on size and kind of major GAG chain moiety

Type	Proteoglycan
Large proteoglycan	
Heparan sulfate proteoglycan (HS-PG)	Perlecan Syndecan Glypican
Chondroitin sulfate proteoglycan (ChS-PG)	Aggrecan Versican Neurocan Brevican
Small proteoglycan	
Dermatan sulfate proteoglycan (DS-PG)	Decorin Biglycan Fibromodulin
Keratan sulfate proteoglycan (KS-PG)	Lumican Keratocan

B. Glycosaminoglycans

GAG chains are generally linked *O*-glycosidically to a core protein of proteoglycan through a GlcUA-Gal-Gal-Xyl-Ser linkage region.[22)] GAG is an acidic linear polysaccharide composed of hundreds of repeating disaccharide units: uronic acid (glucuronic, GlcUA or L-iduronic acid, IduUA) or galactose (Gal) combined with

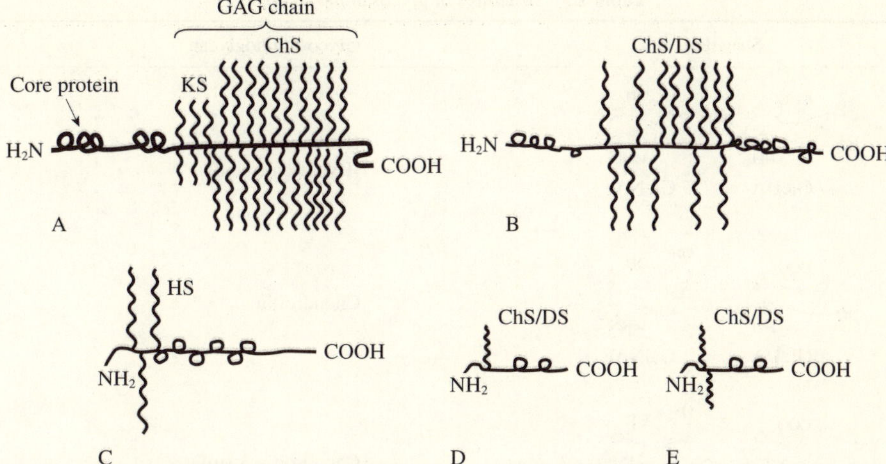

Fig. 1.12 Various structures of proteoglycans. A, aggrecan (monomer); B, versican; C, perlecan; D, decorin; E, biglycan. The left side of all diagrams indicates the N-terminal. GAG, glycosaminoglycan; KS, keratan sulfate; ChS, chondroitin sulfate; DS, dermatan sulfate; HS, heparan sulfate.

hexosamine (*N*-acetylglucosamine, GlcNAc or *N*-acetylgalactosamine, GalNAc or glucosamine, GlcN). GAGs are classified into chondroitin sulfate (ChS), dermatan sulfate (DS), heparan sulfate (HS), heparin (Hep), hyaluronan (HA), keratan sulfate (KS), etc. by the combination of uronic acid and hexosamine that composes the disaccharide units (Table 1.3). Chondroitin (Ch) which has no sulfate group exists in squid skin,[23)] *Caenorhabditis elegans*[24)] and others. Ch is also obtainable by chemical desulfation.

The terms chondroitin sulfate A, B, and C, for, respectively, chondroitin 4-sulfate (Ch4S), DS and chondroitin 6-sulfate (Ch6S), were introduced in 1958.[25)] Oversulfated DS variants such as ChSD, ChSE, ChSH and other homologues of ChS are also identified,[26)] although details are not provided here.

ChS is made up of disaccharide units of GlcUAβ1-3GalNAc and is classified into Ch4S or Ch6S by *O*-sulfation at the C-4 or the C-6 position on a GalNAc, respectively. Ch is a kind of ChS which is not sulfated at all and found in the nematode. HA and Ch are nonsulfated GAGs consisting of disaccharide units of GlcUAβ1-3GlcNAc and GlcUAβ1-3GalNAc, respectively. ChS and DS (also called ChSB) usually exist as hybrid sugar chains in a polymer. DS is an epimer of Ch4S whose GlcUA is converted by epimerization at the C-5 position to IduUA, resulting in an increased content of IduUA in the polymer. HS has GlcNAc as a hexosamine moiety and has both GlcUA and IduUA as a unoric acid moiety.

De-*N*-acetylation at the amino group (-*N*) of GlcNAc and sulfation at the *N*-, C-3, and C-6 positions on a GlcNAc and sulfation at the C-2 position on GlcUA are observed in HS. Hep is distinguished by its many sulfate groups and high content of IduUA compared with HS. The linkages between the uronic acids and hexosamines that compose KS, HS and Hep are always β1-4 links, which are distinguished from the β1-3 links in other GAGs. Gal and GlcNAc are sulfated at the C-

Table 1.3 Structures of glycosaminoglycans

Structure	Glycosaminoglycan
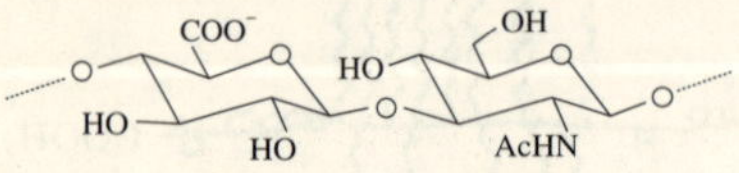 GlcUA GlcNAc	Hyaluronan (Hyaluronic acid)
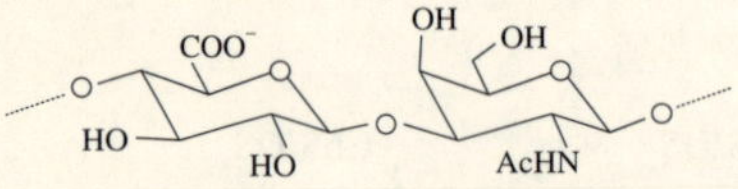 GlcUA GalNAc	Chondroitin
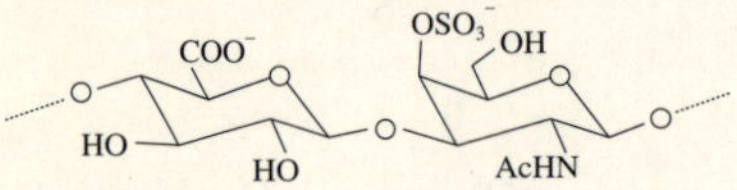 GlcUA GalNAc 4-sulfate	Chondroitin 4-sulfate (Chondroitin sulfate A)
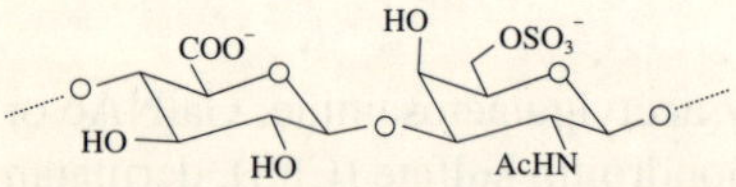 GlcUA GalNAc 6-sulfate	Chondroitin 6-sulfate (Chondroitin sulfate C)
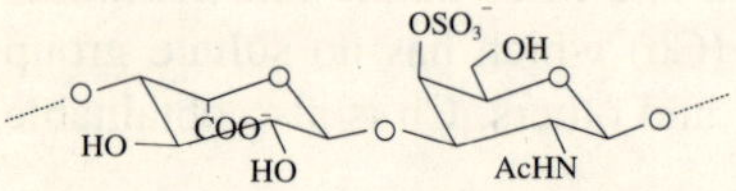 IduUA GalNAc 4-sulfate	Dermatan sulfate (Chondroitin sulfate B)
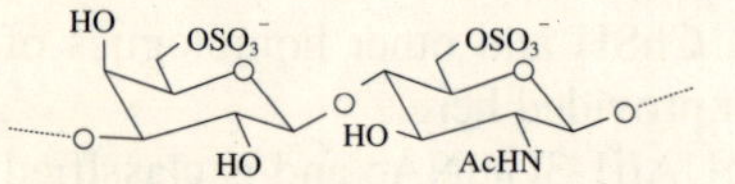 Gal 6-sulfate GlcNAc 6-sulfate	Keratan sulfate
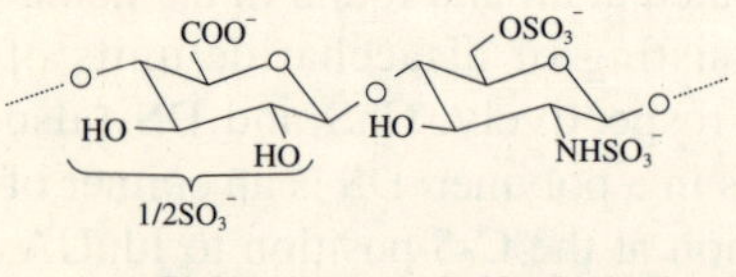 GlcUA (IduUA) *N*-Sulfo-GlcN-6-sulfate (*N*-Sulfo-D-glucosamine 6-sulfate)	Heparan sulfate / Heparin

Disaccharide units of glycosaminoglycans are shown.
GlcUA, glucuronic acid; IduUA, L-iduronic acid; GlcNAc, *N*-acetylglucosamine, GalNAc, *N*-acetylgalactosamine; Gal, galactose.

6 position on KS. All GAGs that exist in nature, with the single exception of HA, exist as sugar components of proteoglycans by linking with protein cores, and have sulfate groups. The molecular weight of GAGs reaches the tens of thousands. Except for KS, the linkage regions between core proteins and GAG chains of proteoglycans are assumed to have a common structure, GlcUA-Gal-Gal-Xyl-Ser (Fig. 1.11).[27)]

The fine structures of GAGs of proteoglycans have considerable variety even on the same sugar chain, since the modification patterns such as epimerization and sulfation are quite different depending on the region. Due to this heterogeneity, structural analysis of the sugar chains that have biological functions is not easy. Currently, the functional domain structures of only a few oligosaccharides, including those of Hep or DS showing inhibition activity against blood coagulation, have been determined.[28, 29)] However, it is expected that the relationships between the structures and functions of GAGs at an oligosaccharide level will be clarified in the near future by the development of analytical techniques. Establishing a foundation of structural analysis of GAGs by glycoengineering using endoglycosidases, as described in the following chapters, is extremely important.

1.1.4 Structures of Glycolipids

A. Classification of Glycolipids

Glycolipids from biological samples are classified into two main categories, glycosphingolipids and glycoglycerolipids. The latter contain a glycerol-based nonpolar portion such as diacylglycerol, dialkylglycerol and alkylacylglycerol and a sugar chain as the hydrophilic portion (Fig. 1.13). Glycoglycerolipids are present in plants, especially abundant in chloroplast, and cell walls of Gram-positive bacteria, mycoplasma and archea. In mammals, this type of lipid is not a major presence except in the testis in which a sulfated glycoglycerolipid called seminolipid (sulfogalactosylalkylacylglycerol)[30)] is present as a major constituent (Fig. 1.13). Seminolipid is also found in myelin of teleost and amphibian. Glycosphingolipids, a general term for glycolipids containing ceramide (*N*-acylsphingosine), are characteristic constituents of the plasma membrane of vertebrates.[31)] Glycosphingolipids are also distributed in plants and some invertebrates.[32)] Sialic acid-containing glycosphingolipids, called gangliosides, are found abundantly in

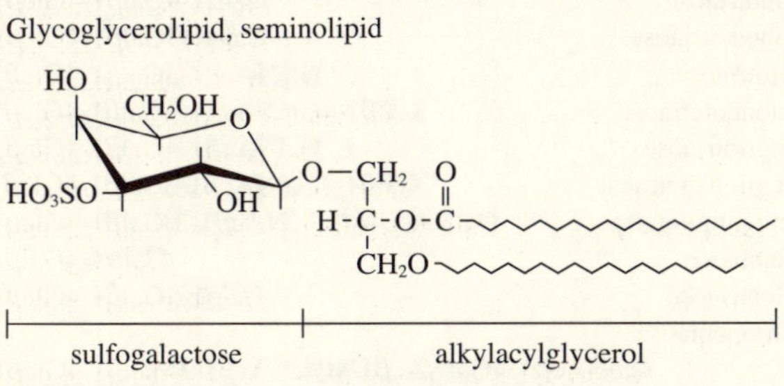

Fig. 1.13 Structure of a sulfated glycoglycerolipid, seminolipid.

neurons and nerve tissues.[33] The structure of a ganglioside, GM1a, is presented in Fig. 1.14. The abbreviations of gangliosides usually follow the nomenclature system of Svennerholm.[34]

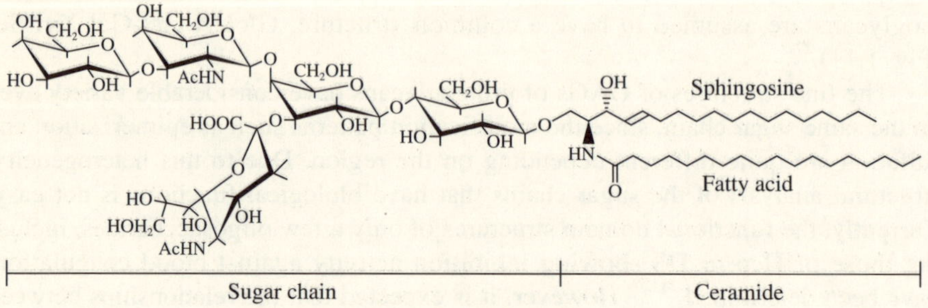

Fig. 1.14 Structure of a glycosphingolipid, GM1a.

B. Chemical Structures of Glycosphingolipids

More than 400 species of glycosphingolipids possessing different sugar structures have been identified, although only seven monosaccharides (glucose, galactose, *N*-acetylgalactosamine, *N*-acetylglucosamine, L-fucose, *N*-acetylneuraminic acid, *N*-glycolylneuraminic acid) have been found in vertebrate glycosphingolipids. In invertebrates, mannose-containing glycosphingolipids are found frequently.[35] The fact that glycosphingolipids containing a novel derivative of sialic acid, deaminated neuraminic acid (KDN),[36] was recently discovered in vertebrates may indicate that the novel structures of glycosphingolipids is likely be revealed in the future through further technological innovation. Glycosphingolipids are currently classified into 11 major groups by the difference in the structures of the sugar moieties bound to ceramide (Table 1.4). The structure of the sugar moiety is determined by the species and the number of constituent monosaccharides, their sequential order, linkages and α-, β-anomeric configurations between monosaccharides. IUPAC-IUB recommends the systematic names and abbreviations of glycosphingolipids (Table 1.4). This table shows 11 basic structures of sugar moiety in glycosphin-

Table 1.4 Systematic nomenclature of glycosphingolipids (GSLs)

Type Names	Basic GSLs	Structures	Abbreviations
Globo	globotriaose	Galα1-4Galβ1-4Glcβ1-1′Cer	Gb3Cer
Isoglobo	globoisotriaose	Galα1-3Galβ1-4Glcβ1-1′Cer	iGb3Cer
Lacto	lactotriaose	GlcNAcβ1-3Galβ1-4Glcβ1-1′Cer	Lc3Cer
Neolacto	lactoneotetraose	Galβ1-4GlcNAcβ1-3Galβ1-4Glcβ1-1′Cer	nLc4Cer
Ganglio	gangliotriaose	GalNAcβ1-4Galβ1-4Glcβ1-1′Cer	Gg3Cer
Isoganglio	ganglioisotetraose	Galβ1-3GalNAcβ1-3Galβ1-4Glcβ1-1′Cer	iGg4Cer
Lactoganglio	lactogangliotetraose	GalNAcβ1-4(GlcNAcβ1-3)Galβ1-4Glcβ1-1′Cer	–
Gala	galabiaose	Galα1-4Galβ1-1′Cer	Ga2Cer
Muco	mucotriaose	Galβ1-4Galβ1-4Glcβ1-1′Cer	Mc3Cer
Arthro	arthropentaose	GalNAcα1-4GalNAcβ1-4GlcNAcβ1-3Manβ1-4Glcβ1-1′Cer	Ar5Cer
Mollu	mollupentaose	Fucα1-4GlcNAcβ1-4Manα1-3Manβ1-4Glcβ1-1′Cer	Ml5Cer

golipids in which sugar chains could further extend to the direction of nonreducing ends. The core structure of these glycosphingolipids is a lactosylceramide (Galβ1-4Glcβ1-1′Cer) except gala-, arthro- and mollu-type glycosphingolipids. Gala-, arthro- and mollu-type glycosphingolipids, all of which are found in invertebrates, contain Galβ1-1′Cer, Manβ1-4Glcβ1-1′Cer and Manβ1-4Glcβ1-1′Cer as the core structure, respectively. Mannose is a constituent of only arthro- and mollu-type glycosphingolipids. For globo- and isoglobo-type glycosphingolipids, galactose is linked to the terminal galactose of lactosylceramide via *α* linkage, forming Galα1-4Gal and Galα1-3Gal, respectively. Likewise, *N*-acetyl-D-glucosamine is linked to the lactosylceramide via *β* linkage to form GlcNAcβ1-3Gal for lacto-type and neolacto-type glycosphingolipids. For basic structure of neolacto-type one, a galactose further extends via *β*1-4 linkage. Instead of *N*-acetylglucosamine, *N*-acetylgalactosamine is linked to the core lactosylceramide for ganglio-type glycosphingolipids. The structures of major neutral and acidic glycosphingolipids found in vertebrates are shown in Table 1.5.

Table 1.5 Structure of major glycosphingolipids (GSLs)

Name	Structure
Neutral glycosphingolipids	
Glucosylceramide (GlcCer)	Glcβ1-1′Cer
Galactosylceramide (GalCer)	Galβ1-1′Cer
Lactosylceamide (LacCer)	Galβ1-4Glcβ1-1′Cer
Globotriaosylceramide (Gb3Cer)	Galα1-4Galβ1-4Glcβ1-1′Cer
Globotetraosylceramide (Gb4Cer), globoside	GalNAcβ1-3Galα1-4Galβ1-4Glcβ1-1′Cer
Globopentaosylceramide (Gb5Cer), Forssman antigen	GalNAcα1-3GalNAcβ1-3Galα1-4Galβ1-4Glcβ1-1′Cer
Lactotetraosylceramide (Lc4Cer), paragloboside	Galβ1-3GlcNAcβ1-3Galβ1-4Glcβ1-1′Cer
Neolactotetraosylceramide (nLc4Cer), neoparagloboside	Galβ1-4GlcNAcβ1-3Galβ1-4Glcβ1-1′Cer
Gangliotetraosylceramide (Gg4Cer), asialoGM1	Galβ1-3GalNAcβ1-4Galβ1-4Glcβ1-1′Cer
Acidic glycosphingolipids	
Sulfatide (HO_3SGalCer)	HO_3S-3Galβ1-1′Cer
GM3	NeuAcα2-3Galβ1-4Glcβ1-1′Cer
GM2	GalNAcβ1-4(NeuAcα2-3)Galβ1-4Glcβ1-1′Cer
GM1a	Galβ1-3GalNAcβ1-4(NeuAcα2-3)Galβ1-4Glcβ1-1′Cer
GD1a	NeuAcα2-3Galβ1-3GalNAcβ1-4(NeuAcα2-3)Galβ1-4Glcβ1-1′Cer
GD1b	Galβ1-3GalNAcβ1-4(NeuAcα2-8NeuAcα2-3)Galβ1-4Glcβ1-1′Cer
GT1b	NeuAcα2-3Galβ1-3GalNAcβ1-4(NeuAcα2-8NeuAcα2-3)Galβ1-4Glcβ1-1′Cer
GQ1b	NeuAcα2-8NeuAcα2-3Galβ1-3GalNAcβ1-4(NeuAcα2-8NeuAcα2-3)Galβ1-4Glcβ1-1′Cer

C. Synthesis and Degradation of Glycosphingolipids *in vivo*

Glycosphingolipid synthesis starts from a reaction in which glucose or galactose is transferred to a ceramide to produce glucosylceramide and galactosylceramide, respectively. This step is catalyzed by UDP-glucose:ceramide glucosyltransferase (GlcT)[37)] and UDP-galactose:ceramide galactosyltransferase (GalT),[38)] respectively. The catalytic domain of GlcT is located on the cytosolic face of the Golgi membrane, while that of GalT is located on the lumen face of the endoplasmic reticulum (ER). The rule of extension of sugar chains of glycosphingolipids is common in all mammals, *i.e.*, a monosaccharide is sequentially transferred to a glucosylceramide or galactosylceramide from a nucleotide sugar by one of a series of specific glycosyltransferases, all of which are localized on the lumen face of the Golgi membrane. The biosynthetic pathway of gangliosides is illustrated in Fig. 1.15.

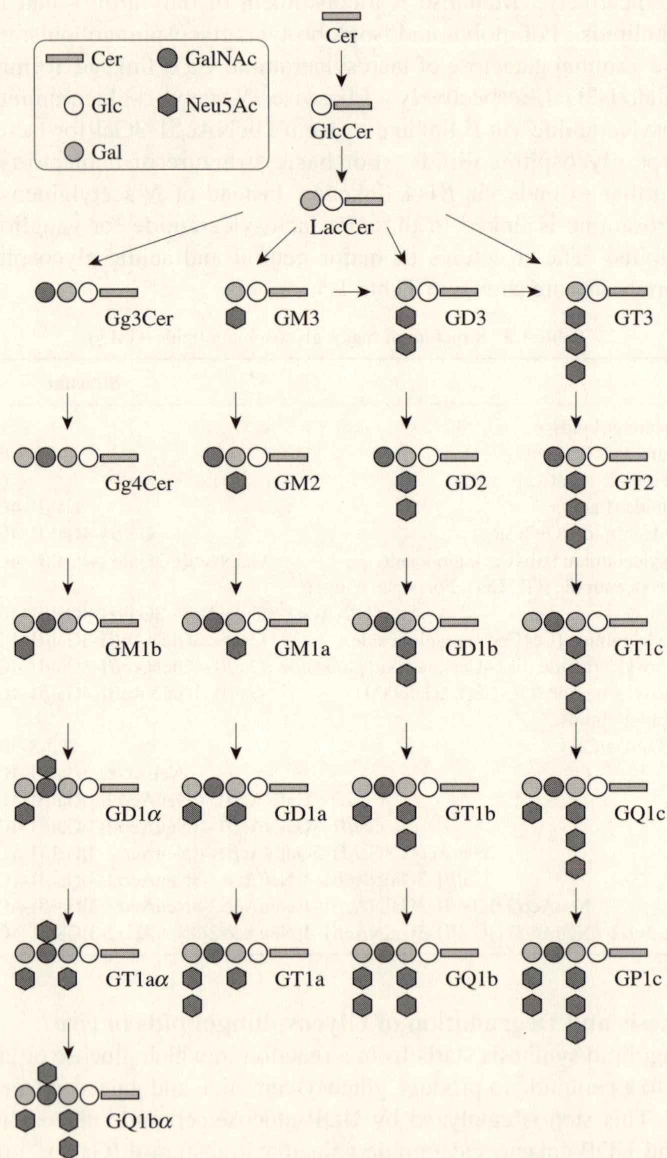

Fig. 1.15 Biosynthetic pathway of gangliosides.

After recycling between the plasma membrane and intracellular organs, vertebrate glycosphingolipids are finally transported to lysosomes where they are all hydrolyzed sequentially from the nonreducing end by exo-type glycoside hydrolases (glycosidases). All the hydrolysis reactions of glycosphingolipids *in vivo* seem to require specific activator proteins, although these can be replaced by certain detergents *in vitro*.[39] The balance of the synthesis and degradation of glycosphingolipids is completely regulated in the cell. If a glycosidase is lacking due to genetic deficiency, a glycosphingolipid accumulates in the lysosomes and causes a serious disease. Several disorders of glycosphingolipid metabolism due to a lack of either a specific glycosidase or an activator protein have been identified. Some invertebrates possess endoglycoceramidase (ceramide glycanase) hydrolyzing the glycosidic linkage between oligosaccharides and ceramides of various glycoshingolipids.[40-42] In the hydra *Hydra magnipapillata*, glycosphingolipids are first hydrolyzed by endoglycoceramidase to produce oligosaccharide and ceramide. The oligosaccharides produced are then sequentially degraded by exoglycosidases.[43] This catabolic pathway is completely different from that of mammals, which are likely to be missing the endoglycoceramidase. It should be stressed that this catalytic pathway seems to be unique but quite ubiquitous in invertebrates except insects since endoglycoceramidase activity is widely distributed in *Cnidaria*, *Mollusca*, *Annelida*, *Porifera*, *Echinodermata*, *etc*.

1.2 Enzyme Activity

1.2.1 Measurement of the Amount of Protein

Approximate amount of proteins can be assessed by the following methods. Except for weight, the amount of proteins is expressed in terms of a standard protein, such as bovine serum albumin.

1) Weight. Although the sample used is a purified protein, the freeze-dried sample contains considerable amounts (10 ~ 20%) of water, salts and other substances.

2) Absorption at 280 nm. Proteins contain chromophores such as tryptophan, phenylalanine, and tyrosine residues that absorb 280 nm light. However, most absorbance is attributable to tryptophan residues.
According to Lambert-Beer's law:

$$\frac{I_t}{I_0} = 10^{-\varepsilon cd} \qquad A = \log\frac{I_0}{I_t} = \varepsilon cd$$

ε, molar absorption coefficient; c, molar concentration; d, length of cell; A, absorbance.
Therefore, absorbance A is proportional to the concentration for a given sample

(ε=constant) using the same wavelength and the cell used (d=constant) (Fig. 1.16). Absorbance depends on the wavelength. A typical UV spectrum is shown in Fig. 1.17.

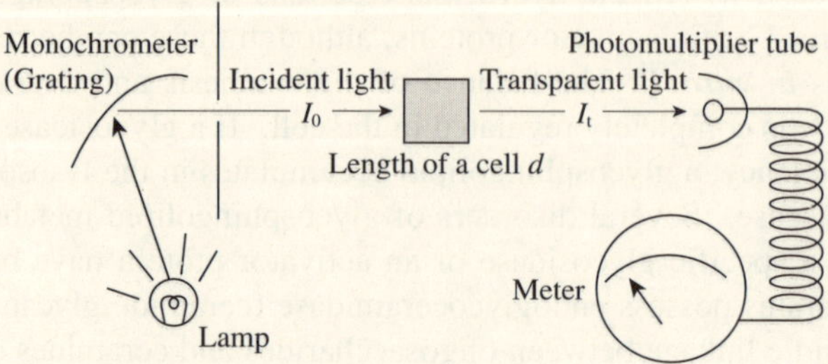

Fig. 1.16 Principle of a spectrophotometer.

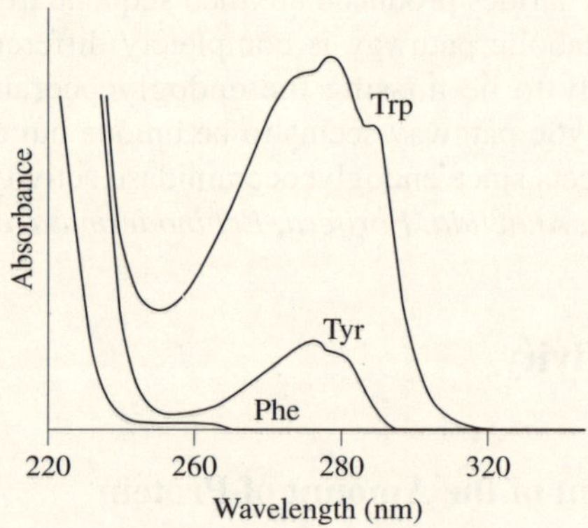

Fig. 1.17 UV spectrum of equimolar amounts of tryptophan, phenylalanine, and tyrosine.

As each protein contains different amounts of these amino acid residues, each protein has its own value (ε). Thus the amount of a protein is usually expressed in terms of a standard protein (bovine serum albumin, for example) as a reference compound. Bovine serum albumin with A=0.63[44] at 280 nm for the concentration of 1 mg/mL using the light path of 1 cm is used for the calculation of protein concentration, since the weight sample contains considerable moisture. Roughly 1% of a protein solution gives 10 ~ 20 absorbance at 280 nm.

3) Bicinchoninic acid (BCA). Cupric ion [Cu (II)] is reduced with peptide bonds, tyrosine, cysteine and tryptophan residues to produce cuprous [Cu (I)] ion in an alkaline solution. The cuprous ion produced is coupled with two molecules of BCA to produce purple color that is measured at 562 nm.

4) Coomassie Brilliant Blue (CBB). The amount of protein is roughly estimated from the color of a band on polyacrylamide gel electrophoresis (PAGE) where proteins are visualized with CBB. Colorization on gels depends on the hydropho-

bic nature of the protein. Usually around 1 μg is visible as blue color. This largely depends on the decolorization process.

5) The amount of protein is determined by quantitative determination of component amino acids such as asparagine, alanine and leucine in the acid hydrolysates if the amino acid sequence and molecular weight of the protein is known.

1.2.2 Measurement of Initial Reaction Velocity

Due to the specific activities of enzymes, the amount of an enzyme is estimated from the quantitative measurement of its specific activity even in a sample contaminated with other materials such as other proteins or carbohydrates. The hydrolysis of sucrose is used as an example.

$$\text{Sucrose} + H_2O \rightarrow \text{Glucose} + \text{Fructose}$$

Sucrose is actually not hydrolyzed in aqueous solution at room temperature and at neutral pH. However, if an enzyme (*e.g.*, invertase) is added to this solution, the reaction takes place much faster. The degree of the reaction is determined by measuring the amount of product formed (*e.g.*, glucose) by glucose oxidase, reducing power, HPLC-PAD, and other methods. The amount of product (glucose) increases with reaction time. The reaction velocity v is obtained by dividing the increment of the product p during the reaction time.

$$v = \frac{\Delta p}{\Delta t}$$

The increase of the product is not linear throughout the reaction time, but deviates from the straight line as the reaction proceeds (Fig. 1.18).

This deviation is partly due to the denaturation of an enzyme, inhibition of the reaction by the product, reverse reaction and decrease in the amount of substrate. Therefore, a simple way to eliminate these factors, often used is the measurement of initial velocity within the straight line region. Usually the initial velocity (product of less than around 20% of the initial substrate concentration) is calculated by quantifying the product p_1 at time t_1 (Fig. 1.18).

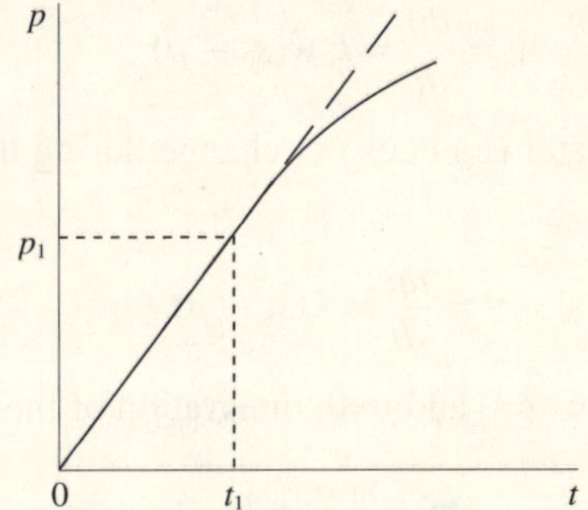

Fig. 1.18 Progress curve of an enzyme reaction.

$$v = \frac{\Delta p}{\Delta t} = \frac{p_1}{t_1}$$

The initial velocity increases with the amount of an enzyme added (Fig. 1.19). The amount of enzyme is obtained by measuring the initial velocity. The range where a linear relationship exists between initial velocity and enzyme concentration should be used. The initial reaction velocity changes by 1) enzyme concentration and 2) substrate concentration.

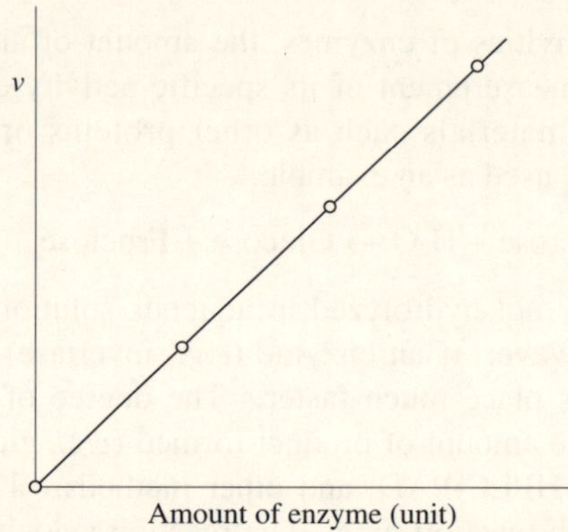

Fig. 1.19 Estimation of the amount of an enzyme.

The amount of enzyme is often expressed in units of enzyme activity. For example one unit of enzyme activity is defined as the amount of the enzyme that releases 1 nmol of glucose from 12 mM sucrose per minute at pH 6.8 and at 37°C under the conditions used. The enzymatic reaction proceeds as described below using the hydrolysis of sucrose as an example.

$$\text{Sucrose} + H_2O \xrightarrow{k'} \text{Glucose} + \text{Fructose}$$

$s_0 - p$ w p at time t

s_0, the initial concentration of a substrate (sucrose); p, concentration of a product (glucose) at time t. Initial velocity v is expressed as

$$v = \frac{dp}{dt} = k'w(s_0 - p)$$

If the concentration of water (w) does not change during the reaction, velocity v is expressed as

$$v = \frac{dp}{dt} = k(s_0 - p)$$

Using the initial conditions t=0 and p=0, integration of the formula gives

$$\ln \frac{s_0}{s_0 - p} = kt$$

Then p is written as

$$p = s_0 (1 - e^{-kt})$$

As seen in Fig. 1.20, the product increases almost linearly for the initial time.

$$p \cong k s_0 t$$

As time proceeds, the product increases not linearly but deviates from the tangent line at time 0 due to the decrease in substrate concentration and finally reaches s_0.

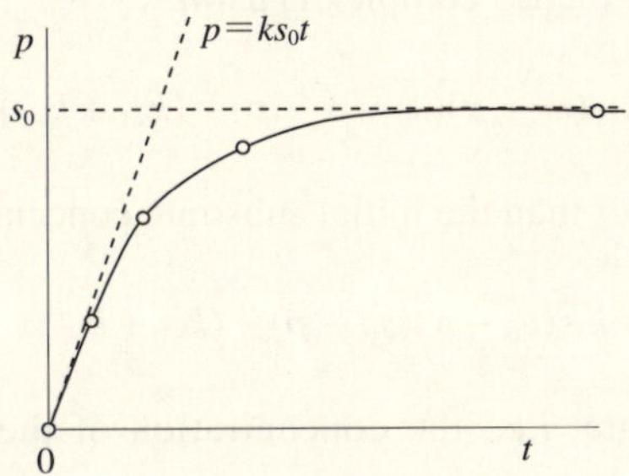

Fig. 1.20 Progress curve of a product by digestion of a substrate with an enzyme.

1.2.3 Kinetics of Enzyme Activity

The initial velocity was found to depend on the substrate concentration. Using invertase Brown observed that the reaction is first-order kinetic.[45] However, when the concentration of sucrose is over 5%, then v becomes constant (Fig. 1.21). When mineral acid is used, the reaction proceeds with first-order kinetics.

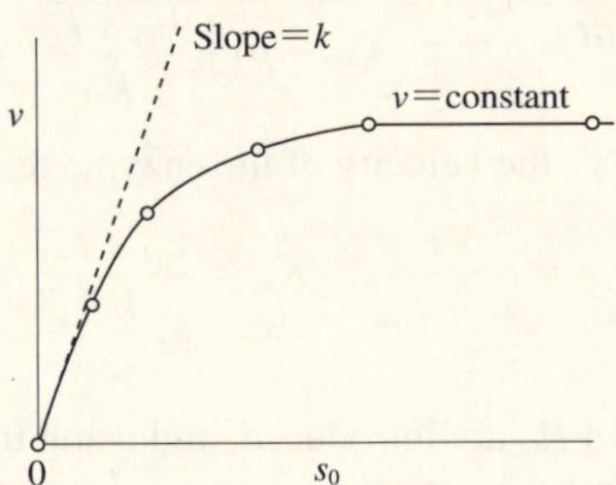

Fig. 1.21 Velocity versus initial substrate concentration for given amount of an enzyme.

Brown assumed the intermediate, enzyme-substrate complex to explain the results.[45] Later Michaelis and Menten proposed the solution assuming equilibrium between an enzyme, substrate, and enzyme-substrate complex (ES complex) in 1913 by introducing initial velocity.[46] Briggs and Haldane later introduced the steady state analysis of the enzyme-substrate complex in 1925.[47]

The following is a steady state analysis of the enzyme reaction assuming the enzyme-substrate complex. As the second substrate H_2O is of sufficiently high concentration to saturate an enzyme, the concentration of the second substrate is included in the velocity constant k_{+2}.

$$\begin{array}{cccccccc} E & + & S & \underset{k_{-1}}{\overset{k_{+1}}{\rightleftarrows}} & ES & ES+(H_2O) & \overset{k_{+2}}{\rightarrow} & P+E \\ e_0-x & & s_0-x-p & & x & x & & p \quad \text{at time } t \end{array}$$

e_0, initial enzyme concentration; x, concentration of the ES complex.
Change of the amount of the ES complex is dx/dt.

$$\frac{dx}{dt} = k_{+1}(e_0 - x)(s_0 - x - p) - (k_{-1} + k_{+2})x$$

Assuming that x is smaller than the initial substrate concentration s_0.

$$\frac{dx}{dt} = k_{+1}(e_0 - x)(s_0 - p) - (k_{-1} + k_{+2})x$$

Applying the steady state, *i.e.*, the concentration of the ES complex does not change, then $dx/dt=0$

$$k_{+1}\,(e_0-x)\,(s_0-p) - (k_{-1} + k_{+2})\,x = 0$$

Then

$$x = \frac{e_0(s_0 - p)}{(s_0 - p) + \dfrac{k_{-1} + k_{+2}}{k_{+1}}}$$

Initial velocity $v=dp/dt$ is expressed as

$$v = \frac{dx}{dt} = k_{+2}x = \frac{k_{+2}e_0(s_0 - p)}{(s_0 - p) + \dfrac{k_{-1} + k_{+2}}{k_{+1}}}$$

At the initial stage , $p << s_0$, the velocity of the enzyme reaction can be written as

$$v = \frac{k_{+2}e_0s_0}{s_0 + \dfrac{k_{-1} + k_{+2}}{k_{+1}}} = \frac{Vs_0}{s_0 + K_m} \qquad K_m = \frac{k_{-1} + k_{+2}}{k_{+1}} \qquad V = k_{+2}e_0$$

V (maximum velocity) and K_m are introduced, and using these values we can speculate the nature of the enzyme reaction.
K_m is called the Michaelis constant and is restricted to the substrate concentration at $v=V/2$ (see Fig. 1.22).
1) If k_{+2} is the rate-limiting step, then $k_{+2} << k_{-1}$, and

$$K_m = \frac{k_{-1}}{k_{+1}} = \frac{[E][S]}{[ES]} = K_s$$

Under these conditions, K_m is K_s (the dissociation constant of the ES complex) indicating affinity between the substrate and the enzyme. The smaller is the K_m val-

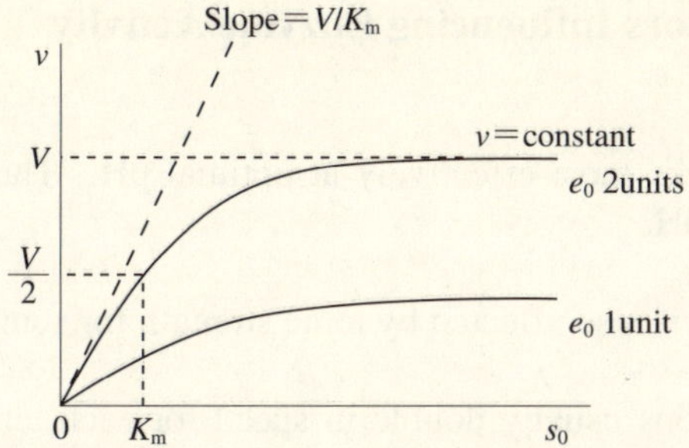

Fig. 1.22 Dependence of the initial substrate concentration on the initial velocity.

ue, the stronger is the affinity. K_m is the reciprocal of the affinity of the enzyme and substrate.

2) If $s_0 >> K_m$, the substrate concentration is larger than K_m, then

$$v = V = k_{+2}e_0$$

V is a measure of k_{+2} (k_{+2} : number of reactions per unit of time). In actual reactions, the k_{+2} reaction step includes a number of intermediate stages; therefore k_{cat} (turnover number) is often used.

3) If $s_0 << K_m$, the substrate concentration is smaller than K_m, then

$$v = \frac{e_0 k_{+2}}{K_m} s_0 \propto \frac{k_{+2}}{K_m}\left(or \frac{k_{cat}}{K_m} \right), \text{ and } \frac{k_{cat}}{K_m} \text{ is referred to as the specificity constant.}$$

The initial velocity is proportional to the k_{+2}/K_m value when e_0 is constant. K_m and V are obtained using the Lineweaver and Burk plot as follows.[48)]

$$v = \frac{Vs_0}{s_0 + K_m} \text{ is written as } \frac{1}{v} = \frac{1}{V} + \frac{K_m}{V}\frac{1}{s_0}$$

The values are obtained graphically (Fig. 1.23). Substrate concentrations (s_0) should be around the K_m value*.

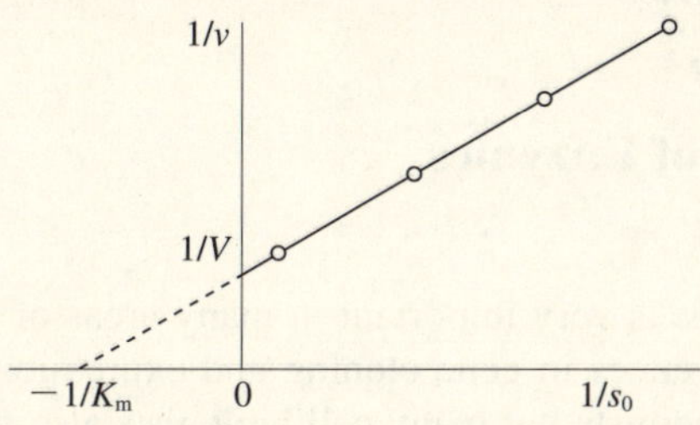

Fig. 1.23 A Lineweaver-Burk plot of the Michaelis-Menten equation.

* For further readings, see Reference 49.

1.2.4 Other Factors Influencing Enzyme Activity

(1) pH

An enzyme works most effectively at optimal pH. The enzyme reaction is usually done at this pH.

(2) Ionic strength

Enzyme activity is also affected by ionic strength for some enzymes.

(3) Temperature

Chemical reactions usually double in speed for each ten-degree rise in temperature. However, for enzyme reactions higher temperatures make the enzyme inactive due to denaturation. Enzyme activity reaches maximum value at optimal temperature. However, this value depends on the reaction time.

(4) Metal ions

Enzymes often require cofactors. These are often metal cations, such as Mn^{2+}, Mg^{2+} , Ca^{2+} and others. Therefore, when a chelating reagent like EDTA and EGTA is added to the reaction mixture, enzyme activity is often lost.

(5) Inhibitors

Enzyme activity is inhibited by inhibitors. For details see Reference.[49)]

(6) Lipid

Proteins having transmembrane domain(s) sometimes require membrane components for enzyme activity. These include sphingolipids, glycerophospholipids and others.

(7) Organic compound

Organic compounds such as acetate sometimes partially inhibit enzyme action when the enzyme recognizes an acetyl group in the structure of substrates.

(8) Stability

Enzymes are usually unstable in alkaline pH and acidic pH. Therefore the enzymes are most stable at pH of around neutral. Some enzymes lose their activity on freezing and thawing. Enzymes are usually unstable at around –5 ~ –10°C. Therefore samples should be kept at temperatures below –30°C. During thawing, the solid should be treated as fast as possible in an water bath at 37°C followed by immediate cooling with ice.

1.3 Purification of Enzymes

Purification of enzymes is very important in many areas of biochemical research. Recently, with the advances in gene cloning and expression, not only enzymologists and protein biochemists but many cell biologists also need to purify enzyme proteins. Although enzyme purification is often considered to be not so interesting, it is absolutely necessary for understanding the function and properties of enzyme in cells. Roughly speaking, the purification of enzyme protein is carried out by the following steps: (1) homogenation and disruption of target cells and tissues, (2) concentration of the extract, (3) separation of the protein by various chromatographic techniques and (4) estimation of purity of the purified protein.

1.3.1 Homogenation and Disruption of Cells and Tissues

The first step in the purification of an enzyme is the preparation of an extract containing the enzyme protein in soluble form. This extract is prepared according to the location of the enzyme in the cell. If the enzyme is membrane-bound or present in an inclusion body, the protein is insoluble, and a method to solubilize the enzyme will be required. For an intracellular enzyme, the cells must be disrupted by any method of disruption. In most cases, pretreatment may be necessary. For example, tissues may require removal of fat and mincing into small pieces. In the case of cells grown in a culture medium, concentration of cells is necessary to carry out efficient disruption. As proteases may be also released by disruption, some protease inhibitors must be added to the extraction mixture. Furthermore, as mechanical cell disruption may cause overheating resulting in denaturation of the enzyme protein, the extract and equipment should be pre-chilled and kept cool. Addition of a sulfhydryl reagent such as 2-mercaptoethanol and dithiothreitol (DTT) and/or chelating agent such as EDTA will prevent enzyme inactivation, because an oxidizing environment can sometimes cause denaturation or aggregation of the protein.

1. Mammal tissues: Most animal proteins are located in specific organs or muscles. The surrounding fatty tissues must be removed. Both organs and tissues should first be homogenized to form a cell paste. Waring blenders, grinders, mincers and mills need to be used for disruption. Animal cells lack rigidity and their relatively large size makes them easy to disrupt. The extract is obtained after centrifugation.

2. Plant tissues: Plants have a rigid cell wall composed of insoluble carbohydrate complexes, lignin or wax, which surrounds the plasma membrane. Cellulosic microfibrils cover the plasma membrane in a random manner and are not easy to disrupt. On the other hand, phenolic compounds, which inactivate proteins, are present in vacuoles and will be released by disruption, in which case prompt extraction is essential.

3. Bacteria: In bacteria, the plasma membrane is surrounded by a cell wall that is rigid and strengthened by peptidoglycan. In Gram-positive bacteria the major component of the cell wall is peptidoglycan, together with teichoic acids, teichuronic acids and other carbohydrates. In contrast, the peptidoglycan layer in the cell wall of Gram-negative bacteria is thick and is surrounded by an outer layer containing a lipopolysaccharide and proteins. Lysozyme is a useful enzyme frequently used to break down peptidoglycan, and it is efficient for disruption of Gram-positive bacteria. Gram-negative bacteria are less susceptible to lysozyme because of the presence of an outer layer. However, their susceptibility can be increased in the presence of EDTA, which destabilizes the lipopolysaccharide structure. If it is used in an isotonic solution, the outer cell wall will be removed and the contents of the periplasmic space are released. It is relatively easy to disrupt bacterial cells by grinder (for coccus) or sonic oscillation (for bacilli).

4. Fungi and yeast: The cell walls of filamentous fungi and yeasts are almost completely composed of polysaccharide. In the filamentous fungi it is mostly composed of chitin linked with glucans, and in yeast it is composed of mannan and glucan. The remainder of the cell wall is composed of protein and lipid. Most fungi and yeast can be disrupted by grinder.

In general, the tissues of higher organisms, mammals and plants are tough. Organs and tissues may be disrupted and simultaneously homogenized in blenders. These techniques are not usually satisfactory for the disruption of microbial cells because of their rigidity. Small volumes of cell paste can be ground in a mortar with alumina or sand to release the intracellular components. Sonication has been employed successfully for the disruption of many microbial cells but coccus is relatively resistant. The release of protein will be almost proportional to the acoustic power input and independent of cell concentration. However, it should not be used for cell pastes greater than about 20% w/v. About 170 watts of acoustic power at 20 kHz is usually required. In some cases, the addition of a detergent (*e.g.*, SDS or Triton) to release membrane-bound enzymes is advantageous. For large volumes, a device such as the Dyno-Mill is preferred. This is a device in which beads are agitated by rapidly rotating discs. The technique is limited to application on a very large scale and is best used for microorganisms that have a strong resistance against the other disruption methods. It is particularly suitable to use for yeast, fungi and coccus such as *Staphylococcus*, *Micrococcus* and *Streptococcus*.

As mechanical methods of disruption mostly generate heat, it is essential to pre-chill the extraction (keep at about 4°C) to prevent heat denaturation of proteins. Once the cell is disrupted, the desired protein may be gradually degraded by intrinsic proteases. Then, prompt extraction, addition of inhibitors and cooling to reduce their effects are necessary. Most commonly, 50 mM phenylmethylsulfonyl fluoride (PMSF) is added to the buffers as the protease inhibitor prior to use. Since disruption is only one of many steps in the isolation of a desired protein, methods which facilitate subsequent purification steps should be chosen.

For an extracellular enzyme, such as one secreted into culture broth or present in serum or urine, cells and other particles must be removed to prevent interference in the purification steps and separated by centrifugation to obtain the supernatant. Most microbial endoglycosidases are extracellular enzymes found in the culture fluid of microorganisms.

1.3.2 Salting Out and Dialysis

In the steps of the purification of an enzyme, the precipitation method is used for concentrating proteins prior to a subsequent purification step. This method is also used for a rough separation of protein during the early stages of a purification procedure and is followed by chromatographic separation.

The solubility of a protein molecule in an aqueous solvent is determined by the distribution of charged hydrophilic and hydrophobic groups on its surface. The charged groups on the surface interact with ionic groups in the solution. The hydrophobic areas on the protein intereact with water molecules causing an ordered matrix of water molecules to form these areas (Fig. 1.24). Precipitates of protein are formed by aggregation of the protein molecule, which is induced by changing pH or ionic strength, or by the addition of an organic solvent. Temperature will affect the degree of aggregation. Precipitate can be recovered by filtration or centrifugation. To remove the precipitating agent that might interfere with subsequent purification steps, the dissolved precipitate should be dialyzed or desalted by gel filtration.

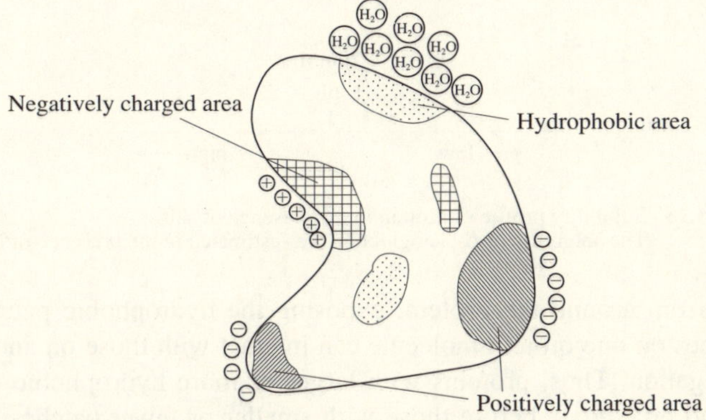

Fig. 1.24 Typical scheme of a protein showing negatively and positively charged areas and interacting with ions in the solution.

One of the easiest methods of precipitating a protein is to adjust the pH of the solution to the isoelectric point (pI) of the protein. Above the pI of the protein the surface is predominantly negatively charged, and charged molecules will be repelled from one another. On the other hand, below the pI the overall charge of the protein molecule will be positive and molecules will repel one another. However, at the pI of the protein the negative and positive charges on the surface of a protein molecule are canceled, and electrostatic attraction between molecules may occur and result in the formation of a precipitate.

Precipitation by the addition of neutral salts is probably the most common method used for fractionating protein. Fig. 1.25 shows the solubility of a protein in the presence of NaCl. At a pH nearly equal to the pI of a protein, the solubility decreases with increase in the concentration of salt added. The precipitated protein is usually not denatured and its activity is recovered by dissolving the pellet. These neutral salts can also stabilize proteins against denaturation, proteolysis or microbial contamination. Such precipitation by increasing the ionic strength is termed salting-out. The salting-out step is a good stage for storing the extract of the enzyme for a long time. Salting-out depends on the hydrophobicity of the surface of the protein, which is different from isoelectric precipitation. Many hydrophobic groups exist inside the protein, but some are located at the surface, in patches. Water is forced into contact with these groups (Fig. 1.24). When salts are added, water dissolves the salt ions, and as the salt concentration increases water

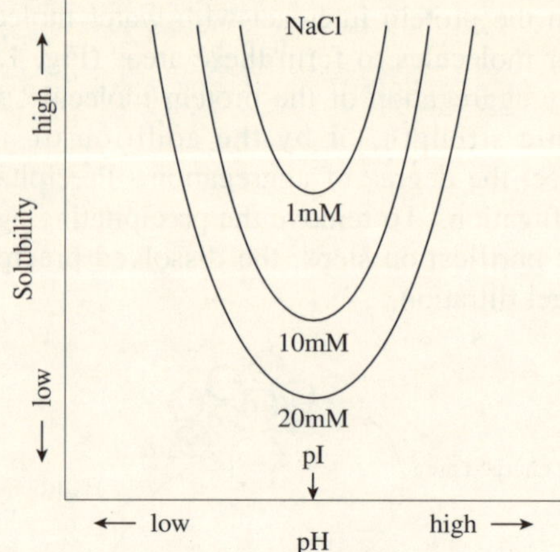

Fig. 1.25 Solubility profile of protein in the presence of salt.
The solubility of β-lactoglobulin was estimated in the presence of NaCl.

is removed from around the protein, exposing the hydrophobic patches. Hydrophobic patches on one protein molecule can interact with those on another, resulting in aggregation. Thus, proteins with larger or more hydrophobic patches will aggregate and precipitate before those with smaller or fewer patches, resulting in fractionation. Increasing the temperature increases the amount of precipitation. However, salting-out is usually performed at 4°C to decrease the risk of inactivation of sample protein.

The effectiveness of the salt is mainly determined by the nature of the anion, and phosphate is the most effective of various salts such as sulfate, acetate and chloride. However, in practice, ammonium sulfate is the most commonly used salt because it is cheap and sufficiently soluble. A saturated ammonium sulfate solution in water is approximately 4 M. The ammonium sulfate concentration is usually quoted as percent saturation, assuming that the solution can be dissolved with the same amount (as percent saturation) of ammonium sulfate. To calculate the grams of ammonium sulfate to add to one liter of solution at 20°C to give a desired percentage, the following equation is used:

$$\text{Ammonium sulfate, g} = 533\ (S_2 - S_1)\ /\ 100 - 0.3\ S_2$$

where S_1 is the starting percentage, and S_2 the final percentage. This equation allows for the increase of volume that occurs on addition of the salt.

Ammonium sulfate will slightly acidify the enzyme solution; therefore diluted ammonia water should be used to maintain the pH at around 7.0. The temperature is also important for precipitation by ammonium sulfate: the higher the temperature the lower the solubility of the protein. Slow addition of the ammonium sulfate which is ground into pieces and moderate stirring are also important. The resulting precipitate is usually collected by centrifugation. The pellet should be dissolved in a small volume of buffer and denatured proteins should be removed by centrifugation.

It is also necessary to remove salts after the precipitation step in order to proceed to the next step. This is achieved by various methods including dialysis, ultrafiltration and gel filtration. 1: Dialysis is carried out as follows. The protein solution is put in a bag of semi-permeable membrane and placed in the required buffer. Small molecules can pass through the membrane while large molecules are retained. The semi-permeable dialysis tube is usually made of cellulose acetate with pores 1-20 nm in diameter (the size of these pores is determined by the minimum molecular weight of the protein molecules required). Dialysis is often carried out overnight, usually at 4°C. 2: An alternative method for desalting is ultrafiltration. This method is suitable for large-scale application. For this ultrafiltration, various equipment is used. First, water or buffer is added to the protein solution, then concentrated by any ultrafiltration equipment: this process is repeated until the ionic strength of the filtrate reaches that of the added buffer or water. 3: Another alternative method is gel filtration. This method is only applicable to small volumes. The maximum volume of sample should not exceed 25-30% of the column volume for adequate resolution. Otherwise the protein cannot be separated from the salts. This method results in dilution of the sample.

1.3.3 Gel Filtration

Gel filtration is a partition chromatography used for separating molecules of different sizes. The basic principle of gel filtration is that molecules are partitioned between solvent and a stationary phase of defined porosity gel. The separation is carried out using a porous gel matrix (in bead form) packed in a column and surrounded by solvent. When a sample containing a mixture of molecules which are smaller and larger than the pores of the matrix is eluted on the column packed with a porous gel matrix, the smaller molecules can enter the matrix pores and move slowly through the column, and the larger molecules are excluded from the porous gel matrix and eluted fast from the column (Fig. 1.26). Therefore, all the molecules are eluted in order of decreasing size. Generally, long and thin proteins elute faster than globular proteins of the same size.

The supports most commonly used are gels consisting of cross-linked polyacrylamide, agarose and dextran. In choosing the most suitable gel matrix for an application, several factors should be considered. If all proteins in a mixture must be separated from any compound having low molecular weight, the proteins must be excluded from the pore matrix. Suitable media include Sephadex G-25 or G-50 and Bio-Gel P-6 or P-10. For the resolution of different proteins that are close in size, a matrix of the correct fractionation range must be chosen. Smaller beads (fine and superfine grades) usually give better resolution because molecular diffusion is lower. Usually, an ordinary type of gel should be used. Fig. 1.27 shows the relationship between elution behavior and molecular properties. As shown in this figure, selectivity curves for each of the gels are different, and the choice of gel type depends on the molecular size of the desired protein. Technical bulletins from the gel manufacturers are available for this purpose.

The choice of column dimension is also important. Generally, the longer the

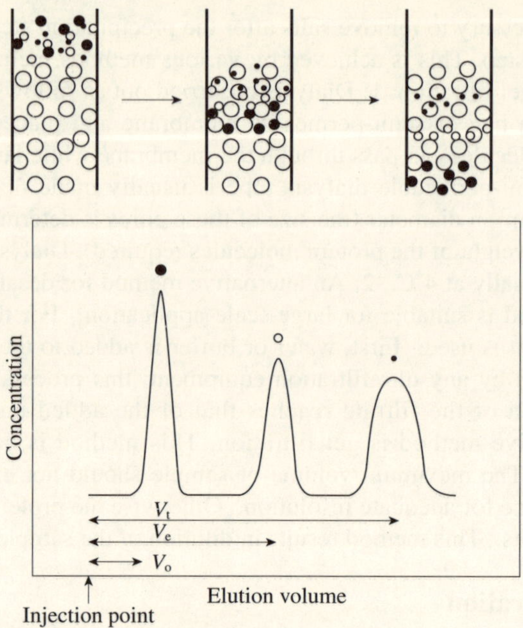

Fig. 1.26 Diagrammatic representation of the principle of gel filtration chromatography and its elution profile.
Upper figures are a diagrammatic representation of gel filtration chromatography. ●, molecules larger than matrix pores; ○, molecules intermediate in size; •, molecules smaller than matrix pores.
Bottom figure is the elution profile of a mixture of molecules of three different sizes.

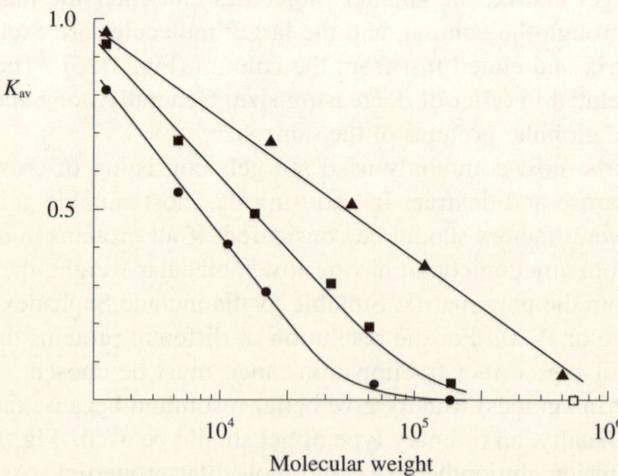

Fig. 1.27 Relationship between elution behavior and molecular properties of protein.
▲, Sephadex G-200; ■, Sephadex G-100; ●, Sephadex G-75.

column, the better the resolution. A column that is up to 100 cm in length with an internal diameter of 1-2.5 cm (20-40 times longer than wider) is usually sufficient for good separation. Typically, the sample volume should be 1-5% of the total bed volume. If the bed consists of hard beads of the same size, the void volume will be approximately 35% of the total bed volume.

The elution volume for a protein depends on both its size and shape. Therefore the gel filtration can be used for determination of the molecular weight of the sample protein. The molecular weight can be expressed in terms of the elution volume from the column. The volume of solvent between the injection point and the maximum peak of the sample protein is known as the elution volume (V_e) (Fig. 1.26). In the case of very large molecules which are excluded from the pore matrix, the elution volume is equal to the void/exclusion volume (V_o). The void volume is usually determined by Blue Dextran. The elution volume of a molecule is dependent on the fraction of pore matrix available for diffusion; this is represented by the constant K_d. Hence,

$$V_e = V_o + K_d\,(V_s)$$

where V_s represents the volume of solvent within the pore matrix. This value is difficult to determine experimentally and is therefore substituted by the term $V_t - V_o$ where V_t represents the total bed volume. Total bed volume can be measured by determining the elution volume of a small molecule, such as acetone. The expression becomes $K_d = (V_e - V_o)/(V_t - V_o)$ where K_d is the coefficient expressing the fraction of gel volume available for diffusion of a given sample protein. A plot of the logarithm of molecular weight versus K_d will give a linear relationship.

1.3.4 Ion-exchange Chromatography

Ion-exchange chromatography is a useful separation method for enzyme proteins based their charges and can be used to resolve enzyme proteins that have different charges. Separation of proteins is achieved by the differences in equilibrium between a buffered mobile phase and stationary phase consisting of a matrix to which charged groups are attached. Ion exchange of protein involves their adsorption to the charged groups of a solid support followed by their elution in a buffer of higher ionic strength (Fig. 1.28). This method is widely applicable and low in cost compared with other enzyme separation methods.

Proteins carry both positively and negatively charged groups on their surfaces (Figs.1.24 and 1.28). These are mostly the side chains of acidic and basic amino acids. Positive charges are derived from histidine, lysine, arginine and the N-terminal amino groups. Negative groups are due to aspartic and glutamic acids, C-terminal carboxyl groups and cysteine residues. The net charge on a protein depends on the relative numbers of positively and negatively charged groups, and this varies with pH. The pH at which a protein has an equal number of positively and negatively charged groups is termed the isoelectric point (pI). Most proteins have a pI of between pH 5 and 9. Above their pI proteins have a net negative charge (use anion exchanger) while below it their overall charge is positive (use

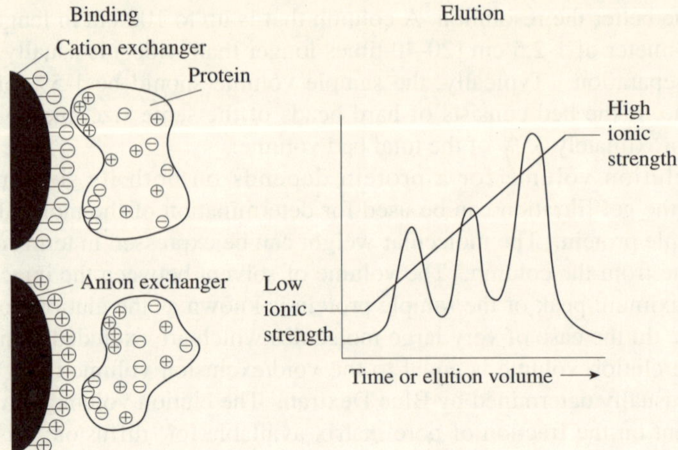

Fig. 1.28 Simple mechanism of binding and elution of protein in ion-exchange chromatography.

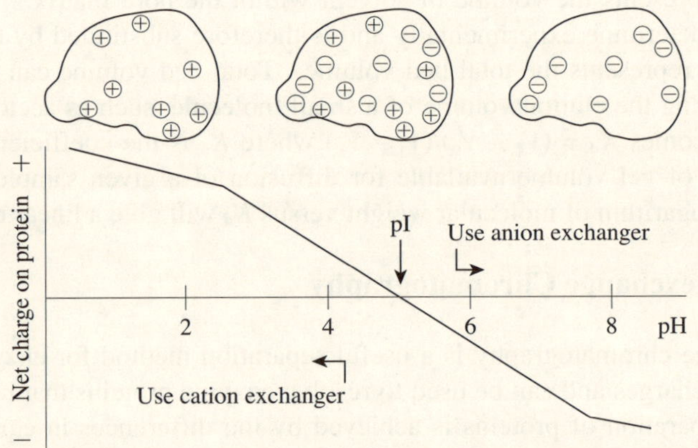

Fig. 1.29 Typical scheme of net charge on a protein.

cation exchanger) (Fig. 1.29).

For the effective use of ion-exchange chromatography in enzyme purification, the stationary phase must be capable of binding either positively-charged or negatively-charged proteins. For this purpose ion-exchange matrices are modified or derivatized with positively charged groups for the adsorption of anionic proteins (anion exchangers) or negatively charged groups for the adsorption of cationic proteins (cation exchangers). In order to bind to the stationary phase, the counter-ions of both groups must become electrolytically dissociated. Counter-ions interrupt the exchanger groups, preventing their binding with a protein. Na^+ and H^+ are the common counter-ions for cation exchangers while Cl^- and OH^- are usually used with anion exchangers (Fig. 1.30). Counter-ions can be arranged according to their strength of interaction with their respective ionogenic groups at equal con-

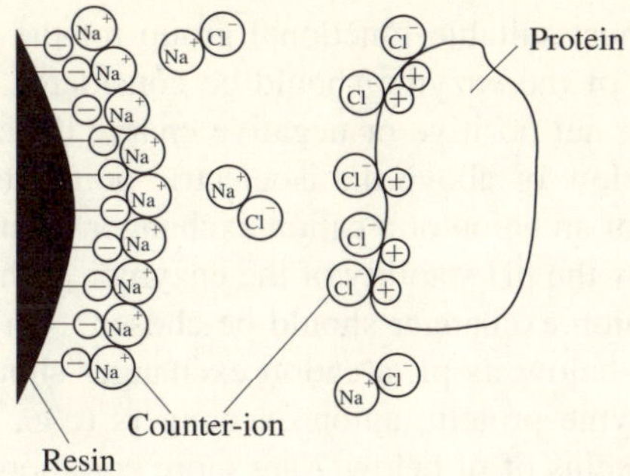

Fig. 1.30 Function of counter-ion on cation-exchange chromatography.

centration.

Cations: $K^+ > NH_4^+ > Na^+ > H^+$

Anions: $NO_3^- > PO_4^- > HSO_3^- > Cl^- > HCO_3^- > HCOO^- > CH_3COO^- > OH^-$

Consequently, chloride will replace hydroxide ions at equal strength as the counter-ion for an anion exchanger.

Ionogenic groups used in both ion exchangers are further divided into strong and weak groups. Strong ionogenic groups are ionized over the wide pH range that is normally used in protein purification, while weak ionogenic groups have a narrower pH range and are only partly ionized. The ionogenic groups used in the ion exchange of proteins are listed in Table 1.6. The most commonly used functionalities are the weak ionogenic groups. The diethylaminoethyl (DEAE) group is usually used in an anion exchanger to purify negatively charged proteins, while the carboxymethyl (CM) group is used in a cation exchanger to purify positively charged proteins. Strong ion exchangers are used based on sulfomethyl and triethylaminoethyl functionalities for strong cation and anion exchanges, respectively.

Table 1.6 Ion-exchange groups used in the purification of proteins

Ion-exchanger	Name	Abbreviation
Strong anion exchanger		
$-CH_2N^+(CH_3)_3$	Trimethylaminoethyl	TAM–
$-C_2H_4N^+(C_2H_5)_3$	Triethylaminoethyl	TEAE–
$-C_2H_4N^+(C_2H_5)_2CH_2CH(OH)CH_3$	Diethyl-2-hydroxypropylaminoethyl	QAE–
Weak anion exchanger		
$-C_2H_4N^+H_3$	Aminoethyl	AE–
$-C_2H_4N^+H(C_2H_5)_2$	Diethylaminoethyl	DEAE–
Strong cation exchanger		
$-SO_3^-$	Sulfo	S–
$-CH_2SO_3^-$	Sulfomethyl	SM–
$-C_3H_6SO_3^-$	Sulfopropyl	SP–
Weak cation exchanger		
$-COO^-$	Carboxy	C–
$-CH_2COO^-$	Carboxymethyl	CM–

In selecting the most suitable functional group for the purification of an enzyme the pH stability of the enzyme should be considered. Enzyme proteins are amphoteric and have a net positive or negative charge depending on whether the pH of the buffer is below or above the isoelectric point, respectively (Fig.1.29). Therefore, the choice of an anion or a cation exchanger must be made. In fact, the decision is restricted by the pH stability of the enzyme. If the enzyme is more stable above its pI, an anion exchanger should be chosen. On the other hand, if the protein is more stable below its pI, a cation exchanger should be chosen. In the purification of an enzyme protein, anion exchangers (*e.g.*, DEAE) are most frequently used since proteins of pI below 7 are more common. With strong ion exchangers their ionogenic groups are all charged over a wide pH range. However, weak ion exchangers are normally chosen for the ion exchange of proteins of between pI 6 and 9. Since weak cation exchangers lose their charge below pH 6 and weak anion exchangers above pH 9, strong ion exchangers may be used for proteins having pI outside these values.

The mobile phase used in ion-exchange chomatography is aqueous since the electrolyte properties of water contribute to the dissociation of the ionogenic groups and matrix. In an aqueous solution the surface charge groups of a protein are associated with counter-ions in a way similar to the ionogenic groups on the exchanger. Commonly used buffers in the ion exchange of proteins are shown in Table 1.7. There are three elution methods for ion-exchange column chromatography of enzyme purification: the isoclatic elution method, the stepwise elution method and the gradient elution method.

Table 1.7 Buffers commonly used in the ion exchange of proteins

Ion-exchanger	Buffer	pK_a	Buffering range
Cation	Acetate	4.8	4.8 – 5.2
	Citrate	4.8	4.2 – 5.2
	Mes*	6.2	5.5 – 6.7
	Phosphate	7.2	6.7 – 7.6
	Hepes**	7.6	7.6 – 8.2
Anion	L-Histidine	6.1	5.5 – 6.0
	Imidazole	7.0	6.6 – 7.1
	Triethanolamine	7.7	7.3 – 7.7
	Tris***	8.1	7.5 – 8.0
	Diethanolamine	8.8	8.4 – 8.8
	Ammonium	9.3	–

*Mes, 2-(*N*-morpholino)-ethanesulfonic acid;
**Hepes, *N*-2-hydroxyethylpiperazine-*N*′-2-ethanesulfonic acid;
***Tris, tris(hydroxymethyl)aminomethane.

1. The isoclatic elution method: The elution is carried out using a single solution. The sample volume should be between 1% and 5% of the bed volume. A long column is used with a diameter-to-length ratio of around 1:20.
2. The stepwise elution method: This method is carried out using a sequential, discontinuous change in pH and salt concentration. The sample should adsorb to 5-

10% of the total bed capacity. The column is usually shorter (20-40 cm).
3. The gradient elution method: The composition of the eluent (pH or ionic strength) is changed continuously. The sample protein should be 5-10% of the bed capacity. The column is usually 20-40 cm with a diameter-to-length ratio of less than 1:5. The total volume of eluent should be about five times the bed volume. If the gradient is too steep, the resolution will be not good, but a too shallow gradient will result in dilution and take a long time. An increasing pH gradient can be used for cation exchangers while a decreasing pH gradient is used for anion exchangers.

1.3.5 Affinity Chromatography

Proteins contain binding sites for interaction with other biomolecules called ligands. Ligands may be small molecules such as substrates for enzymes or larger molecules such as peptide hormones. The interaction of a binding site of protein with a ligand is determined by the overall size and shape of the ligand as well as by the number and distribution of the complementary surfaces of proteins. The complementary surfaces may involve a combination of charged and hydrophobic moieties and exhibit molecular interactions such as hydrogen bonding. This binding activity of a protein, which is stereoselective and of high affinity, can be used for the purification of the protein in a technique known as affinity chromatography. Affinity chromatography is the most powerful technique for purification since its high selectivity can purify a single protein from a crude mixture of proteins even with a low amount of the protein. If the affinity of the ligand for the protein is sufficiently high, the technique offers simultaneously the concentration of a protein from a large volume of solution. The principle of this chromatography is shown in Fig. 1.31.

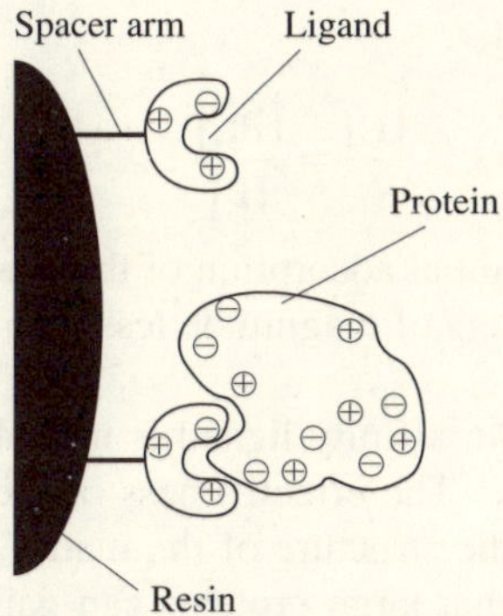

Fig. 1.31 Simple principle of binding of protein in affinity chromatography.

The construction of an affinity adsorbent for the purification of a particular protein involves three major factors. These are choice of a suitable ligand, selection of a support matrix and spacer, and attachment of the ligand to the support

matrix.
1. Choice of a ligand: Suitable pairs of protein and ligand combinations for affinity chromatography are, for example, antigen-antibody, hormone-receptor, glycoprotein-lectin or enzyme-substrate/cofactor/effector. In practice, the natural biological ligand may be difficult to obtain, in which case analogues or mimics are preferentially employed. The following factors must be considered in selecting a ligand for enzyme purification.
(i) Specificity of the enzyme: Ideally, the ligand should recognize only the desired enzyme protein to be purified. The choice may be made from the known biological properties of the enzyme. If this is not possible a ligand known to recognize a group of enzymes should be selected (*e.g.*, nucleotide for nucleotide-requiring enzymes and protease inhibitor for proteases).
(ii) Reversibility: The ligand should form a reversible complex with the enzyme. The complex should be resistant to the washing buffer but is easily dissociable without denaturing the enzyme.
(iii) Stability: The ligand should be stable under the conditions to be used for immobilization as well as the conditions of use (this includes resistance to proteolysis and to denaturation by eluents).
(iv) Size: The ligand should be large enough because it contains several groups that are able to interact with the protein resulting in sufficient stereoselectivity and affinity.
(v) Affinity: The interaction of a protein P and a ligand L can be described by the equation:

$$\mathrm{PL} \rightleftarrows \mathrm{P} + \mathrm{L}$$

where the equilibrium dissociation constant, K_L, is defined by:

$$K_{\mathrm{L}} = \frac{[\mathrm{P}][\mathrm{L}]}{[\mathrm{PL}]}$$

A simple rearrangement gives:

$$\frac{[\mathrm{L}]}{K_{\mathrm{L}}} = \frac{[\mathrm{PL}]}{[\mathrm{P}]}$$

which implies that for substantial adsorption of the protein from solution the value of K_L must be about two orders of magnitude less than the concentration of immobilized ligand.
2. Selection of the matrix: An affinity ligand is immobilized to a solid matrix via one or more covalent bonds. The effectiveness of the immobilized ligand in purifications is dependent on the structure of the matrix. The matrix should have a high degree of porosity so that large proteins can gain access to an immobilized ligand, and should also be chemically stable under the conditions used for coupling, operation, regeneration and a range of pH and temperature in the enzyme reaction. Its properties should not be substantially altered on functionalization. Agarose or cross-linked agarose is one of the better matrix materials advisable for use.
3. Choice of spacer: A spacer molecule may be employed to keep the ligand a suf-

ficient distance from the matrix backbone. It may also be effective if the ligand is immobilized through a site near enough to the protein-binding surface to interfere with protein binding. The spacer should be hydrophilic and not bind proteins itself . The methylene groups are usually used and the number most often successful is 6 - 8. Although cyanogen bromide (CNBr) is extremely toxic, CNBr activation is one of the most widely used processes in the preparation of affinity adsorbents. It reacts with hydroxyls in agarose matrix to produce a support which can subsequently be derivatized with spacer molecules or ligands containing primary amines. Commercially activated agarose is recommended and the details of coupling procedures are provided in manufacturer's instructions.

If the desired enzyme is a glycoprotein, lectin affinity chromatography is a very useful means of purification.[50)] Lectins are a family of carbohydrate binding proteins that are produced by slime molds, plants and animals. They are recognized for their ability to agglutinate human erythrocytes of different types (A, B and O) by binding to specific receptors on their surface and are often called agglutinins. The specificity of lectins was determined by the ability of particular sugars to inhibit these biological responses. Mostly, the *N*-linked sugar chain is bound covalently to the enzyme protein. This group can be divided on the basis of oligosaccharide composition into high-mannose type, complex type and hybrid type [51)] (see section 1.1.2). For example, castor bean agglutinin (RCA), a galactose-binding lectin, primarily binds to complex type and hybrid type oligosaccharides in glycoproteins. Concanavalin A (Con A) mainly binds to high-mannose type oligosaccharide in glycoproteins. A number of the more common lectins can be purchased immobilized on agarose from commercial sources. Some important structural characteristics of the common lectins are summarized in Table 1.8. Various eluents can be applied either stepwise or by gradient. Most glycoproteins are eluted with the sugar or an analogue for which the lectin has an affinity. For example, glycoproteins binding to Con A are frequently eluted with methyl α-D-mannoside (0-0.5 M) in equilibration buffer and those binding to soybean agglutinin eluted with *N*-acetyl-D-galactosamine (0-0.5 M) in equilibration buffer.

1.3.6 Other Chromatographic Techniques

1. Hydroxylapatite: Calcium phosphate gels have been used frequently in protein purification. Crystalline hydroxylapatite [$Ca_{10}(PO_4)_6(OH)_2$] is prepared by slowly mixing calcium chloride and sodium phosphate to form a precipitate of brushite ($Ca_2HPO_4 \cdot 2H_2O$). The brushite is then boiled with sodium hydroxide to convert it to hydroxylapatite. The mechanism of protein adsorption onto hydroxylapatite is thought to involve both Ca^{2+} and PO_4^{3-} groups on the crystal surface.[52)] As these charged groups are closely arranged on the crystal it is likely that dipole-dipole interactions exist between adsorbent and protein, although purely electrostatic interactions cannot be ruled out. Since hydroxylapatite adsorbs proteins by a mechanism different from other separation techniques, it is a useful additional chromatographic procedure which can be used to separate protein mixtures that cannot be achieved by alternative methods such as ion exchange and hydrophobic interac-

Table 1.8 Biochemical characteristics and specificities of various lectins

Lectin (abbreviation)	Sugar specificity	Metal ions required
Concanavalin A (Con A)	α-D-Man > α-D-Glc > α-D-GlcNAc	Ca^{2+} Mn^{2+}
Lentil lectin (LCA)	α-D-Man > α-D-Glc > α-D-GlcNAc	
Soybean lectin (SBA)	α-D-GalNAc > β-D-GalNAc > α-D-Gal	Ca^{2+} Mn^{2+}
Castor bean lectins		
(RCA60)	D-GalNAc	
(RCA120)	β-D-Gal > D-Gal	
Wheat germ agglutinin (WGA)	(β-D-GlcNAc)$_3$ > (β-D-GlcNAc)$_2$	Ca^{2+} Mn^{2+} Zn^{2+}
Peanut agglutinin (PNA)	β-D-Gal-(1-3)-D-GalNAc > D-GalNH$_2$ = α-D-Gal	
Pea lectin	α-D-Man	
Phytohemagglutinin (PHA)	D-GalNAc	Ca^{2+} Mn^{2+}
Bandeirea simplicifolia		
BS-I	α-D-Gal	Ca^{2+} Mg^{2+}
BS-II	α-D-GlcNAc	Ca^{2+} Mg^{2+}
Ulex europeus agglutinin (UEA-I)	α-L-Fucose	
Dolichos biflorus agglutinin (DBA)	α-D-GalNAc	
Jacalin	α-D-Gal	
Helix pomatia lectin	α-D-Gal >> α-D-GlcNAc > α-D-GalNAc	

tion chromatography. The adsorption of proteins can be carried out using batch or column method. In the case of large volume samples the former is preferred. Adsorption is carried out in a low concentration of sodium or phosphate buffer (< 20 mM) at a neutral pH. The hydroxylapatite column is washed in a low concentration of phosphate buffer (pH 6.5-7.0) before application of the sample. Typical flow rates are 10-25 mL cm^{-2} h^{-1}. Protein elution is usually achieved with a stepwise or continuous gradient of increasing concentration up to 500 mM. Potassium phosphate is preferred to sodium phosphate at higher concentrations at 4°C.

2. Hydrophobic interaction chromatography: The simple tertiary structure of protein shows a hydrophobic core surrounded by a hydrophilic outer part. However, surface hydrophobicity is caused by the presence of nonpolar amino acids such as alanine, methionine, tryptophan and phenylalanine at the surface of the protein.[53) It seems that the hydrophobicity of the surface of the protein not only contributes to stabilize protein conformation but also forms the basis of specific interactions connected with the biological function of the protein. The surface hydrophobic amino acids are usually arranged in patches. The number, size and distribution of these non-ionic regions is a characteristic of each protein and can be used for its separation. The principle of this chromatography is indicated in Fig. 1.32. It can be used directly after precipitation of proteins because the ionic strength of the protein will enhance hydrophobic interaction. In purification the required enzyme is eluted in a gradient of decreasing ionic strength. The choice of hydrophobic ligand is important for ease of elution. The most commonly used non-ionic adsorbents are based on agarose, and octyl or phenyl functionalities are bound to it. Phenyl Sepharose is less hydrophobic than Octyl Sepharose. Extremely hydrophobic enzymes (*e.g.*, membrane-bound enzyme) may adsorb too strongly to Octyl Sepharose and require very strong eluting conditions. Thus Phenyl Sepharose should be used. In contrast, mildly hydrophobic enzymes may not adsorb to Phenyl Sepharose and will require the use of more hydrophobic matrices. The ionic strength of elution will also change the protein hydrophobicity and therefore its degree of adsorption. Mildly hydrophobic proteins may not be adsorbed until the ionic strength is increased to below that required for precipitation. Prior to the chromatography, the ionic strength of the sample should be adjusted with salt, and buffered preferably near the pI of the required enzyme to enhance non-ionic adsorption. The packed column should be similarly equilibrated. The sample is then applied to the matrix and washed with the same buffer. The capacities of hydrophobic matrices are usually as high as those of the ion exchangers.

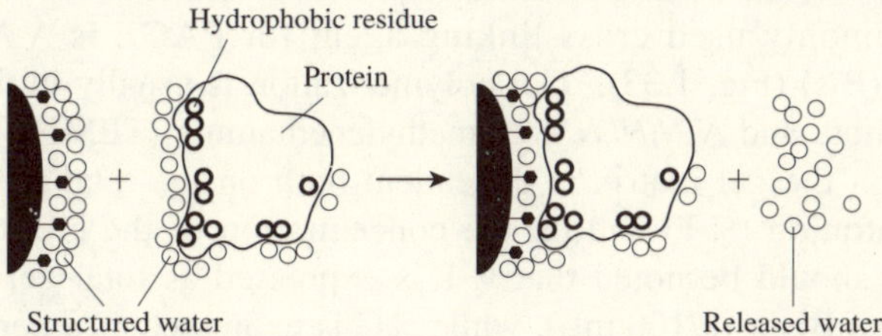

Fig. 1.32 Principle of hydrophobic interaction chromatography.

3. HPLC and FPLC: HPLC gives high resolution in a very short time (usually within 30 min). It uses small rigid uniform beads (10-40 μm diameter) to give higher separation of proteins. However, only small amounts (1-10 mg) can be loaded onto the columns, hence this is mostly used as an analytical technique. Silica-based packings are unstable above pH 7.5, so alkaline buffers should not be used. The FPLC system can be used for large-scale purification of proteins. The

systems will operate under moderate pressure, giving higher speed and resolution than conventional chromatography. The hydrophobic matrices are used in this system. The packings are based on 10-μm agarose particles for use at flow rates of 150 mL cm^{-2} h^{-1} at a pressure of up to 30,390 hPa (30 atm). The matrix has a capacity of 10 mg/mL, allowing purification of up to 80 mg of protein at a time.

1.3.7 Gel Electrophoresis

Proteins are charged at pHs other than their pIs and will migrate in an electric field according to their net charges. Electrophoresis is an ideal analytical technique to resolve and separate the individual components of protein mixtures. The separation occurs in a support medium to counteract the effects of convection and diffusion that occur during electrophoresis. A variety of matrices including starch, agarose and cellulose acetate can be used for electrophoresis. However, polyacrylamide gel electrophoresis (PAGE) is the best choice for most applications because of its high resolution. The electrophoretic techniques are available for separation of proteins on the basis of a combination of their three major properties: size, net charge and relative hydrophobicity. Electrophoresis is often used to analyze soluble proteins while retaining their enzymatic and biological properties. An important procedure is PAGE in the presence of the anionic detergent, sodium dodecyl sulfate (SDS), by which proteins can be characterized depending on the molecular size of their constituent polypeptides. PAGE is an indispensable analytical procedure for determining protein purity. It is also routinely used for monitoring the progress of a purification procedure and for the identification of fractions containing the desired protein. In addition, the electrophoretic method can be used for characterization of purified proteins in terms of microheterogeneity, degradation and subunit structure.

Polyacrylamide gels are formed by the copolymerization of acrylamide monomers with a cross-linking agent to form a three-dimensional lattice. The best and most commonly used cross-linking agent for PAGE is *N,N'*-methylenebisacrylamide (Bis) (Fig. 1.33). Gel polymerization is usually initiated with ammonium persulfate and *N,N,N',N'*-tetramethylenediamine (TEMED). The size of the pores within the gel matrix is dependent both on the total monomer (acrylamide) concentration (%T) and on the concentration of the cross-linking agent (Bis) (%C). It should be noted that %T is expressed as total gel concentration (acrylamide plus Bis as g/100 mL), while %C is expressed as a percentage of the total gel concentration (*i.e.*, g of Bis/mL). Pore size can be increased by reduction of %T at a fixed %C. The alternative approach is to increase %C at a fixed %T. Consequently, the increase in pore size is thought to be due to the formation of a bead-like structure within the gel rather than a three-dimensional network.

Theoretically, there is an optimal gel concentration for the resolution of a desired protein. But there is almost no gel concentration which will give the best separation of a more complex protein mixture. Therefore, gels containing a linear or nonlinear concentration gradient of polyacrylamide are often used. The pore radius of these gels decreases with increasing gel concentration, so effective sepa-

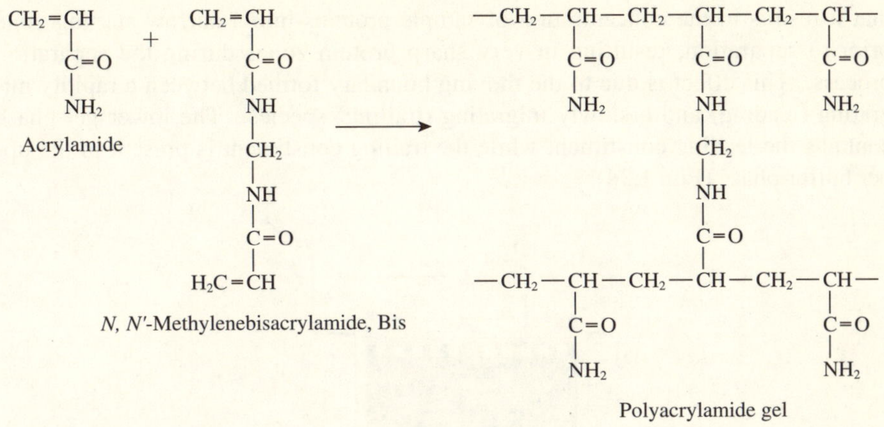

$CH_2{=}CH{-}C({=}O){-}NH_2$ + $CH_2{=}CH{-}C({=}O){-}NH{-}CH_2{-}NH{-}C({=}O){-}CH{=}CH_2$ ⟶ Polyacrylamide gel

Acrylamide

N, N'-Methylenebisacrylamide, Bis

Polyacrylamide gel

Fig. 1.33 Copolymerization of acrylamide with *N,N'*-methylenebisacrylamide to form polyacrylamide gel.

ration of proteins with a wider range of molecular masses can be carried out by this electrophoresis. The resulting separation profiles are stable and highly resolved. Most routine separation is carried out by linear polyacrylamide gradients. However, it may be necessary to use more complicated gradient shapes for a particularly difficult separation of proteins. Therefore, various procedures have been developed for the preparation of gradient polyacrylamide gels. The simplest procedure is to use a two-chamber gradient-forming apparatus.

The anionic detergent, SDS, is a very effective solubilizing agent for a wide range of proteins. Most proteins bind 1.4 g of SDS per 1 g of protein,[54)] effectively masking the intrinsic charge of the polypeptide chains, so that net charge per mass becomes approximately constant. Subsequent electrophoretic separation is dependent only on the molecular radius of the protein, which is roughly approximate to molecular size, and occurs as a result of molecular sieving through the gel matrix. The concentration of polyacrylamide gel determines the effective separation range of SDS-PAGE. For example, a 5%T gel will separate proteins in the range of 20-350 kDa, while a 10%T gel is suitable for proteins in the range of 15-200 kDa. If the sample to be analyzed contains proteins with a wide range of molecular weights, it is advantageous to use a gradient SDS-PAGE system. For example, a 3-30%T gradient gel can separate proteins in the range of 10 kDa to nearly 500 kDa. However, small polypeptides and peptides (<10 kDa) cannot be separated by SDS-PAGE.

SDS-PAGE is the most popular technique of electrophoretic separation for the majority of protein samples. For optimal reaction with SDS, samples should be boiled for 3-5 min in the presence of 2-5% SDS and reagents such as 0.1 M 2-mercaptoethanol or 20 mM dithiothreitol to cleave disulfide bonds. The most commonly used buffer system for SDS-PAGE is that devised by Laemmli[55)] based on the discontinuous buffer system of Davis[56)] with the addition of 0.1% SDS. The advantage of this buffer system compared with a homogeneous buffer system is that it results in the concentration of sample proteins into a narrow starting zone prior to separation, resulting in very sharp protein zones during the separation process. This effect is due to the moving boundary formed between a rapidly migrating (leading) and a slowly migrating (trailing) species. The lower gel phase contains the leading constituent while the trailing constituent is present in the upper buffer phase (Fig. 1.34).

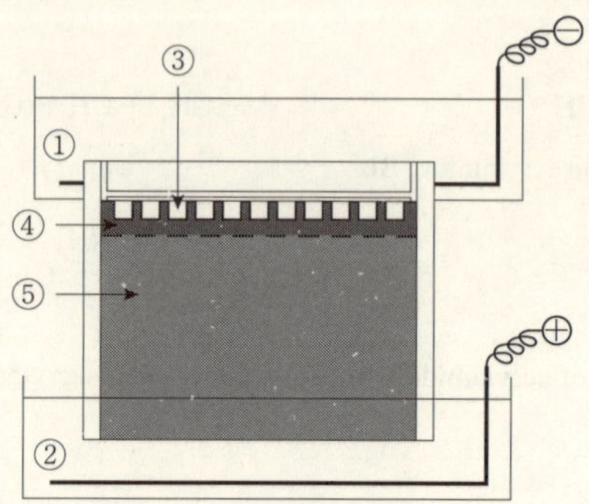

Fig. 1.34 Diagram of the apparatus of SDS-PAGE.
① Cathode chamber including 0.025 M Tris-0.19 M glycine (pH 8.3)-0.1% SDS, ② anode chamber including 0.025 M Tris-0.19 M glycine (pH 8.3)-0.1% SDS, ③ sample well including 0.063 M Tris-0.06 M HCl (pH 6.8)-2% SDS, ④ stacking gel including 0.063 M Tris-0.06 M HCl (pH 6.8)-0.1% SDS, ⑤ separating gel including 0.38 M Tris-0.06 M HCl (pH8.9)-0.1% SDS

PAGE of proteins under denaturing conditions such as SDS-PAGE results in the complete loss of biological activity of the proteins being separated. Therefore, electrophoresis under native conditions is used to analyze soluble proteins because of the advantage of retention of biological and enzymatic properties. To estimate the characterization of protein purity, both electrophoretic methods of native PAGE and SDS-PAGE should be done.

The most general protein staining methods after electrophoresis are based on the use of the nonpolar, sulfated triphenylamine dye Coomassie Brilliant Blue R–250 (CBB). Staining is usually carried out using 0.1% CBB in 45% methanol, 45% distilled water and 10% acetic acid. The time required for staining depends on the thickness of the gel and its polyacrylamide concentration. A 10%T gel of 0.5-1 mm thickness should be stained for about 2 h. Destaining is accomplished by gentle agitation in the same acid-methanol solution.

1.4 Cloning of Endoglycosidases

Glycoside hydrolases (EC 3.2.1.X) are a remarkably diverse group of enzymes and catalyze the degradation of a huge variety of naturally occurring carbohydrates and glycoconjugates. The methodologies for cloning and sequencing DNA have become simpler yet more powerful, so that today, almost any DNA desired can be cloned. The requirements for a gene cloning laboratory cannot be exactly formulated because they depend greatly on the sources of the genes being cloned. Moreover, new methods are constantly being invented and existing techniques are altered in response to changing needs. This section focuses on the strategy of cloning of one of the endoglycosidases, endo-β-*N*-acetylglucosaminidase, from *Arthrobacter protophormiae* (Endo-A), which exhibits strong transglycosylation activity (cf. Chapter 2, section 2.1.1). Methods for expressing large amounts of protein from a cloned gene introduced into *Escherichia coli* have proven invaluable in the purification and functional analysis of glycosidases. The determination of amino acid residues essential for the transglycosylation activity of Endo-A by random and site-directed mutagenesis is also described. Gene cloning and functional analysis of the Endo-A gene is summarized in Fig. 1.35.

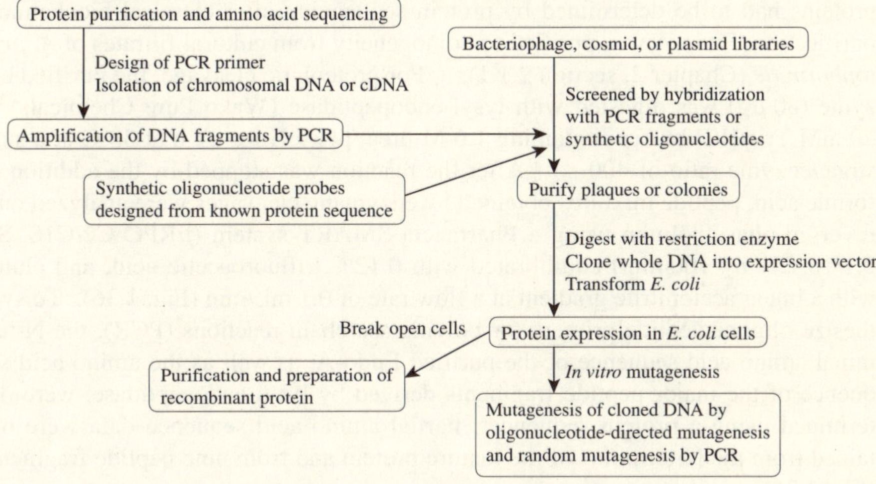

Fig. 1.35 A summary flowchart of cloning and functional analysis of endoglycosidase.

In 1984, the primary amino acid sequence of endo-β-*N*-acetylglucosaminidase from *Streptomyces plicatus* (Endo-H) was reported.[57] The complete amino acid sequences of endo-β-*N*-acetylglucosaminidase from *Flavobacterium meningosepticum* (Endo-F_1, -F_2, and -F_3) and *Flavobacterium* sp. (Endo-Flavo) were also determined by molecular cloning and analysis of the nucleotide sequences,[58-61] and these proteins have similar molecular weight (267-299 amino acids). Recently, the three-dimensional structures of Endo-H, Endo-F_1 and Endo-F_3 have been solved by X-ray crystallography.[62-64] These enzymes have highly conserved amino acid residues, especially the active site residues Asp-130 and Glu-132 in Endo-H.[65] The glutamic acid residue has been identified as the proton donor, and the aspartic acid residue, stabilizing the intermediate in a substrate-assisted hydrolysis mechanism[66] in which the *N*-acetyl group of the substrate acts as the nucleophile.[67] However, these enzymes do not transfer oligosaccharides to GlcNAc-peptides.[68] In contrast, the molecular weight of Endo-A (approximately 72,000) is quite different from that of other endo-β-*N*-acetylglucosaminidases. Endo-A, unlike these other enzymes, is shown to have powerful transglycosylation activity.[69] It is of particular interest to understand why Endo-A has strong transglycosylation activity. To obtain such information, the Endo-A gene was cloned and its nucleotide sequence determined.[70] Random and site-directed mutagenesis coupled with a genetic screen to isolate Endo-A mutants with altered hydrolysis and transglycosylation activity was used.[71]

1.4.1 Peptide Purification and Determination of a Partial Amino Acid Sequence of Endo-A

Before the availability of recombinant DNA techniques, the primary structure of proteins had to be determined by protein sequencing after classical biochemical purification. Endo-A was purified to homogeneity from cultural filtrates of *A. protophormiae* (Chapter 2, section 2.1.1).[72] For proteolytic cleavage, the purified enzyme (80 μg) was digested with lysyl-endopeptidase (Wako Pure Chemicals) in 20 mM Tris/HCl buffer containing 1.6 M urea, pH 9.1, for 21 h at 30°C at a substrate/enzyme ratio of 400 : 1. After the reaction was stopped by the addition of formic acid, peptide mixtures obtained by enzymatic cleavages were analyzed on a reversed-phase column using a Pharmacia SMART system (μRPC C2/C18, SC 2.1/10, 2.1 by 100 mm) equilibrated with 0.12% trifluoroacetic acid, and eluted with a linear acetonitrile gradient at a flow rate of 0.1 mL/min (Fig. 1.36). To synthesize oligonucleotide primers for polymerase chain reactions (PCR), the N-terminal amino acid sequence of the purified Endo-A, as well as the amino acid sequence of the major peptide fragments derived by lysyl-endopeptidase, were determined using a protein sequencer. Partial amino acid sequence data were obtained from the N-terminus of the mature protein and from nine peptide fragments (Fig. 1.36).

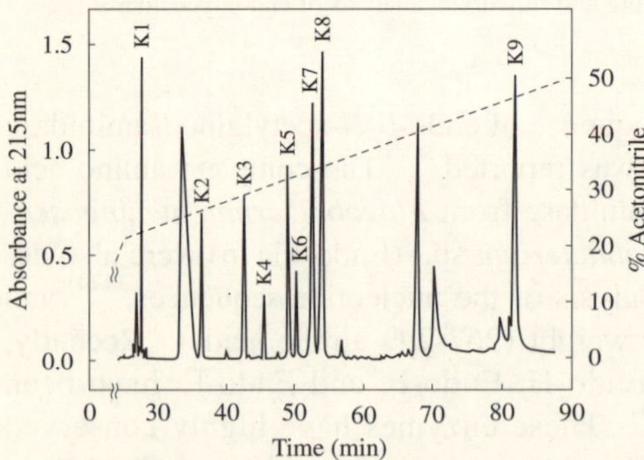

K1. QSNVHHYRVYK
K2. STATSVPFVTHFNTGSGAQFSAEGK
K3. PAAPNVNVRQYDPDPSGIQLVXEK
K4. ELIGTSAGDRIYLEGLVEESK
K5. AHTSLGLYRPDWAFQSSETMEAFYE
K6. QNDVRLHIEALSETFVPSDARMIDIK
K7. STYNGPLSSHXFPEELAQXEPD
K8. XXQFWVGSTGNPAETDGQSNXPG MAHHFPA
K9. DAHLVSLSALNRHTSGVPSQGAP VFYENTFSY

Fig. 1.36 Separation and amino acid sequence of lysyl-endopeptidase peptides. The reverse-phase HPLC column was equilibrated with 0.12% trifluoreacetic acid and developed with a liner acetonitrile gradient at a flow rate of 0.1 mL/min.
[Reprinted from *Arch. Biochem. Biophys.* **338**, Takegawa, K. *et al.*, 24, Copyright (1997), with permission from Elsevier]

1.4.2 Polymerase Chain Reaction and Primer Selection

The polymerase chain reaction (PCR) is a rapid procedure for the *in vitro* enzymatic amplification of a specific segment of DNA. The gene of interest must usually be amplified from genomic DNA or cDNA by PCR before it can be cloned into an expression vector (Fig. 1.37). The first step is the design of the necessary primers. Usually a primer length of 18-30 bases is optimal for most PCR applications. Shorter primers can lead to amplification of nonspecific PCR products. The GC content of a primer should be between 40 and 60%. The 3′-end of the primer molecule is especially critical for the specificity and sensitivity of PCR. The degeneracy of the genetic code for the selected amino acids of the region targeted for amplification must be examined. Although methionine and tryptophan are encoded by a single codon, the other amino acids may be encoded by two to six different triplets (Table 1.9). Selection of amino acids with minimal degeneracy is desired. Therefore, given a choice, the use of peptide regions containing amino acids requiring four or six codons should be avoided.

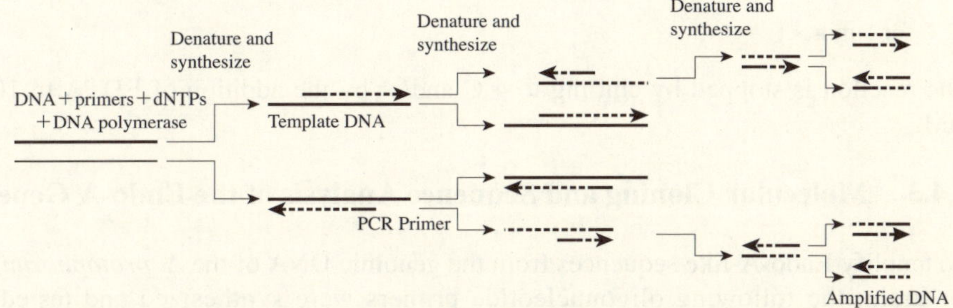

Fig. 1.37 The polymerase chain reaction. In the presence of DNA polymerase and excess dNTPs, oligonucleotides that hybridize specifically to the target sequence can prime new DNA synthesis. The first cycle is characterized by a product of indeterminate length, and the second cycle produces the discrete short product which accumulates exponentially with each successive round of amplification. This process can lead to the many million-fold amplification of the discrete fragment over the course of 25 to 35 cycles.

Table 1.9 Codon degeneracy

Amino Acid	Codons
Met, Trp	1
Cys, Asp, Glu, Phe, His, Lys, Asn, Gln, Tyr	2
Ile	3
Ala, Gly, Pro, Thr, Val	4
Leu, Arg, Ser	6

A typical PCR experiment is carried out under the following conditions:

– Reagents –

Template DNA	100-250 ng
Primer 1	0.2 μM
Primer 2	0.2 μM
dNTPs	200 μM of each dNTP
10 X polymerase buffer	1 X
DNA polymerase	1-2.5 units
Total volume	20-100 μL

– Methods–

92-94°C for 10-30 sec (depending on PCR machine)	25-35 cycles
TM (melting temperature) –5°C for 10-30 sec (adjust according to annealing temperatures of primers)	
72°C for 30 sec (long PCR products (>1 kb) require longer)	

↓

72°C for 5 min

The reaction is stopped by chilling to 4°C and/or by the addition of EDTA to 10 mM.

1.4.3 Molecular Cloning and Sequence Analysis of the Endo-A Gene

To amplify Endo-A-like sequences from the genomic DNA of the *A. protophormiae* strain, the following oligonucleotide primers were synthesized and tested: primer 1, 5′GTTGGATCCTTT (C) CCA (T/G/C) GAA(G)GAA (G) T (C) TA (T/G/C) GCA (T/G/C) CA3′ and primer 2, 5′GTTGAATTCGTA (G) TTA (G) AAA (G) TGA (T/G/C) GTA (T/G/C) ACA (G) AA3′. The primers included restriction sites at their 5′-ends to facilitate cloning of the resulting amplified products (*Bam*HI and *Eco*RI). Primer 1 encodes amino acids FPEELAQ while primer 2 encodes FVTHFNT. The PCR mixture contained 3 μg *A. protophormiae* genomic DNA, 5 mM primers, 50 mM KCl, 10 mM Tris/HCl (pH 8.3), 1.5 mM $MgSO_4$, 0.2 mM dNTPs, and 3 units Takara ExTaq DNA polymerase (Takara Shuzo Co.), in a total volume of 50 μL. PCR was performed in an Eppendorf tube using 35 repetitions of the following temperature cycle: denaturation, 94°C for 60 seconds, annealing, 49°C for 90 seconds, and extension, 72°C for 90 seconds. The reaction product was resolved by electrophoresis in a 1.5% agarose gel. A fragment of approximately 900 bp was recovered and ligated into the pBluescript KS vector.

The complete sequence of the PCR product was determined, and five of the nine peptides were found in the PCR fragment. As this PCR product is therefore part of the Endo-A gene, this gene fragment was used as a probe to recover the Endo-A gene from the *A. protophormiae* genomic library. In order to clone the entire Endo-A gene code, chromosomal libraries were constructed using λEMBL3. *A. protophormiae* chromosomal DNA was isolated and digested with

*Sau*3AI, and fragments of 8 to 15 kb were ligated into *Bam*HI-digested λEMBL3, then packaged into phage particles using the Gigapack III packaging kit (Stratagene), and used to infect the *E. coli* P2393 strain. Screening of the chromosomal phage libraries resulted in the isolation of clones containing the entire Endo-A gene by plaque hybridization. To determine the location of the Endo-A gene within the fragment, various subclones in which *Cla*I, *Pst*I, *Kpn*I and *Hin*dIII digests of the fragment had been ligated at appropriate sites in pBluescript KS cloning vectors were prepared. Finally, the plasmid (4.7-kb *Cla*I-*Pst*I fragment containing the Endo-A gene) was sequenced.[70)]

1.4.4 Gene Probes from Conserved Sequences of Endo-β-*N*-acetylglucosaminidases

With the advent of molecular cloning and the ability to deduce amino acid sequences from gene sequences, the number of new sequences has increased dramatically. This has resulted in a huge database of proteins that can be grouped into families by common structural and functional properties. Endo-β-*N*-acetylglucosaminidases are classified into two glycoside hydrolase (GH) families based on their amino acid sequence homologies (Carbohydrate-Active Enzymes server at URL: http://afmb.cnrs-mrs.fr/CAZY/).[73)] Endo-H, Endo-Flavo and Endo-F belong to the GH family 18 containing chitinases. Recently, Endo-A has been classified into GH family 85 including endo-β-*N*-acetylglucosaminidase from *Mucor hiemalis*,[71)] *Caenorhabditis elegans*,[74)] *Diplococcus* (*Streptococcus*) *pneumoniae*,[75)] *Bacillus halodurans*,[76)] and human.[77)] Furthermore, apparent orthologues for Endo-A protein are found in mammals, various plants, yeasts and intestinal bacteria (Chapter 2 section 2.1.1). Analysis of the sequences of protein within GH family 85 revealed that these proteins include a (FY) (DEH) G (WY) (LF) (ILVF) NXE segment comprising the amino acid residues at positions 165 to 173 in Endo-A (see Fig. 2.5). The glutamic acid residue at position 173 is predicted to be a common catalytic residue in family 85 enzymes on the basis of a computational analysis of glycoside hydrolase families.[78)] This may be an appropriate size to construct an oligonucleotide probe to clone cDNAs encoding endo-β-*N*-acetylglucosaminidases.

1.4.5 Expression and Purification of Endo-A from *Escherichia coli*

The Endo-A fragment was subcloned into pBluescript. Endo-A activity was detected mainly in the periplasmic space of cells. The amount of Endo-A in the periplasmic space increased in the exponential phase and maximized in the early stationary phase. The transformed *E. coli* XL1-Blue strain carrying a recombinant plasmid was grown aerobically in 3% peptone and 1% NaCl at 37°C for 72 h. The cells (a 4-L-cultivation) were centrifuged and extract from the periplasmic space of the cells obtained by the osmotic shock procedure.[79)] This extract was applied to a DEAE-Toyopearl 650 M (2.5 × 10 cm) previously equilibrated with 10 mM phosphate buffer (pH 7.0). The column was washed with the same buffer and the enzyme eluted in 4-mL fractions with a gradient formed between 10 mM phosphate buffer and 0.5 M NaCl in 10 mM phosphate buffer (pH 7.0). The active fractions were combined, and the enzyme solution was applied to a Phenyl-Sepharose CL-4B column (0.8 × 22 cm) equilibrated with 1 M ammonium sulfate in 10 mM phosphate buffer (pH 7.0). The enzyme was eluted in 2-mL fractions with a gradient formed between 1 M ammonium sulfate and 40% ethyleneglycol in 10 mM phosphate buffer (pH 7.0). The active fractions were combined and concentrated by the addition of ammonium sulfate to 90% saturation. Recombinant Endo-A migrated as a single protein band by SDS-PAGE analysis. The recombinant Endo-A was purified to homogeneity by relatively simple procedures, and the recombinant and native Endo-A showed very similar optimum pH profiles and transglycosylation activity.[70)]

1.4.6 Random Mutagenesis of the Endo-A Gene

The classical approach to genetics is to create *in vivo* mutations randomly throughout the genome, then isolate those that display a particular phenotype. It is now possible to create mutations such as deletions, insertions and specific base changes at predetermined sites in a DNA molecule. Oligonucleotide-directed mutagenesis is proving valuable for the study of protein structure and function. If the DNA sequence of the protein coding region of a gene is known, then a synthetic oligonucleotide can be used to specifically change one amino acid to another. However, there are no well-characterized GH family 85 enzymes, and important amino acid residue(s) for the transglycosylation have yet to be identified. Therefore, random and site-directed mutagenesis coupled with a genetic screen to define the amino acid residues critical for the hydrolysis and transglycosylation activity of Endo-A was used. Random mutagenesis of the Endo-A gene was carried out by error-prone PCR amplification (Fig. 1.38).[78)] The template DNA for the mutagenic amplification was the pET-Endo-A plasmid, which contains the Endo-A gene cloned in the pET32a vector. S-Tag primer and T7 terminator primer were used for the amplification. PCR was performed using 39 repetitions of the following temperature cycle: 94˚C for 1 min, 53˚C for 1.5 min, 72˚C for 2 min, and Takara Taq DNA polymerase and 0.2 mM of $MnCl_2$ were added for random mutagenesis. The amplified material was digested with the appropriate restriction enzyme and ligated into a pET32a vector. The ligation mixtures were transformed into *E. coli* BL21 strain. The transformants were characterized by determining protein expression and enzyme activity (Fig. 1.38). The mutagenized inserts were confirmed by DNA sequencing using the Li-Cor 4200 sequencer (ALOKA).

Based on the results of the random mutation analysis, amino acids for site-directed mutagenesis was selected. One of the Endo-A mutants, E173Q, was found to have lost enzyme acitivity, indicating that Glu-173 is essential for the enzymatic function of Endo-A as expected. In the course of characterizing Endo-A mutants, one mutant for which no transglycosylation product could be detected but which maintained substantial hydrolysis activity was isolated. Site-directed mutagenesis revealed that the transglycosylation product was not detected in W216R

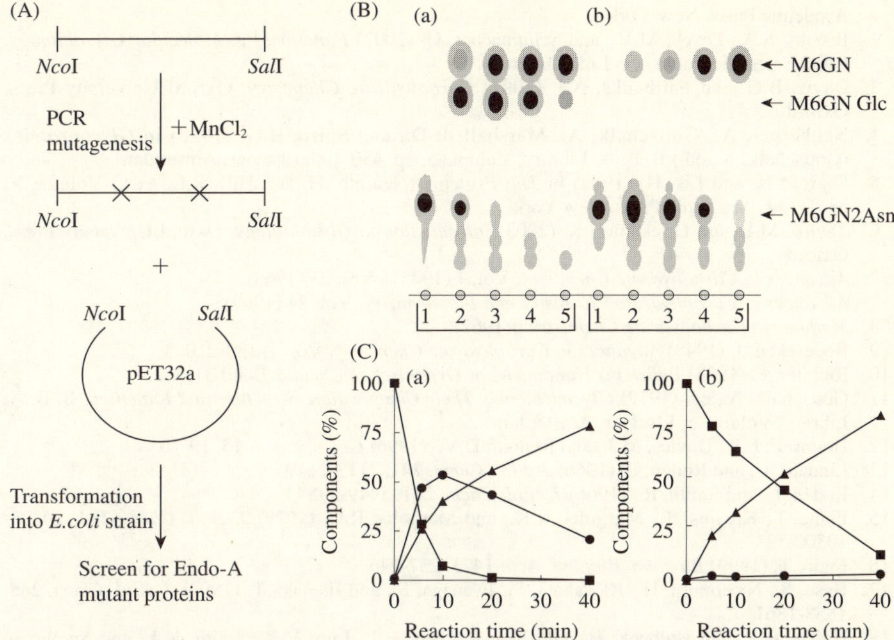

Fig. 1.38 (A) Analysis of mutant of Endo-A gene. Endo-A gene was mutagenized by amplification using PCR performed under error-prone conditions. The pET32a plasmid and the mutagenized DNA were ligated and transformed into *E. coli* BL21 strain. The transformants were screened for hydrolysis and transglycosylation activity. The mutagenized DNA were transformed into *E. coli* strain. The transformants were screened.
(B) TLC analysis of transglycosylation activity of Endo-A W216R mutant. $(Man)_6(GlcNAc)_2Asn$ was incubated with wild-type Endo-A (a) or mutant W216R (b) in the presence of 0.1 M glucose at 37°C for 0 (lane 1), 5 (lane 2), 10 (lane 3), 20 (lane 4), and 40 (lane 5) min.
(C) Relative rates of each spot. The percentage of these spots was evaluated with NIH image software. ▲, hydrolysis product ($Man_6GlcNAc$); ●, transglycosylation product ($Man_6GlcNAc$-Glc); ■, remaining substrate ($Man_6GlcNAc_2Asn$).
[Reprinted from *Biochem. Biophys. Res. Commun.* **283**, Fujita, K. and Takegawa, K., 683, Copyright (2001), with permission from Elsevier]

mutant. The specific activity and K_m value of the W216R mutant was nearly identical to that of the wild-type. Wild-type Endo-A produced transglycosylation products (53.1%) and hydrolysis products (40.1%) for 10 min (Fig. 1.38). However, the W216R mutant did not produce transglycosylation products but did produce hydrolysis products (33.7%) (Fig. 1.38). Mutation to a positively charged residue (W216K) abrogated the transglycosylation activity the same as W216R.[71)] These results indicate that the tryptophan residue at 216 of Endo-A plays a key role in the transglycosylation. The strategy described here should be generally applicable to the cloning and characterization of many desirable glycosidases.

References

1. Pigman, W. and Horton, D., eds. (1972) *The Carbohydrates (Chemistry and Biochemistry)*

Academic Press, New York
2. Brooks, S.A., Dwek, M.V., and Schumacher, U. (2002) *Functional & Molecular Glycobiology*, BIOS Scientific Publishers Ltd., Oxford
3. Davis, B.G. and Fairbanks, A.J. (2002) *Carbohydrate Chemistry*, Oxford University Press, Oxford
4. Neuberger, A., Gottschalk, A., Marshall, R.D., and Spiro, R.G. (1972) in *Glycoproteins* (Gottschalk, A., ed.) B. B. A. Library, Volume 5, pp. 450-490, Elsevier, Amsterdam
5. Sharon, N. and Lis, H. (1982) in *The Proteins* (Neurath, H. and Hill, R.L., eds.) Volume V, pp. 1-144, Academic Press, New York
6. Taylor, M.E. and Drickamer, K. (2003) *Introduction to Glycobiology*, Oxford University Press, Oxford
7. *Advances in Carbohydrate Chemistry*, Vol. 1 (1945) - Vol. 23 (1968)
Advances in Carbohydrate Chemistry and Biochemistry, Vol. 24 (1969) -
8. *Methods in Carbohydrate Chemistry* I (1962) -
9. Boeeseken, J. (1949) *Advances in Carbohydrate Chemistry*, Vol. 4, 189-210
10. Richiter, F. (1938) *Beilsteins Handbuch der Organischen Chemie*, Band 31, 1-2
11. Gottschalk, A., ed. (1972) *Glycoproteins. Their Composition, Structure and Function*. B. B. A. Library, Volume 5, Elsevier, Amsterdam
12. Hounsell, E.F., Davies, M.J., and Renouf, D.V. (1996) *Glycoconj. J.* **13**, 19-26
13. Lindahl, U. and Rodén, L. (1966) *J. Biol. Chem.* **241**, 2113-2119
14. Rodén, L. and Smith, R. (1966) *J. Biol. Chem.* **241**, 5949-5954
15. Finne, J., Krusius, T., Margolis, R.K., and Margolis, R.U. (1979) *J. Biol. Chem.* **254**, 10295-10300
16. Endo, T. (1999) *Biochim. Biophys. Acta* **1473**, 237-246
17. Hase, S., Nishimura, H., Kawabata, S., Iwanaga, S., and Ikenaka T. (1990) *J. Biol. Chem.* **265**, 1858-1861
18. Harris, R.J., van Halbeek, H., Glushka, J., Basa, L.J., Ling V.T., Smith, K.J., and Spellman, M.W. (1993) *Biochemistry* **32**, 6539-6547
19. Kuraya, N., Omichi, K., Nishimura, H., Iwanaga, S., and Hase, S. (1993) *J. Biochem.* **114**, 763-765
20. Spiro, R.J. (1967) *J. Biol. Chem.* **242**, 4813-4823
21. Torres, C.R. and Hart, G.W. (1984) *J. Biol. Chem.* **259**, 3308-3317
22. Lindahl, U. and Rodén, L. (1972) in *Glycoproteins* (Gottschalk, A., ed.) B. B. A. Library, Volume 5, Part A, p. 491, Elsevier, Amsterdam
23. Anno, K., Kawai, Y., and Seno, N. (1964) *Biochim. Biophys. Acta* **1**, 348-349
24. Yamada, S., Van Die, I., Van den Eijnden, D.H., Yokota, A., Kitagawa, H., and Sugahara, K. (1999) *FEBS Lett.* **459**, 327-331
25. Hoffman, P., Linke, A., and Meyer, K. (1958) *Fed. Proc.* **17**, 1078-1082
26. Suzuki, S., Saito, H., Yamagata, T., Anno, K., Seno, N., Kawai,Y., and Furuhashi, T. (1968) *J. Biol. Chem.* **243**, 1543-1550
27. Hassell, J.R., Kimura, J.H., and Hascall, V.C. (1986) *Annu. Rev. Biochem.* **55**, 539-567
28. Petitou, M., Driguez, P.A., Duchaussoy, P., Herault, J.P., Lormeau, J.C., and Herbert, J.M. (1999) *Bioorg. Med. Chem. Lett.* **9**, 1161-1166
29. Preissner, K.T., Koyama, T., Muller, D., Tschopp, J., and Muller-Berghaus, G. (1990) *J. Biol. Chem.* **265**, 4915-4922
30. Ishizuka, I., Suzuki, M., and Yamakawa, T. (1973) *J. Biochem.* **73**, 77-87
31. Hakomori, S. (1981) *Annu. Rev. Biochem.* **50**, 733-764
32. Sugita, M., Fujii, H., Inagaki, F., Suzuki, M., Hayata, C., and Hori, T. (1992) *J. Biol. Chem.* **267**, 22595-22598
33. Ledeen, R. W. and Wu, G. (1991) *J. Neurochem.* **56**, 95-104
34. Svennerholm, L. (1963) *J. Neurochem.* **10**, 613-623
35. Sugita, M., Nakae, H., Yamamura, T., Takamiya, Y., Itasaka, O., and Hori, T. (1985) *J. Biochem.* **98**, 27-34
36. Song, Y., Kitajima, K., Inoue, S., and Inoue, Y. (1991) *J. Biol. Chem.* **266**, 21929-21935
37. Ichikawa, S. and Hirabayashi, Y. (1998) *Trends Cell Biol.* **8**, 198-202
38. Coetzee, T., Fujita, N., Dupree, J., Shi, R., Blight, A., Suzuki, K., and Popko, B. (1996) *Cell* **86**, 209-219
39. Sandhoff, K. and Kolter, T. (2003) *Philos. Trans. R. Soc. Lond. Biol. Sci.* **358**, 847-861
40. Li, Y.-T., Ishikawa, Y., and Li, S.-C. (1987) *Biochem. Biophys. Res. Commun.* **149**, 167-172
41. Basu, S.S., Dastgheib-Hosseini, S., Hoover, G., Li, Z., and Basu, S. (1994) *Anal. Biochem.* **222**,

270-274
42. Horibata, Y., Okino, N., Ichinose, S., Omori, A., and Ito, M. (2000) *J. Biol. Chem.* **275**, 31297-31304
43. Horibata, Y., Sakaguchi, K., Okino, N., Iida, H., Inagaki, M., Fujisawa, T., Hama, Y., and Ito, M. (2004) *J. Biol. Chem.* **279**, 33379-33389
44. Webster, G.C. (1970) *Biochim. Biophys. Acta* **207**, 371-373
45. Brown, A.J. (1902) *J. Chem. Soc. (Trans.)* **81**, 373-388
46. Michaelis, L. and Menten, M.L. (1913) *Biochem. Z.* **49**, 333-369
47. Briggs, G.E. and Haldane, J.B.S. (1925) *Biochem. J.* **19**, 338-339
48. Lineweaver, H. and Burk, D. (1934) *J. Am. Chem. Soc.* **56**, 658-666
49. Dixon, M., Webb, E.C., Throne, C.J.R., and Tipton, K.F. (1979) *Enzymes*, Longman Group Ltd., London
50. Lis, H. and Sharon, N. (1973) *Annu. Rev. Biochem.* **42**, 541-574
51. Osawa, T. and Tsuji, T. (1987) *Annu. Rev. Biochem.* **56**, 21-42
52. Bernardi, G., Giro, M.-G., and Gaillard, C. (1972) *Biochim. Biophys. Acta* **278**, 409-420
53. Hofstee, B.H.J. (1975) *Biochem. Biophys. Res. Commun.* **63**, 618-624
54. Reynolds, J.A. and Tanford, C. (1970) *J. Biol. Chem.* **245**, 5161-5165
55. Laemmli, U.K. (1970) *Nature* **227**, 680-685
56. Davis, B.J. (1964) *Ann. N.Y. Acad. Sci.* **121**, 404-427
57. Robbins, P.W., Trimble, R.B., Wirth, D.F., Hering, C., Maley, F., Maley, G.F., Das, R., Gibson, B.W., Royal, N., and Biemann, K. (1984) *J. Biol. Chem.* **259**, 7577-7583
58. Takegawa, K., Mikami, B., Iwahara, S., Morita, Y., Yamamoto, K., and Tochikura, T. (1991) *Eur. J. Biochem.* **202**, 175-180
59. Fujita, K., Nakatake, R., Yamabe, K., Watanabe, A., Asada, Y., and Takegawa, K. (2001) *Biosci. Biotechnol. Biochem.* **65**, 1542-1548
60. Tarentino, A.L., Quinones, G., Schrader, W.P., Changchien, L., and Plummer, T.H., Jr. (1992) *J. Biol. Chem.* **267**, 3868-3872
61. Tarentino, A.L., Quinones, G., Changchien, L., and Plummer, T.H., Jr. (1993) *J. Biol. Chem.* **268**, 9702-9708
62. Rao, V., Cui, T., Guan, C., and Van Roey, P. (1995) *Structure* **3**, 449-457
63. Van Roey, P., Rao, V., Plummer, T.H., Jr., and Tarentino, A.L. (1994) *Biochemistry* **33**, 13989-13996
64. Wadding, C.A., Plummer, T.H., Jr., Tarentino, A.L., and Van Roey, P. (2000) *Biochemistry* **39**, 7878-7885
65. Watanabe, T., Kobori, K., Miyashita, K., Fujii, T., Sakai, H., Uchida, M., and Tanaka, K. (1993) *J. Biol. Chem.* **268**, 18567-18572
66. Tews, I., Terwisscha van Scheltinga, A.C., Perrakis, A., Wilson, K.S., and Dijkstra, B.W. (1997) *J. Am. Chem. Soc.* **119**, 7954-7959
67. Rao, V., Cui, T., Guan, C., and Van Roey, P. (1999) *Protein Sci.* **8**, 2338-2346
68. Takegawa, K., Tabuchi, M., Yamaguchi, S., Kondo, A., Kato, I., and Iwahara, S. (1995) *J. Biol. Chem.* **270**, 3094-3099
69. Takegawa, K., Yamaguchi, S., Kondo, A., Iwamoto, H., Nakoshi, M., Kato, I., and Iwahara, S. (1991) *Biochem. Int.* **25**, 829-835
70. Takegawa, K., Yamabe, K., Fujita, K., Tabuchi, M., Mita, M., Izu, H., Watanabe, A., Asada, Y., Sano, M., Kondo, A., Kato, I., and Iwahara, S. (1997) *Arch. Biochem. Biophys.* **338**, 22-28
71. Fujita, K. and Takegawa, K. (2001) *Biochem. Biophys. Res. Commun.* **283**, 680-686
72. Takegawa, K., Nakoshi, M., Iwahara, S., Yamamoto, K., and Tochikura, T. (1989) *Appl. Environ. Microbiol.* **55**, 3107-3112
73. Henrissat, B. (1991) *Biochem. J.* **280**, 309-316.
74. Kato, T., Fujita, K., Takeuchi, M., Kobayashi, K., Natsuka, S., Ikura, K., Kumagai, H., and Yamamoto, K. (2002) *Glycobiology* **12**, 581-587
75. Muramatsu, H., Tachikui, H., Ushida, H., Song, X., Qiu, Y., Yamamoto, S., and Muramatsu, T. (2001) *J. Biochem.* **129**, 923-928
76. Fujita, K., Takami, H., Yamamoto, K., and Takegawa, K. (2004) *Biosci. Biotechnol. Biochem.* **68**, 1059-1066
77. Suzuki, T., Yano, K., Sugimoto, S., Kitajima, K., Lennarz, W.J., Inoue, S., Inoue, Y., and Emori, Y. (2002) *Proc. Natl. Acad. Sci. USA* **99**, 9691-9696
78. Rigden, D.J., Jedrzejas, M.J., and de Mello, L.V. (2003) *FEBS Lett.* **544**, 103-111
79. Suzuki, H., Kumagai, H., and Tochikura, T. (1986) *J. Bacteriol.* **168**, 1332-1335

2

General Introduction of Various Endoglycosidases

2.1 Endoglycosidases That Relate to *N*-Glycans

2.1.1 Endo-β-*N*-acetylglucosaminidase

Two types of glycosidic linkages are generally present in glycoproteins. One comprises GlcNAc linked *N*-glycosidically to asparaginyl residues, and the other GalNAc or monosaccharide such as Glc, Gal, Man, Fuc and GlcNAc linked glycosidically to the hydroxyl groups of Ser or Thr residues of the protein.

Endo-β-*N*-acetylglucosaminidase (Endo-β-GlcNAc-ase, EC 3.2.1.96) is an endoglycosidase that hydrolyzes the *N,N'*-diacetylchitobiose moiety in oligosaccharides bound to the Asn residue of various glycoproteins or glycopeptides through *N*-glycosidic linkage. This enzyme is unique with regard to the enzymatic action by which one GlcNAc residue remains bound to the protein/peptide as it cleaves the *N,N'*-diacetylchitobiose moiety (Fig. 2.1). The enzyme is useful in elucidating the structure and function of the oligosaccharide of glycoproteins because it can release oligosaccharide without causing damage to either the oligosaccharide or protein/peptide moieties. For this purpose, peptide-N^4-(*N*-acetyl-β-D-glucosaminyl) asparagine amidase (peptide *N*-glycanase) is also used. This enzyme is a kind of amidase that acts on the glycosylamide linkages of Asn-linked

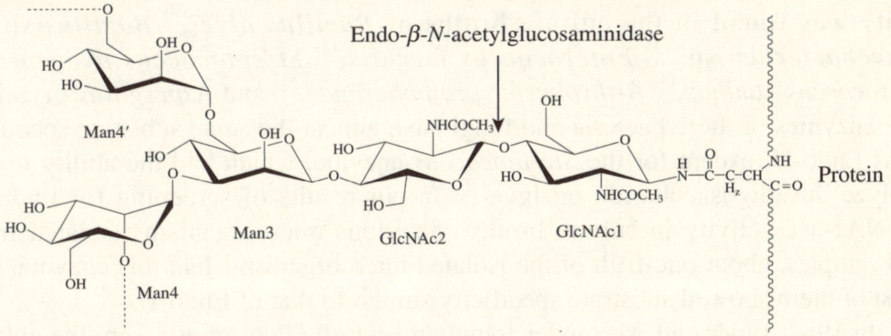

Fig. 2.1 Enzymatic action of Endo-β-GlcNAc-ase. Man, mannose; GlcNAc, *N*-acetylglucosamine.

oligosaccharides and liberates native oligosaccharides.

Endo-β-GlcNAc-ase was first found in the culture fluid of *Diplococcus pneumoniae* (now named *Streptococcus pneumoniae*) by Muramatsu in 1971.[1] Muramatsu discovered it in the enzyme activity releasing an oligosaccharide from mouse myeloma γ-globulin. Similar activity was subsequently found in the culture filtrate of *Streptomyces griseus* by Tarentino *et al.* in 1972.[2] The enzyme could release oligosaccharides from thyroglobulin B and ovalbumin. These enzymes were purified to homogeneity[3,4] and termed Endo-D and Endo-H, respectively. Since both enzymes were found, they have been used as a valuable tool in glycoprotein research, and the enzyme activity has been found from various sources.

A. Microbial Endo-β-GlcNAc-ases

Endo-D of *Diplococcus (Streptococcus) pneumoniae* and Endo-H of *Streptomyces griseus* have different specificities. Endo-D can hydrolyze a complex type oligosaccharide chain of "side-chain-free" glycopeptides and a few high-mannose type oligosaccharide chains in many glycopeptides.[5] The glycan specificity of the enzyme is the trisaccharide structure Manα1-3Manβ1-4GlcNAc, which is linked to proximal GlcNAc.[6] On the other hand, Endo-H generally hydrolyzes high-mannose type oligosaccharide chains of glycopeptides. This enzyme requires a tetrasaccharide structure Manα1-3Manβ1-6Manβ1-4GlcNAc as its glycan specificity.[7] As Endo-H has properties that enable it to act readily on high-mannose type oligosaccharide chains, the enzyme is useful for determining the type of oligosaccharide chain. This type of specificity of Endo-H is very popular in the Endo-β-GlcNAc-ases of microorganisms.

Endo-CII is one of two Endo-β-GlcNAc-ases found in the culture broth of *Clostridium perfringens* by Ito *et al.* in 1975.[8] This enzyme showed a glycan specificity similar to that of Endo-H except that most of the hybrid type oligosaccharide chains which are easily hydrolyzed by Endo-H cannot be released by this enzyme. This enzyme was reported to require a branched pentasaccharide structure Manα1-6 (Manα1-3) Manβ1-4GlcNAc.[7] On the other hand, Endo-CI, which is another Endo-β-GlcNAc-ase of *C. perfringens*, shows exactly the same substrate specificity as that of Endo-D.[8] Besides these enzymes, similar enzyme activity was found in the culture broths of *Bacillus alvei*,[9] *Bacillus* sp.,[10] *Flavobacterium* sp.,[11] *Enterococcus faecalis*,[12] *Streptococcus pyogenes*,[13] *Myxococcus xanthus*,[14] *Arthrobacter protophormiae*,[15] and *Aspergillus oryzae*.[16] The enzymes of these bacteria and fungi have almost the same substrate specificity as Endo-H, except for the *Streptococcus* enzyme, which had the ability to hydrolyze the oligosaccharide on IgG.[13] In our results of screening for Endo-β-GlcNAc-ase activity in culture broths of various microorganisms isolated from soil samples, about one-fifth of the isolated microorganisms had this enzyme, and most of them showed substrate specificity similar to that of Endo-H.

In 1982, Elder and Alexander found an Endo-β-GlcNAc-ase from the culture broth of *Flavobacterium meningosepticum* that could hydrolyze both high-mannose and complex type oligosaccharides of the Asn-linked oligosaccharide chains.[17] This enzyme, named Endo-F, cleaved glycoproteins of retrovirus, lymphocytic choriomeningitis virus and Pichinde virus in the presence of a non-ionic detergent such as Nonidet P-40. However, the Endo-F preparation contained a deglycosylating amidase, identified as peptide-N^4-(*N*-acetyl-β-D-glucosaminyl) asparagine amidase (PNGase), which hydrolyzed the glycosylamide linkage of a broad range of the Asn-linked oligosaccharides bound to protein.[18] Later, two enzymes were separated by hydrophobic-interaction chromatography.[19] Then Endo-F was shown to preferentially cleave high-mannose type oligosaccharides and also biantennary complex type oligosaccharides at a slower rate.[20] Endo-F has been reported to cleave complex type oligosaccharides of various glycoproteins such as thyroid-stimulating hormone,[21] human granulocyte/macrophage colony-stimulating factor (hGM-CSF),[22] herpes simplex virus,[23] *etc*. Later Trimble and Tarentino showed that the Endo-F preparation consisted of three distinct endoglycosidases which have been designated Endo-F_1, F_2 and F_3.[24] Endo-F_1, which was the predominant enzyme in the preparation, was very similar in substrate specificity to well-characterized Endo-H, which hydrolyzed typical high-mannose type oligosaccharides.[24] Endo-F_2 preferentially hydrolyzes the biantennary complex type oligosaccharides, and also high-mannose type oligosaccharides at a greatly diminished rate.[25] Endo-F_3 hydrolyzes both the bi- and triantennary complex type oligosaccharides, but the rates of hydrolysis were slower than those of Endo-F_1 and Endo-F_2.[25] It is interesting that bacteria have some Endo-β-GlcNAc-ases showing different substrate specificities. They are fully active in the presence of non-ionic detergents such as Nonidet P-40 and Triton X-100. A similar enzyme capable of acting on complex type oligosaccharides of glycoproteins was found in the culture broth of a fungus which was isolated from soil and identified as *Mucor hiemalis*.[26] Table 2.1 shows the substrate specificity of various microbial enzymes.

Table 2.1 Substrate specificity of various microbial Endo-β-GlcNAc-ases

Enzyme source	Types of Asn-linked oligosaccharides acting by enzyme
Endo-H (*Streptomyces plicatus*)	high-mannose, hybrid
Endo-D (*Diplococcus pneumoniae*)	high-mannose, complex (monoantennary)
Endo-CII (*Clostridium perfringens*)	high-mannose, hybrid
Endo-F_1 (*Flavobacterium meningosepticum*)	high-mannose, hybrid
Endo-F_2 (*Flavobacterium meningosepticum*)	high-mannose, complex (biantennary)
Endo-F_3 (*Flavobacterium meningosepticum*)	high-mannose, complex (biantennary, triantennary)
Endo-A (*Arthrobacter protophormiae*)	high-mannose
Endo-M (*Mucor hiemalis*)	high-mannose, hybrid, complex (biantennary)

B. Animal Endo-β-GlcNAc-ases

The first discovery of animal Endo-β-GlcNAc-ase was reported by Nishigaki *et al.* in 1974.[27] The enzyme activity was found to be present in rat and porcine organs. Evidence for the presence of this enzyme came from the study of urinary oligosaccharides excreted by patients with lysosomal-enzyme-deficiency diseases. Urinary oligosaccharides with reducing end GlcNAc residues have been found in several different glycosidosis,[28] leading to the hypothesis that the Endo-β-GlcNAc-ase acts in the absence of the particular exoglycosidases and that its action is an early step in glycoprotein catabolism. However, no evidence for a lysosomal form of the activity was found, and on examination of the subcellular distribution of the enzyme, its cytoplasmic localization was confirmed. This cytosolic enzyme activity has been found in a wide variety of mammalian sources such as rat liver,[29] human skin fibroblasts,[30] human saliva,[31] hen oviduct,[32] *Caenorhabditis elegans*[33] and human cells.[34]

Most animal Endo-β-GlcNAc-ases hydrolyze high-mannose type oligosaccharide chains faster than complex and hybrid type sugar chains. Endo-β-GlcNAc-ase from rat liver hydrolytically cleaves Manα1-6(Manα1-3)Manβ1-4GlcNAcβ1-4*R* in which *R* represents either GlcNAc-Asn or GlcNAc.[35] However, the enzyme can barely act on the oligosaccharide chains with Fucα1-3 or 6GlcNAc-Asn as their *R* residues. The fact may support the hypothesis proposed for the mechanism of urinary excretion of Asn-linked oligosaccharides in fucosidosis patients. Because of the lack of lysosomal α-L-fucosidase, Asn-linked oligosaccharides with an α-L-Fuc on the proximal GlcNAc residue of the oligosaccharide chain cannot be hydrolyzed by Endo-β-GlcNAc-ase and excreted in urine. The Endo-β-GlcNAc-ase from hen oviduct (Endo-HO)[32] was purified to homogeneity, and its substrate specificity revealed that oligosaccharide chains containing the Manα1-2Manα1-3Manβ1-4GlcNAcβ1-4GlcNAc structure are good substrates. The Endo-β-GlcNAc-ase from *C. elegans* (Endo-CE)[33] was cloned and the recombinant enzyme expressed in *Escherichia coli* exhibited substrate specificity mainly for high-mannose type oligosaccharides, and $Man_8GlcNAc_2$ oligosaccharide was the best substrate. Suzuki *et al.* reported that both Endo-β-GlcNAc-ase and α-mannosidase are required to form a free oligosaccharide with the structure $Man_5GlcNAc$, which is then transported into the lysosomes by an unknown mechanism.[34] On the other hand, the human salivary enzyme (Endo-HS) is specific for complex type oligosaccharide chains and can release the oligosaccharide chains from native glycoproteins regardless of the existence of an L-Fuc residue on the proximal GlcNAc of the *N,N'*-diacetylchitobiose core of their sugar chains.[31] The enzyme was less active on the incomplete complex type oligosaccharide chains.

Accumulating evidence indicates that the *N*-glycosylation process plays critical roles in protein folding, intracellular transport and degradation of proteins. In recent years, it has been found that during the *N*-glycosylation of proteins in the endoplasmic reticulum (ER), a significant number of free oligosaccharides are formed.[36] When the free oligosaccharides are formed in the lumen of the ER, they are transported out of the ER to the cytosol.[37] This export process is important for cells because of the potential competition between glycoproteins and free oligosaccharides in the glycan-dependent folding/degradation process that operates in the ER.[38] The combined actions of Endo-β-GlcNAc-ase and a cytosolic α-mannosidase were found to be essential for free oligosaccharides to be transferred into lysosome, where they are further degraded.[39]

C. Plant Endo-β-GlcNAc-ases

The occurrence of Endo-β-GlcNAc-ases has been demonstrated in several plants such as fig,[40] jack bean,[41] pea,[42] radish,[43] barley,[44] *Ginkgo biloba*,[45] soybean,[46] and tomato.[46] Two types of *N*-glycans can be found in plants. The oligomannoside-type presents an inner core substituted (on Man4 and 4′) by 2-6 additional Man residues and a complex type glycans which also contain other sugars such as additional GlcNAc, Gal, Xyl (β1,2 linked to Man3) or Fuc (α1,3 linked to GlcNAc1). Two enzymes may be involved in the removal of *N*-glycan: Endo-β-GlcNAc-ase and peptide-N^4-(*N*-acetyl-β-D-glucosaminyl) asparagine amidase (PNGase). Due to their specificity of substrate and to the nature of the unconjugated *N*-glycans (free oligosaccharides) found in plants, Endo-β-GlcNAc-ase is probably involved in the de-*N*-glycosylation of oligomannoside-type glycans linked to proteins whereas the complex type glycans of glycoproteins could be hydrolyzed by PNGase. Kimura *et al.* reported that the plant Endo-β-GlcNAc-ase was highly active towards the high-mannose type *N*-glycans bearing the Manα1-2Manα1-3Manβ1-structural unit,[47] although the high-mannose type free *N*-glycans in plant cells have a common core structural unit: Manα1-6Manα1-3Manα1-6(Manα1-3)Manβ1-4GlcNAc. Optimum pH of plant Endo-β-GlcNAc-ases is 6-7, and the enzyme may not reside in acidic organelles such as the vacuole in plant cells. It is speculated that the enzyme may be localized in the cytosolic fraction. Alternatively, the reason for de-*N*-glycosylation may be found in the physiological functions of the *N*-glycans released, considered as signaling molecules.[48]

D. Structure and Gene of Endo-β-GlcNAc-ase

The structure of Endo-β-GlcNAc-ase has been well studied for microbial enzymes. The complete amino acid sequences of some microbial Endo-β-GlcNAc-ases have been determined. The enzymes are classified into two glycoside hydrolase (GH) families of 18 and 85 from their structures. GH family 18 enzymes include Endo-H,[49] Endo-F_1,[50] Endo-F_2,[51] Endo-F_3,[51] *Flavobacterium* sp. enzyme[52] and others including Endo-S.[13] The primary sequences of first four enzymes and Endo-S were determined from the nucleotide sequence and that of *Flavobacterium* sp. enzyme was from analyses of enzyme-digested peptides. The proteins of Endo-H, Endo-F_1, Endo-F_2, Endo-F_3 and *Flavobacterium* sp. enzyme consist of 271, 289, 290, 290 and 267 amino acids, respectively.[49-52] There is 60% structural identity between Endo-H and *Flavobacterium* sp. enzyme, and 32% identity between Endo-F_1 and *Flavobacterium* sp. enzyme, although both enzymes are derived from the same genus and have similar substrate specificity. Endo-F_2 and Endo-F_3, which can hydrolyze complex type oligosaccharides, share 32% amino acid identity, but only 15% with that of Endo-F_1 or Endo-H. However, there are a few amino acid residues that are common to all five enzymes. They are distributed in five clusters. Kuranda and Robbins showed two highly conserved regions common to a number of chitinases.[53] They also found two similar regions in the amino acid sequence of Endo-H. Later, the presence of such conserved regions was confirmed in four other Endo-β-GlcNAc-ases.[50] The relative spacing between the two regions remained fairly constant at about 40 amino acids in five Endo-β-GlcNAc-ases and many chitinases (Fig. 2.2). This suggests that they represent the catalytic domain cleaving the β1,4 glycosidic bond between adjacent GlcNAc residues. Many chitinases are known to belong to GH family 18 and are named family 18 chitinases. The phylogenetic relationship of Endo-β-GlcNAc-ase in GH family 18 is shown in Fig. 2.3.

Endo-H	^{78}PLQQG - - IKVLLSVLGN	^{123}GLDGVDFDDEYA
Endo-F_1	^{79}PLQDKG - - IKVILSILGN	^{123}NLDGVFFDDEYS
Endo-F_2	^{72}YLHKRG - - TKVIIT-LGD	^{117}NLDGIDIDIESS
Endo-F_3	^{74}SLQSRG - - IKVLQNIDDD	^{119}KLDGISLDIEHS
Flavobacterium sp. Endo-β-GlcNAc-ase	^{74}PLQAKG - - IKVSLSILGN	^{119}GLDGVDLDDEYS
Bacillus chitinase A1	^{147}KLKQTNPNLKTIISVGGW	^{195}NFDGVDLDWEYP
Bacillus chitinase D	^{251}YLQSQG - - KKVLISMGGA	^{294}GFNGLDIDLEGS
Serratia chitinase A	^{258}ALKQAHPDLKILPSIGGW	^{306}FFDGVDIDWEFP
Serratia chitinase B	^{80}ALKAHNPSLRIMFSIGGW	^{135}GFDGVDIDWEYP

Fig. 2.2 Protein sequence similarities between various microbial Endo-β-GlcNAc-ases and chitinases in GH family 18.
The numbers in each sequence represent the position relative to the N-terminal of that protein. Identical residues are shaded.

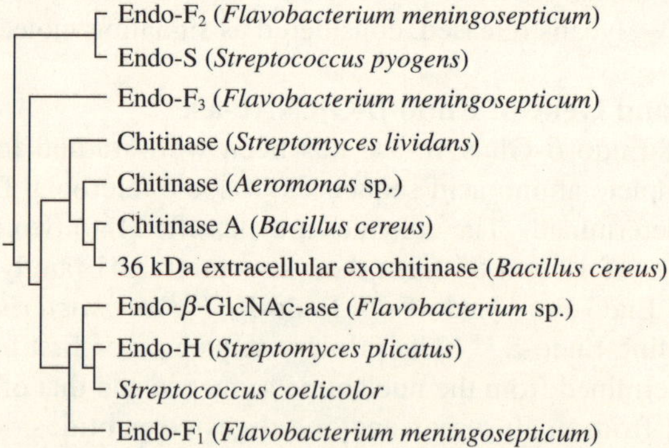

Fig. 2.3 Phylogenetic relationship of GH family 18 enzymes. A phylogenetic tree was constructed by means of ClustalW using the neighbor-joining method.

On the other hand, Endo-A from *Arthrobacter protophormiae*,[54] Endo-D from *Streptococcus pneumoniae*,[55] and Endo-M from *Mucor hiemalis*[56] belong to GH family 85. Their amino acid sequences that were designated by base sequences were very different from those of GH family 18 enzymes. Endo-CE from

C. elegans[33)] and Endo-β-GlcNAc-ase from human cells[34)] also belong to this family. Interestingly, the corresponding homologues encoding putative Endo-β-GlcNAc-ase were found in the genomic sequences of *Drosophila melanogaster* (GenBank accession number AY075262) and *Arabidopsis thaliana* (GenBank accession number AC009991). GH family 85 enzymes may be divided into two groups, *i.e.* bacterial and eukaryotic enzymes, as shown in Fig. 2.4, that describe the phylogenetic relationship of this family of enzymes. The bacterial enzymes are secreted ones, and the eukaryotic ones are predicted to be involved in the processing of free oligosaccharide in the cytosol. GH family 85 enzymes possess a consensus segment in which the Glu residue was identified as the catalytic residue (Fig. 2.5). However, the corresponding Glu residue was also found in the enzymes of GH family 18. Therefore, both enzymes of GH families 18 and 85 may possess a consensus motif in which Glu is a critical active residue.

Many GH family 85 homologues were recently cloned. The genes of GH family 85 Endo-β-GlcNAc-ases were identified as follows (GenBank accession numbers are shown in parentheses): *Arthrobacter protophormiae* AKU 0647 (U59168), *Bacillus halodurans* C-125 (AP001509), *Bifidobacterium longum* NCC2705 (AE014762), *Lactobacillus plantarum* WCFS1 (AL935252), *Clostridium perfringens* 13 (AP003186), *Lactococcus lactis* subsp. *lactis* IL1403 (AE006379), *Streptococcus pneumoniae* TIGR4, *Mucor hiemalis* (AB060586), *Pichia anomala* (AJ306295), *Anopheles gambiae* str. (AAAB01008846), *Drosophila melanogaster* (AY075262), *Arabidopsis thaliana* ORF At3g11040 (AC009991), ORF At5g05460 (AB010692), *Homo sapiens* (AF512564), *Mus musculus* (AK048667), and *Caenorhabditis elegans* (AB079783).

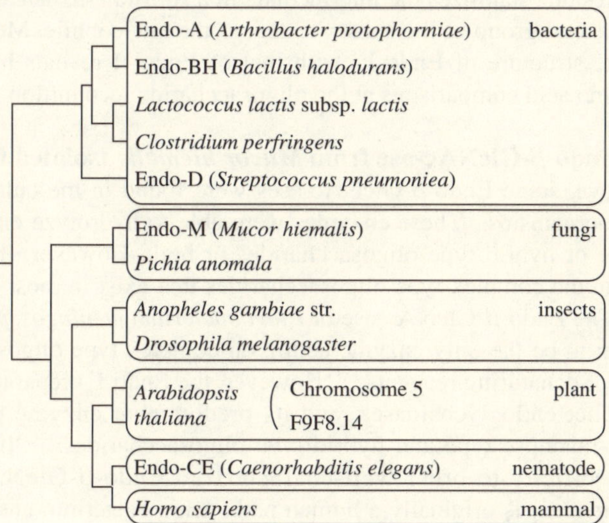

Fig. 2.4 Phylogenetic relationship of GH family 85 enzymes. A phylogenetic tree was constructed by means of ClustalW using the neighbor-joining method.

Organism	Sequence		Sequence		Group
Arthrobacter protophormiae (Endo-A)	YGFDGWFINQE	173	IAWQNHLTDRN	224	bacteria
Bacillus halodurans	YGFDGWFINQE	196	IRWQNYLTDEN	247	
Bifidobacterium longum	YGFDGWFINQE	232	IDWQNALTDEN	287	
Lactobacillus plantarum	YGFDGWFINQE	222	MDWQNALTKEN	294	
Clostridium perfringens	YGFDGWFINQE	230	MDWQNALTDKN	286	
Lactococcus lactis subsp. *lactis*	YGFEGWFINEE	225	VDWQNQLNDQN	275	
Streptococcus pneumoniae (Endo-D)	YGYDGYFINQE	324	RYHQDGLGEYN	379	
Mucor hiemalis (Endo-M)	CQFHGWLLNVE	177	LEWQNELNSAN	236	fungi
Pichia anomala	FGFEGWLVNVE	179	LRWQNELNSRN	234	
Anopheles gambiae str.	YGFDGWLFNIE	183	IHWQNQLTWKN	236	insects
Drosophila melanogaster	YGFDGWLINME	251	VFYQNAVNEWN	304	
Arabidopsis thaliana (At3g11040)	LGFDGWLINIE	181	LQWQDQLTELN	234	plants
Arabidopsis thaliana (At5g05460)	LGFDGWLINIE	176	LAWQDQLTENN	248	
Homo sapiens (HsENGase)	FRFDGWLINIE	237	LKWQDELNQHN	290	mammals
Mus musculus	FRFDGWLINIE	229	LKWQDELNDQN	282	
Caenorhabditis elegans (Endo-CE)	FGFDGWLINIE	154	LHWQNWLNEMN	207	nematode

Fig. 2.5 Amino acid sequence similarities of GH family 85 enzymes. Completely conserved residues and conserved substitutions are highlighted in black and dark gray, respectively. Filled and open triangles indicate the residues corresponding to the important amino acids for the hydrolysis and transglycosylation activity of Endo-A, respectively.

The three-dimensional structure of Endo-H and Endo-F_1 have been resolved by X-ray crystallography.[57)] As shown in Fig. 2.2, these enzymes have highly conserved amino acid residues in the active site of Endo-H that correspond with Asp-130 and Glu-132. The Glu residue has been identified as the proton donor, and the Asp residue stabilizes the intermediate in a substrate-assisted hydrolysis in which the *N*-acetyl group of the substrate acts as the nucleophile. Moreover, comparison of the structure of Endo-F_3 with that of Endo-H reveals highly distinct folds and amino acid comparisons at the oligosaccharide recognition sites.[58)]

E. Novel Endo-β-GlcNAc-ase from *Mucor hiemalis* Isolated from Soil

As noted above, some Endo-β-GlcNAc-ases were found in the culture broths of various microorganisms. These enzymes were able to hydrolyze either the high-mannose type or hybrid type oligosaccharides or both. However, they could not entirely act on the complex type oligosaccharides that exist in most glycoproteins of animals. The Endo-β-GlcNAc-ase of *Flavobacterium meningosepticum* (Endo-F) was shown to be the only enzyme acting on complex type oligosaccharides in the presence of denaturing reagents.[17)] However, the Endo-F preparation consisted of three distinct endoglycosidases, and its predominant enzyme preferentially cleaved high-mannose type and hybrid type oligosaccharides.[25)] It is similar in substrate specificity to other well-characterized Endo-β-GlcNAc-ases. As *F. meningosepticum* is originally a human pathogenic bacterium and its synthesis of the enzyme is very poor, it is difficult to handle the bacterium and obtain the enzyme in sufficient quantities. However three different enzymes have been cloned.[50-51)]

1) Discovery of Endo-M

We attempted to find a microorganism that produced Endo-β-GlcNAc-ase to release complex type oligosaccharides from glycoproteins.[26] We searched for an enzyme source in microorganisms from soil samples. The screening for the enzyme was carried out as follows. The culture fluid of microorganism isolated from soil was incubated in 100 mM potassium phosphate buffer (pH 6.0), with dansyl (DNS) transferrin glycopeptide prepared as the substrate by exhaustive pronase-digestion of human transferrin containing only complex type oligosaccharides followed by its dansylation. After overnight incubation at 37°C, the reaction mixture was put on a silica gel sheet and thin-layer chromatography (TLC) was done with a solvent system of 1-butanol/acetic acid/water (3/1/1). If the enzyme activity was found, the product which was released from the substrate might be detected at a position identical to that of DNS-Asn-GlcNAc on exposure to ultraviolet light. The screening work was very tedious. However, after screening about 800 strains, we found a novel Endo-β-GlcNAc-ase acting on complex type oligosaccharide in the culture broth of a fungus isolated from soil. These soil samples were collected at a picnic welcoming newcomers to our laboratory in the suburbs of Kyoto, Japan. We identified this fungus as *Mucor hiemalis* on the basis of its morphological properties, and named the enzyme Endo-M after its source.[59]

2) Purification of Endo-M

The purification of Endo-M was started by collecting the filtrated broth (8.6 L) obtained by filtration of culture broth (9 L) of *Mucor hiemalis* cultivated for three days. The filtrated broth collected was added to carboxymethyl cellulose (200 g wet weight) and stirred for 30 min. After filtration, ammonium sulfate was added to 70% saturation and left standing for two days. The resulting precipitate was collected, dissolved with 10 mM potassium phosphate buffer (pH 7.0), then dialyzed against the same buffer. The dialyzed solution was put on a DEAE-Sepharose CL-6B column and Endo-M was eluted with 10 mM potassium phosphate buffer (pH 7.0) including 0.2 M NaCl. The active fractions were added to ammonium sulfate (85% saturation), followed by standing overnight. The resulting precipitate was dissolved and put on a column of Sephadex G-200, and the active fractions were concentrated by ultrafiltration. The concentrated enzyme solution was applied to a hydroxylapatite column previously equilibrated with 10 mM potassium phosphate buffer (pH 7.0), and the enzyme was eluted with 100 mM potassium phosphate buffer (pH 7.0). To the concentrated enzyme solution, an equal volume of 100 mM potassium phosphate buffer (pH 7.0) 60% saturated with ammonium sulfate was added, and the viscous solution was applied to a Toyopearl TSK-gel HW-65F column equilibrated with 100 mM potassium phosphate buffer (pH 7.0) saturated with 30% ammonium sulfate. The enzyme was eluted with 100 mM potassium phosphate buffer (pH 7.0) 15% saturated with ammonium sulfate. The enzyme solution was concentrated and desalted by ultrafiltration then put on a column of Con A-Sepharose 4B previously equilibrated with 50 mM potassium phosphate buffer (pH 7.0). The enzyme was eluted with 50 mM potassium phosphate buffer (pH 7.0) containing 0.1 M NaCl and 1% methyl α-D-mannoside. The results of enzyme purification are summarized in Table 2.2.[59)] Although a minor contaminant was still apparent, the final enzyme preparation was shown to be almost homogeneous on PAGE. The molecular weight of the enzyme was approximately 90,000. The enzyme showed maximum activity in the pH range of 6.0 to 6.5, and was stable from pH 7 to 8. The enzyme could be stored (pH 7.0) at –20°C for over three months without loss of activity.

Table 2.2 Purification of Endo-β-GlcNAc-ase from *Mucor hiemalis*

	Total protein (mg)	Total activity (milliunits)	Specific activity (milliunits/mg)	Yield (%)
Culture broth	11,700	1,160	0.099	100
CM-cellulose treatment	10,800	800	0.074	69
Ammonium sulfate precipitate (70% saturation)	200	290	1.45	25
DEAE-Sepharose CL-6B column chromatography	15.6	224	14.3	19
Sephadex G-200 column chromatography	2.74	82.8	30.2	7.1
Hydroxylapatite column chromatography	0.345	14.7	42.6	1.2
TSK-gel HW-65F column chromatography	0.055	4.3	78.1	0.37
ConA-Sepharose column chromatography	0.03	2.8	93.3	0.24

3) Substrate Specificity of Endo-M

The activity of Endo-M was measured using DNS-sialotransferrin glycopeptide or DNS-asialotransferrin glycopeptide as the substrate.[59)] The assay mixture was composed of 30 nmol of substrate, 100 mM potassium phosphate buffer (pH 6.0) and the enzyme solution, in a total volume of 35 μL. After incubation for an appropriate time at 37°C, the reaction was terminated by the addition of 10 μL of 20% trichloroacetic acid or boiling. An aliquot was applied to HPLC using a reversed-phase column (0.46×25 cm, Unicil QC-18: GL-Science). The reaction mixture was analyzed with a fluorescence spectrophotometer. The column was eluted with 11% acetonitrile in 25 mM sodium borate buffer (pH 7.5) at 0.5 mL/min. For detection of DNS-compounds, an excitation wavelength of 320 nm and emission wavelength of 540 nm were used. Another analysis method was carried out as follows: An aliquot of the reaction mixture was applied to filter paper and paper chromatography was performed by the solvent system of 1-butanol/acetic acid/water (3/1/1). The reaction product, DNS-Asn-GlcNAc, which was detected on the chromatogram from its strong fluorescence upon exposure to ultraviolet light, was extracted with water and its amount determined by fluorescence spectrometry (excitation, 313 nm; emission 540 nm). One unit of the enzyme was defined as the amount of enzyme that yielded 1 μmol of DNS-Asn-GlcNAc per min at 37°C under the conditions used.

This Endo-β-GlcNAc-ase was found to cleave not only the high-mannose type and hybrid type of Asn-linked oligosaccharide but also the complex type of biantennary oligosaccharides in glycoproteins,[59] unlike other well-characterized microbial Endo-β-GlcNAc-ases which can act on only high-mannose type and hybrid type oligosaccharides. Table 2.3 shows the substrate specificity of Endo-M in comparison with that of the enzyme from *Flavobacterium* sp.,[11] a typical microbial enzyme. Both enzymes could rapidly hydrolyze ovalbumin glycopeptide GP-IV and GP-V, which have a high-mannose type oligosaccharide, although the activity of Endo-M for GP-V was half that for GP-IV.[60] This substrate specificity of Endo-M is different from those of other microbial Endo-β-GlcNAc-ases that have the highest activity for GP-V. On the other hand, GP-I, GP-II and GP-III, which have hybrid type oligosaccharides, were slowly hydrolyzed. Endo-M activities for hybrid type oligosaccharides appeared to be slightly lower than that for complex type asialobiantennary oligosaccharides. The enzyme was also active on complex type sialobiantennary oligosaccharides. However, fucosyl biantennary oligosaccharide with the L-Fuc residue bound to core-proximal GlcNAc of sialobiantennary oligosaccharide was not effective at all. The Asn residue may not be entirely necessary for Endo-β-GlcNAc-ase activity because the enzyme could act on pyridylaminated (PA)-oligosaccharides having no Asn residue.

4) Transglycosylation Activity of Endo-M

Endo-M has a specific characteristic in that it possesses transglycosylation activity. It was able to transfer the intact complex type of sialo- and asialobiantennary oligosaccharides from glycopeptide to suitable acceptors during hydrolysis of the glycopeptide.[61] When asialotransferrin glycopeptide of human serum was incubated with Endo-M in the absence or presence of GlcNAc, and the oligosaccharides in the supernatant were pyridylaminated, HPLC of the PA-oligosaccharides obtained revealed two separate peaks in the reaction mixture to which GlcNAc was added, although a single peak was found in the reaction mixture without GlcNAc. A new peak was found at the elution position coinciding with PA-biantennary complex type oligosaccharide. These findings indicated that the biantennary complex type oligosaccharide of asialotransferrin glycopeptide was transferred to free GlcNAc by Endo-M. *N,N'*-Diacetylchitobiose was also an effective acceptor of the biantennary complex type oligosaccharide. Glc and Man were also effective as an acceptor. Table 2.4 shows the acceptor specificity of Endo-M transglycosylation using sialotransferrin glycopeptide as the oligosaccharide donor. It revealed that the enzyme requires the 3- and 4-OHs of an acceptor sugar residue to be equatorial. Both α-linked and β-linked sugar residues can act as the acceptor for oligosaccharide. However, the latter was a better acceptor than the former. GlcUA did not act as the acceptor although the OHs of C-3 and C-4 positions are equatorial. The difference between GlcUA and Glc is the residue at the C-6 position: the former is a COOH group and the latter a CH_2OH group. 6-OH is relatively important. Cyclohexanol did not act as an acceptor.

Besides complex type oligosaccharides, the enzyme was also capable of transferring high-mannose type oligosaccharide to an acceptor such as GlcNAc and

Table 2.3 Substrate specificities of *Flavobacterium* sp. Endo-β-GlcNAc-ase and Endo-M for various glycopetides

DNS-glycopetides	Hydrolysis (%)	
	Flavobacterium sp. Endo-β-GlcNAc-ase	Endo-M
Ovalbumin		
GP-V	84	40
Mα1 ↘ 6 Mα1 ↗ 3 Mα1 ↘ 6 Mα1 ↗ 3 Mβ1-4GNβ1-4GN-Asn		
GP-IV	68	81
Mα1 ↘ 6 Mα1 ↗ 3 Mα1 ↘ 6 Mα1-2Mα1 ↗ 3 Mβ1-4GNβ1-4GN-Asn		
GP-III	39	17
(Mα1-2)Mα1 ↘ 6 Mα1 ↗ 3 Mα1 ↘ 6 (GNβ1-4) (GNβ1-4) Mα1 ↗ 3 Mβ1-4GNβ1-4GN-Asn (2 ← GNβ1)		
GP-II	9	10
(Mα1) ↘ 6 Mα1 ↗ 3 Mα1 ↘ 6 (GNβ1-4) (Gβ1-4) GNβ1-4Mα1 ↗ 3 Mβ1-4GNβ1-4GN-Asn (2 ← GNβ1)		
GP-I	5	7
(Mα1) ↘ 6 Mα1 ↗ 3 Mα1 ↘ 6 (GNβ1-4) Gβ1-4GNβ1-4Mα1 ↗ 3 Mβ1-4GNβ1-4GN-Asn (2 ← GNβ1)		
Asialotransferrin GP	0	19
Gβ1-4GNβ1-2Mα1 ↘ 6 Gβ1-4GNβ1-2Mα1 ↗ 3 Mβ1-4GNβ1-4GN-Asn		
Sialotransferrin GP	0	9
NeuAcα2-6Gβ1-4GNβ1-2Mα1 ↘ 6 NeuAcα2-6Gβ1-4GNβ1-2Mα1 ↗ 3 Mβ1-4GNβ1-4GN-Asn		

Each DNS-glycopeptide (33 nmoles) was incubated with enzyme (3 munits) in a total volume of 25 μL for 15 min and an aliquot mixture was analyzed by HPLC. GP, glycopeptide; M, mannose; G, galactose; GN, *N*-acetylglucosamine; NeuAc, *N*-acetylneuraminic acid.

Table 2.4 Acceptor specificity of Endo-M in transglycosylation reaction

Acceptor		Relative yield of transglycosylation products*(%)
*p*NP-*β*-GlcNAc	HO, OH, O, HO, O– *p*NP, NHAc	100
*p*NP-*α*-GlcNAc	OH, HO, O, HO, AcHN, O– *p*NP	86
*p*NP-*β*-Glc	OH, HO, O, HO, OH, O– *p*NP	112
*p*NP-*α*-Glc	OH, HO, O, HO, OH, O– *p*NP	60
*p*NP-*β*-Man	OH, O, HO, OH, HO, O– *p*NP	98
*p*NP-*β*-Gal	HO, OH, O, HO, OH, O– *p*NP	0
*p*NP-*β*-L-Fuc	H_3C, O, O– *p*NP, OH, HO, OH	0
*p*NP-*β*-Xyl	O, HO, HO, OH, O– *p*NP	16
*p*NP-*β*-GlcUA	COOH, O, HO, HO, OH, O– *p*NP	0
Cyclohexanol	HO	0
Cyclohexanediol	HO, HO	0

* Glycoside donor: Sialotransferrin glycopeptide.

N,N'-diacetylchitobiose. However, the yield of the transglycosylation product having a high-mannose type oligosaccharide was lower than that of the product having an asialo- or sialocomplex type oligosaccharide. When the yield of each transglycosylation product from sialotransferrin glycopeptide having a sialocomplex type oligosaccharide (NeuAc-Gal-GlcNAc-Man)$_2$ManGlcNAc$_2$, asialotransferrin glycopeptide having asialocomplex type oligosaccharide (Gal-GlcNAc-Man)$_2$ManGlcNAc$_2$ and Man$_6$GlcNAc$_2$-Asn-peptide from ovalbumin was examined using Fmoc-Asn-GlcNAc as the acceptor, the enzyme transferred the complex type oligosaccharides to acceptors more effectively than the high-mannose type ones.[62] Furthermore, it was of interest that the effectiveness of the transfer reactions from various donor glycopeptides is the reverse of their hydrolysis reactions. $Man_6GlcNAc_2$-Asn-peptide was most easily hydrolyzed by Endo-M, but it was least effective in the transglycosylation reaction (Fig. 2.6). The yield of the transglycosylation product of the sialocomplex type oligosaccharide reached almost 20% and that of $Man_6GlcNAc_2$-Asn-peptide was at most one-third (about 8%) that from sialocomplex type oligosaccharide. The optimum conditions for transglycosylation activity appeared to be almost the same as those of its hydrolytic activity. These findings suggest that even though the oligosaccharide was once bound to the acceptor by transglycosylation reaction, it can be easily hydrolyzed. This may be one of the reasons why the yield of transglycosylation product is low.

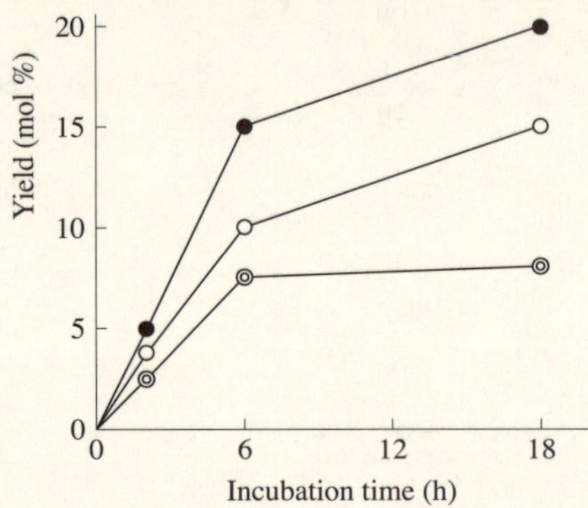

Fig. 2.6 Time course of formation of transglycosylation product from various oligosaccharide donors to Fmoc-Asn-GlcNAc by Endo-M.
●, sialotransferrin glycopeptide; ○, asialotransferrin glycopeptide; ◎, M6-glycopeptide.

The transglycosylation activity of Endo-M can be effectively used to synthesize useful neoglycoproteins and neoglycopeptides containing complex type oligosaccharides. We reported the synthesis of neoglycopeptides by the transglycosylation reaction of sialocomplex type oligosaccharides to the synthetic peptides containing a GlcNAc moiety using Endo-M.[63] The strategies for the syntheses of neoglycopeptides are described in Chapter 3, section 3.1.1.

The oligosaccharide-tranferring reaction by Endo-M can be used for the remodeling of oligosaccharides of glycoproteins/glycopeptides: high-mannose type oligosaccharides in glycoproteins/glycopeptides can be changed to complex types. We transferred the complex type oligosaccharides from hen egg yolk glycopeptides to GlcNAc-remaining protein prepared from bovine pancreas ribonuclease by *Flavobacterium* sp. Endo-β-GlcNAc-ase treatment. Using the transglycosylation activity of Endo-M, the high-mannose type of oligosaccharides in glycoproteins/glycopeptides that are produced by yeasts or molds could be exchanged or remodeled to the complex types which are not synthesized by these microorganisms. This technique is described in section 3.1.2.

5) Molecular Cloning of Endo-M

The gene of Endo-M was cloned and expressed in protease-deficient *Candida boidinii* with a molecular mass of 85 kDa as the monomeric form.[56] The gene encodes a putative 744 amino acids. Although native Endo-M had been isolated from the culture broth,[26] no amino acid sequence corresponding to a signal peptide was found, and its N-terminal peptide sequence started with a proline. Site-directed mutagenesis showed Glu-177 and Trp-228 to be important for the hydrolytic activity and the transglycosylation activity of Endo-M, respectively (Fig.2.5).

F. Production of Novel Endo-β-GlcNAc-ase from *Arthrobacter protophormiae* (Endo-A)

Endo-β-GlcNAc-ase (EC 3.2.1.96) hydrolyzes the glycosidic bond in the *N,N'*-diacetylchitobiose moiety of *N*-linked oligosaccharides of various glycoproteins. Although many bacterial strains produce Endo-β-GlcNAc-ase, few studies on the physiological roles of the enzyme in microorganisms have been described to date. It is unclear why bacteria produce and secrete such highly specialized glycohydrolases because the *N*-linked oligosaccharide substrates are not normally found in their environment. To elucidate the distribution and physiological roles of the enzyme, the author and his coworkers investigated culture conditions favoring the production of bacterial Endo-β-GlcNAc-ase. It was found that a Gram-positive bacterium, *Arthrobacter protophormiae*, highly induced Endo-β-GlcNAc-ase (Endo-A) in medium containing ovalbumin.[15] Endo-β-GlcNAc-ase activity in the culture fluid with 0.3% ovalbumin was over ten times higher than in culture fluid without ovalbumin. Production of the enzyme was also induced by glycoproteins such as yeast invertase and bovine ribonuclease B but not by monosacchrides such as mannose, *N*-acetylglucosamine and galactose. When *A. protophormiae* was grown in medium containing ovalbumin, we found that ovalbumin oligosaccharides accumulated in the culture fluid. This bacterium does not produce α-mannosidase or α-mannanase at all in culture fluid, suggesting that *A. protophormiae* does not utilize ovalbumin oligosaccharides as a carbon source.

1) Purification and enzymatic properties of Endo-A

Endo-A activity was assayed with $Man_6GlcNAc_2$-Asn-DNS (dansyl) as the substrate. The reaction mixture was composed of 50 nmol of Man_6 $GlcNAc_2$Asn-DNS, 1 μmol of sodium acetate buffer (pH 6.0), and the enzyme solution in a total volume of 35 μL. After incubation for an appropriate time at 37°C, the reaction was terminated by the addition of 5 μL of 10% trichloroacetic acid. A portion of the reaction mixture was directly analyzed by high-performance liquid chromatography (HPLC) using a reversed-phase column. The column was eluted with 9% acetonitrile in 25 mM sodium borate buffer, pH 7.5, at 1 mL/min. The product, GlcNAc-Asn-DNS, was determined by fluorescence (excitation, 313 nm; emission, 540 nm). One unit was defined as the amount of enzyme yielding 1 μmol of GlcNAc-Asn-DNS per min at 37°C under the assay conditions.[15] The purification of Endo-A from the culture fluid of *A. protophormiae* grown in 1% ovalbumin is described as follows.[15] (1) Solid ammonium sulfate was added to 90% saturation to the culture fluid (1,600 mL). The precipitate was collected by centrifugation and dissolved in 10 mM potassium phosphate buffer (pH 7.0). This solution was dialyzed against the same buffer. (2) The dialyzed solution was applied to a column (3.5×20 cm) of DEAE-Toyopearl 650M equilibrated with 10 mM phosphate buffer (pH 7.0). The column was washed with the same buffer (500 mL) and then with the buffer containing 0.2 M NaCl (300 mL), and the enzyme was eluted with buffer supplemented with 0.3 M NaCl. The active fractions (100 mL) were combined and dialyzed against 10 mM phosphate buffer (pH 7.0) containing 0.1 M NaCl. (3) The dialyzed solution was applied to a column of DEAE-Toyopearl (1.0×5.0 cm) equilibrated with 10 mM phosphate buffer (pH 7.0) containing 0.1 M NaCl. The column was washed with the same buffer then eluted with a linear gradient of 0.1 to 0.3 M NaCl in the same buffer (300 mL). The active fractions (120 mL) were combined and dialyzed against 10 mM phosphate buffer (pH 7.0) including 1.0 M ammonium sulfate with two changes of the buffer. (4) The enzyme solution was applied to a Phenyl-Sepharose CL-4B column (1.0×30 cm) equilibrated with 1.0 M ammonium sulfate in 10 mM phosphate buffer (pH 7.0). The enzyme was eluted with a linear gradient formed between 100 mL of 1.0 M ammonium sulfate and 100 mL of 40% ethylene glycol. The active fractions (40 mL) were combined and concentrated by pressure dialysis using a collodion bag. (5) The concentrated enzyme solution was applied to a Sephadex G-100 column (1.6×104 cm) previously equilibrated with 10 mM phosphate buffer (pH 7.0). The enzyme was eluted with the same buffer in 3-mL fractions. The active fractions (33 mL) were combined and concentrated using a collodion bag. The purified enzyme preparation showed a single protein band with an apparent molecular weight of about 72,000 by SDS-PAGE.

The amino acid sequence of Endo-A is quite different from those of other Endo-β-GlcNAc-ases such as Endo-H from *Streptomyces griseus*,[49] Endo-Fsp from *Flavobacterium* sp.[52] and Endo-F_1 from *F. meningosepticum*,[50] as described in Chapter 1 section 1.4 and Chapter 2 section 2.1.1.D. The enzyme showed a broad optimum pH in the range of pH 5.0 to 11.0.[15] The substrate specificity of Endo-A was examined using various *N*-linked oligosaccharides. The enzyme completely hydrolyzed high-mannose type oligosaccharides $Man_{5-9}GlcNAc_2Asn$. However, Endo-A does not act on complex-type oligosaccharide, $Gal_2GlcNAc_2Man_3GlcNAc_2Asn$, derived from human transferrin. The substrate specificity of Endo-A was very similar to that of Endo-CII from *Clostridium perfringens*;[64] the enzyme specifically hydrolyzes high-mannose type oligosaccharides, but does not act on complex-type oligosaccharides.[15]

2) Discovery of oligosaccharide-transferring activity of Endo-A

During the study of enzymatic kinetics, Masanao Nakoshi, a graduate student, noticed that treatment of Endo-A with a mixture of *N*-acetylglucosamine, glucose and mannose enhanced the enzyme activity, unlike Endo-β-GlcNAc-ase of other microbial origins.[65] Endo-D was competitively inhibited by the addition of mannose, methyl α-mannoside or *p*-nitrophenyl α-mannopyranoside. In contrast, Endo-H was inhibited not by mannose but by yeast mannan.[66] Endo-A activity was assayed with $Man_6GlcNAc_2Asn$-DNS as the substrate, and the hydrolytic product, GlcNAc-Asn-DNS, was analyzed by HPLC. Therefore, the amount of

GlcNAc-Asn was apparently increased by the addition of monosaccharides in the initial stage of the enzyme reaction. The author's group is very interested in the effect of monosaccharides on Endo-A activity and tried to explain the activation mechanism. First, the enzyme products released by Endo-A with or without *N*-acetylglucosamine were analyzed. Treatment of $Man_6GlcNAc_2Asn$ with the enzyme in the presence of free *N*-acetylglucosamine gave a mixture of $Man_6GlcNAc$ (hydrolysis product) and $Man_6GlcNAc$-GlcNAc mixture.[67] Analysis by ^{1}H-NMR of the product revealed that transglycosylation occurred at the C-4 of *N*-acetylglucosamine and the linkage was of the β-configuration. These results suggest that Endo-A has a transglycosylation (oligosaccharide-transferring) activity, and *N*-acetylglucosamine is transferred to the oligosaccharide during *N*,*N'*-diacetylchitobiose core cleavage by the enzyme. This is the story of our discovery of transglycosylation activity of Endo-A.

Many exoglycosidases and endoglycosidases have been shown to have transglycosylation or transfer reaction activities in addition to hydrolytic activity. These transglycosylation or transfer reaction activities of glycosidases have been used for synthesis of various glycosides.[68, 69] Exoglycosidases, although known to possess transglycosylation activity, are hardly suitable for building complex oligosaccharides because of their relative inefficiency. In contrast, there is little information available on the enzymatic synthesis of neoglycoproteins by endoglycosidases. Trimble *et al.* have reported that *N*-linked oligosaccharides, when hydrolyzed by Endo-β-GlcNAc-ase from *F. meningosepticum* (Endo-F),[17] have glycerol attached to their reducing ends.[70] We examined the oligosaccharide-transferring activity of commercial preparations of Endo-F and Endo-H. These enzymes did not transfer oligosaccharides to *N*-acetylglucosamine or GlcNAc-peptides.

3) Principle of transglycosylation activity

Figure 2.7A illustrates the hydrolytic and transglycosylation activities of Endo-A. After forming the enzyme-substrate complex, the enzyme could transfer the donor oligosaccharides to either water (leading to hydrolysis) or an acceptor (leading to transglycosylation). Since the transglycosylation product can be the substrate for the same enzyme, the process will be repeated until all substrates are transferred to water (complete hydrolysis).

The changes with time in the hydrolysis of $Man_9GlcNAc_2$-Asn and transfer of *N*-acetylglucosamine to the oligosaccharide by Endo-A were examined.[71] In the initial stage of the reaction, the substrate, $Man_9GlcNAc_2$-Asn, decreased quickly, and totally disappeared after 15 min. The transfer product, $Man_9GlcNAc_2$, was formed predominantly in the initial stage of the reaction, but decreased after 10 min, and dropped to 11% at 60 min. The hydrolysis product, $Man_9GlcNAc$, slowly increased during the reaction. As the reaction neared completion, almost all of the reducing ends of *N*-acetylglucosamine were removed from the oligosaccharide (Fig. 2.7).

The transglycosylation product was examined by mass spectrometry and NMR. The NMR analyses revealed that the new glycosidic bond formed by trans-

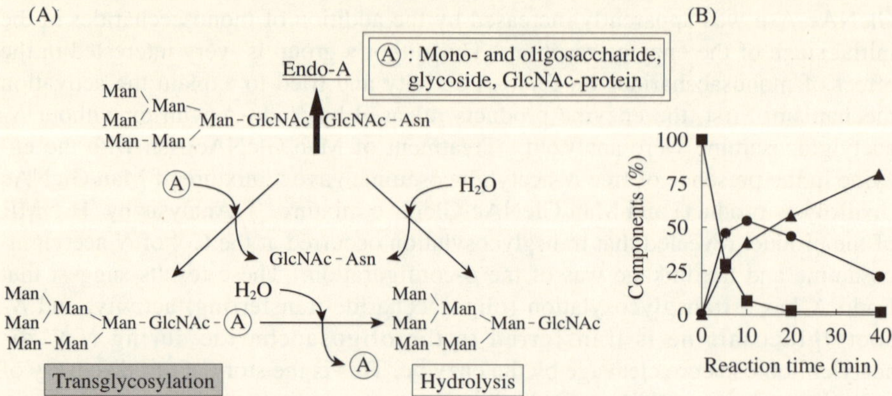

Fig. 2.7 (A) A putative mechanism of Endo-A transglycosylation activity.
(B) Time course of Endo-A action in aqueous solution. The enzymatic reactions were performed in a mixture of 15 nmol of $Man_9GlcNAc_2Asn$, 10 μmol of GlcNAc, and 2.4 milliunits of Endo-A in a total volume of 20 μL with 50 mM ammonium acetate buffer (pH 6.0) at 37°C. The enzyme reaction was stopped by 3-min boiling in a water bath. ●, transglycosylation product; ▲, hydrolysis product; ■, remaining substrate.
[Reprinted with permission from Fan, J.-Q. *et al.* (1995) *J. Biol. Chem.* **270**, 17724]

glycosylation is indeed the GlcNAcβ1-4GlcNAc linkage, the same as that which existed in the donor substrate.[67, 71] Thus, the stereoselectivity and regioselectivity of the Endo-A transglycosylation are not influenced by the anomeric configuration of the acceptor glycoside when transglycosylation is to a *N*-acetylglucosamine residue.[72]

4) The acceptor specificity of Endo-A transglycosylation

The acceptor specificity of Endo-A transglycosylation toward various monosaccharides has been examined. Among the monosaccharides and their derivatives tested, Endo-A could transfer oligosaccharide to *N*-acetylglucosamine, mannose, 2-deoxyglucose and methyl α-D-*N*-acetylglucosaminide. These results indicate that C-1 and C-2 substituents of *N*-acetylglucosamine are not important for the transglycosylation activity. Since 6-deoxyglucose was a fair acceptor, it suggests that 6-OH was relatively unimportant for the transglycosylation. The equatorial 4-OH is essential because *N*-acetylgalactosamine and galactose could not serve as acceptors for transglycosylation. Since 3-deoxyglucose could not be an acceptor at all, and allose, the 3-epimer of glucose, failed to serve as an acceptor, the 3-OH is required to be in equatorial orientation. Fructose could also serve as an acceptor, although the transglycosylation yield was relatively low.[73] Taken together, the enzyme requires the 3- and 4-OHs of an acceptor sugar residue to be equatorial in a normal conformation, but 1-, 2-, and 6-OHs are relatively unimportant. Endo-A can also transfer released $Man_9GlcNAc$ to L-fucose using $Man_9GlcNAc_2Asn$ as a donor substrate. The structure of the oligosaccharide was determined to be $Man_9GlcNAc\beta$1-2-L-Fuc.[73]

The acceptor specificity of Endo-A-catalyzed transglycosylation toward various disaccharides was also investigated.[74)] The enzyme transferred high-mannose type oligosaccharides more efficiently to β-linked disaccharides (cellobiose, gentiobiose, sophorose and laminaribiose) than to α-linked disaccharides (isomaltose, maltose, nigerose, kojibiose and trehalose) as acceptor substrates. These results indicate that Endo-A recognizes the anomeric configuration of the acceptor substrates, and β-linked glycosides are suitable for the synthesis of transglycosylation products.[74)]

5) Transglycosylation activity of Endo-A in media containing organic solvents

We found the hydrolytic activity of Endo-A to be suppressed and the transglycosylation activity enhanced in a reaction mixture containing organic solvent such as acetone, dimethyl sulfoxide (DMSO), and *N,N*-dimethylformamide (DMF) (Table 2.5). Endo-A exerted exclusive transglycosylation activity in media containing acetone, and a hydrolysis product was undetectable under these conditions.[72)] In general, alcohols and acetonitrile are used for the suppression of hydrolysis. The effectiveness of acetone in suppression of hydrolysis is not common, and the mechanism is not well known. However, these hydrophilic organic solvents can be included in the enzyme reaction mixture to decrease water content. The transglycosylation activity of Endo-A toward *N*-acetylglucosamine as an acceptor substrate in media containing organic solvents was further investigated by varying their concentrations.[72)] In acetone-containing media, suppression of the hydrolytic activity increased as acetone concentration was increased to 15-20%, where transglycosylation was maximum. The ratio of transglycosylation to hydrolysis activity rose more sharply between 20 and 30% acetone (Fig. 2.8).

Table 2.5 Transglycosylation activity of Endo-A in organic solvents[a]

Solvent	Remaining Substrate	Yield	
		Hydrolysis	Transglycosylation
	%	%	
None (H_20)	2.0	66.6	32.1
Acetone	8.1	0.8	89.4
Acetonitrile	99.8	0.0	1.4
Dioxane	85.6	0.4	12.6
N,N-Dimethylformamide	71.7	0.0	27.5
Dimethyl sulfoxide	24.8	1.5	74.6
Ethanol	47.1	0.9	55.8
Methanol	2.5	5.9	74.9
Tetrahydrofuran	98.6	0.0	0.0

[a] The enzymatic reaction was carried out with 3 nmol of $Man_9GlcNAc_2Asn$, 10 μmol of GlcNAc, and 2 milliunits of Endo-A in a total volume of 20 μL 25 mM ammonium acetate buffer (pH 6.0) containing 30% (v/v) organic solvent. After incubation at 37°C for 15 min, the hydrolysis and transglycosylation products were quantified with an HPAEC-PAD system.

[Reprinted with permission from Fan, J.-Q. *et al.* (1995) *J. Biol. Chem.* **270**, 17725]

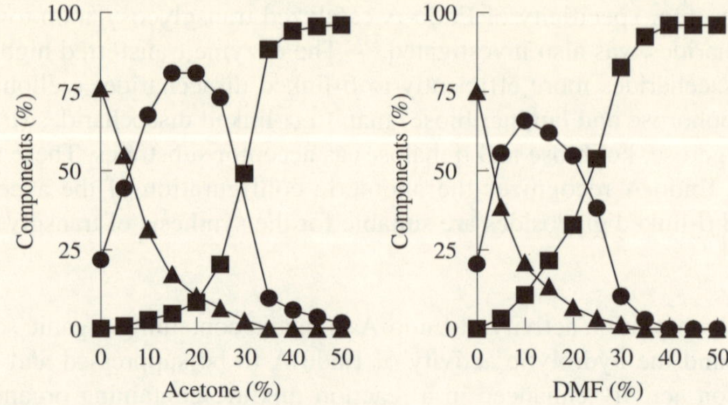

Fig. 2.8 Effect of organic solvents on the transglycosylation activity of Endo-A. The enzymatic reaction was carried out with 15 nmol of $Man_6GlcNAc_2$ Asn, 10 μmol of GlcNAc and 2 milliunits of Endo-A in 20 μL of 25 mM ammonium acetate buffer (pH 6.0) at 37˚C containing different amounts of acetone (left) and *N,N*-dimethylformamide (DMF) (right). The hydrolysis and transglycosylation products were quantified with an HPAEC-PAD system.[73] ●, transglycosylation product; ▲, hydrolysis product; ■, remaining substrate.
[Reprinted with permission from Fan, J.-Q. *et al.* (1995) *J. Biol. Chem.* **270**, 17724]

The stability of Endo-A in organic solvent-containing media was examined. The enzyme was stable in a medium containing up to 30% acetone (v/v) incubated at 37˚C for 30 min. However, after a 30-min preincubation in the medium containing 40% acetone, the enzyme activity was reduced to 46% of the original activity.[72]

The proportion of the hydrolytic products in 30% acetone diminished rapidly when GlcNAc concentration was increased. The transglycosylation and hydrolytic products were found as 86% and 5%, respectively, with 0.2 M *N*-acetylglucosamine. No hydrolytic product could be detected when 0.5 M *N*-acetylglucosamine was used under these conditions. Taken together, the reaction mixture containing 30% acetone achieves the maximum transglycosylation yield.[72]

2.1.2 Endo-β-mannosidase

A. Endo-β-mannosidase Acting on *N*-Linked Sugar Chains

Two kinds of enzymes that can hydrolyze the vicinity of the reducing end of *N*-linked sugar chains of glycoproteins to liberate sugar chains, have so far been identified: the peptide *N*-glycanase[75] that hydrolyzes the amide bond between an aspargine residue and GlcNAc residue and Endo-β-GlcNAc-ase,[1] which hydrolyzes the glycosidic linkage in the *N,N'*-diacetylchitobiose residue of the reducing end of *N*-linked sugar chains (Fig. 2.9). This latter enzyme has been described in a previous section. β-Mannosidases from several mammal tissues have been identified as the enzyme that hydrolyzes the Manβ-linkage.[76] However, an enzyme that hydrolyzes the Manβ-linkage by endo-manner has not yet been reported.

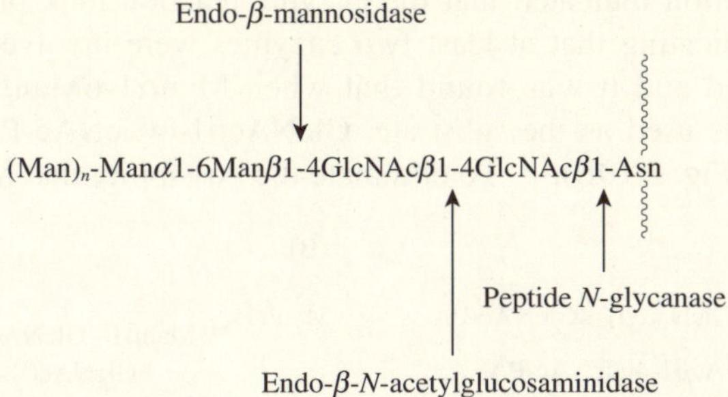

Fig. 2.9 Enzymes acting on the vicinity of the reducing end of *N*-linked sugar chain.

Recently, endo-*β*-mannosidase (E.C. 3.2.1.152), which hydrolyzes the Man*β*1-4GlcNAc linkage of *N*-linked sugar chains by endo-manner, has been identified from lily flowers[77] (Fig. 2.9). In this section, the background of its discovery, its purification procedure, substrate specificity of the enzyme, structure of the endoglycosidase, and its possible role in the degradation and/or processing of *N*-linked sugar chains is described. Availability of this enzyme for the synthesis of oligosaccharides containing the *β*-mannoside structure is described in section 3.3.

B. Discovery of Endo-*β*-mannosidase

The authors have been developing a fluorescence labeling method to analyze sugar chains.[78, 79] In addition to its high sensitivity, small oligosaccharides such as monosaccharides liberated from glycoconjugates can be detected. These are often over looked using conventional methods. In the course of structural studies on the sugar chains of various glycoproteins using the method, GlcNAc*β*1-4GlcNAc as a major (62 %) *N*-linked sugar chain was found by chance in S_4-RNase from the flower of Japanese pear.[80, 81] This is the first example of an *N,N'*-diacetylchitobiose structure found as an *N*-linked sugar chain. In addition to GlcNAc*β*1-4GlcNAc, S_4-RNase also has several high-mannose type and complex type *N*-linked sugar chains.[81] Careful investigation revealed that the sugar chains of intermediate size between GlcNAc*β*1-4GlcNAc and Man*α*1-6(Man*α*1-3)Man*β*1-4GlcNAc*β*1-4GlcNAc (trimannosyl core) structure are missing in S_4-RNase.[81] Taking this observation and the processing mechanism of *N*-linked sugar chains into account, this *N,N'*-diacetylchitobiose structure appeared to be generated by degradation of the trimannosyl core structure. This gave rise to the idea of the presence of endo-*β*-mannosidase, an endoglycosidase that catalyzes the hydrolysis of the Man*β*1-4GlcNAc linkage in the trimannnosyl core structure by endo-manner.

In view of this idea, attempts were made to detect endo-*β*-mannosidase activity in several plant tissues. In an assay for the enzyme activity using Man*α*1-6(Man*α*1-3)Man*β*1-4GlcNAc*β*1-4GlcNAc-PA (PA-; pyridylaminated) as a substrate, a homogenate of plant tissues increasingly produced GlcNAc*β*1-4GlcNAc-PA, which is the expected product obtained by the action of endo-*β*-mannosidase.

Careful investigation indicated that the enzyme reaction took place biphasically (Fig. 2.10A), indicating that at least two enzymes were involved. Several substrates were used and it was found that when Manα1-6Manβ1-4GlcNAcβ1-4GlcNAc-PA was used as the substrate, GlcNAcβ1-4GlcNAc-PA increased linearly with time (Fig. 2.10B).[77] To eliminate the possibility that Manα1-6Manβ1-

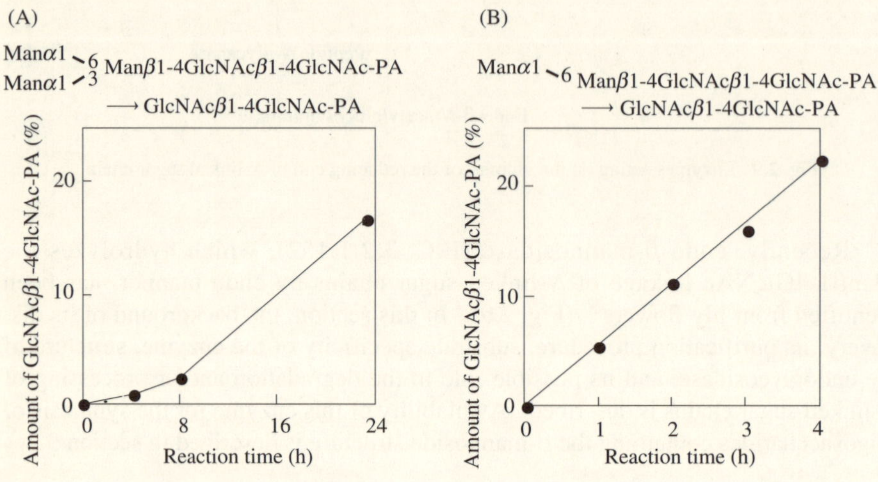

Fig. 2.10 Time course of hydrolysis of Manα1-6(Manα1-3)Manβ1-4GlcNAcβ1-4GlcNAc-PA and Manα1-6Manβ1-4GlcNAcβ1-4GlcNAc-PA with the crude enzyme preparation obtained from lily flowers.
[Reprinted with permission from Sasaki, A. *et al.* (1999) *J. Biochem.* **125**, 365]

4GlcNAcβ1-4GlcNAc-PA was hydrolyzed by a combination of α- and β-mannosidases to produce GlcNAcβ1-4GlcNAc-PA, the hydrolyzates of Manα1-6Manβ1-4GlcNAcβ1-4GlcNAc-PA with a partially purified enzyme were analyzed by high-performance anion-exchange chromatography equipped with a pulsed amperometric detector (PAD).[77] Equimolar amount of GlcNAcβ1-4GlcNAc-PA and Manα1-6Man were detected, but neither Manβ1-4GlcNAcβ1-4GlcNAc-PA nor Man were detected in this analysis (Fig. 2.11), suggesting that the enzyme hydrolyzed Manα1-6Manβ1-4GlcNAcβ1-4GlcNAc-PA to Manα1-6Man and GlcNAcβ1-4GlcNAc-PA (Fig. 2.11).[77] In a similar manner, the partially purified enzyme hydrolyzed larger substrates such as Manα1-3Manα1-6Manβ1-4GlcNAcβ1-4GlcNAc-PA and Manα1-6(Manα1-3)Manα1-6Manβ1-4GlcNAcβ1-4GlcNAc-PA. These results indicated that the enzyme hydrolyzed only the Manβ1-4GlcNAc linkage and that the enzyme is an endoglycosidase.

C. Assay for Endo-β-mannosidase

The enzyme solution and 12.5 μM of Manα1-6Manβ1-4GlcNAcβ1-4GlcNAc-PA in 16 μL of 0.16 M ammonium acetate buffer, pH 5.0, were incubated at 37°C for 30 min. The endo-β-mannosidase activity was measured by quantifying

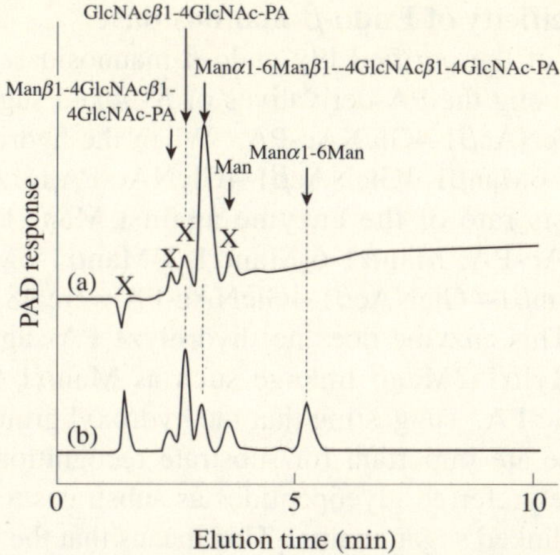

Fig. 2.11 Analysis of hydrolysis products from Manα1-6Manβ1-4GlcNAcβ1-4GlcNAc-PA with the lily endo-β-mannosidase. Manα1-6Manβ1-4GlcNAcβ1-4GlcNAc-PA was incubated with the enzyme for (a) 0 h and (b) 2 h. Products were analyzed by high-performance anion exchange chromatography equipped with a pulsed amperometric detector (PAD). Arrows indicate the elution positions of standard sugar chains. Peaks indicated by x are contaminating materials.

GlcNAcβ1-4GlcNAc-PA generated on size-fractionation HPLC. It is important that Manα1-6Man, the other product from the nonreducing end, is separately confirmed to be detected quantitatively on HPLC equipped with PAD. One unit of the enzyme activity was defined as the amount of enzyme that released 1 nmol of GlcNAcβ1-4GlcNAc-PA from 12.5 μM of Manα1-6Manβ1-4GlcNAcβ1-4GlcNAc-PA per minute under the conditions used.

D. Purification of Endo-β-mannosidase

Endo-β-mannosidase from lily flower was purified to homogeneity.[82)] The hydrolysis activity was fractionated to the soluble fraction and completely precipitated in a solution of 40% saturated ammonium sulfate. The solubilized precipitate was successively applied to DEAE-sephacel, Superdex 200 gel filtration, HA 1000 hydroxyapatite, MonoQ, and PorosHS chromatography.[82)] In the first step of chromatography (DEAE-sephacel), two peaks of enzyme activity were observed. The former major peak was further purified. Finally, this soluble enzyme was purified to 1,200-purification-fold homogeneity. The purified enzyme gave a single band on a native-PAGE and a single peak on a Superdex 200 gel filtration. The molecular weight of the enzyme was estimated to be 78 k judging from the elution volume on Superdex 200 gel filtration, but it gave three protein bands on SDS-PAGE with molecular weight of 28 k, 31 k and 42 k. These observations concluded that the enzyme composed of three polypeptides with mutual noncovalent interaction. The sum of the molecular weights of the three polypeptides (101 k) was not comparable to the one estimated by gel filtration (78 k). This difference may derive from the interaction between the enzyme and dextran-resin of Superdex 200.

E. Substrate Specificity of Endo-β-mannosidase

Substrate specificity of the purified lily endo-β-mannosidase was investigated.[83] The best substrate among the PA-derivatives of *N*-linked sugar chains tested was Manα1-6Manβ1-4GlcNAcβ1-4GlcNAc-PA. When the hydrolysis rate of the enzyme against Manα1-6Manβ1-4GlcNAcβ1-4GlcNAc-PA is taken as 100, the relative initial hydrolysis rate of the enzyme against Manα1-3Manα1-6Manβ1-4GlcNAcβ1-4GlcNAc-PA, Manα1-6(Manα1-3)Manα1-6Manβ1-4GlcNAcβ1-4GlcNAc-PA and Manβ1-4GlcNAcβ1-4GlcNAc-PA were 48, 42, and 4.0, respectively (Table 2.6). This enzyme does not hydrolyze PA-sugar chains containing Manα1-3Manβ or Xylα1-2Manβ linkage such as Manα1-6(Manα1-3)Manβ1-4GlcNAcβ1-4GlcNAc-PA, suggesting that the hydroxyl groups at C-2 and C-3 of the β-linked mannose are important for substrate recognition of the enzyme. In addition, this enzyme preferred glycopeptides as substrates to their corresponding PA-derivatives of *N*-linked sugar chains. This means that the peptide portion to be attached to the sugar chain is also recognized by the enzyme. In fact, K_m values for Manα1-6Manβ1-4GlcNAcβ1-4GlcNAc-PA and Manα1-6Manβ1-4GlcNAcβ1-4GlcNAc-peptide are 1.2 mM and 75 μM, respectively, showing that the affinity of this enzyme to glycopeptides is much higher than to the corresponding PA-sugar chains. Thus the enzyme recognizes a wide area of substrates including the Manβ1-4GlcNAc linkage and the peptide portion. Endo-β-mannosidase was clearly distinguished from other so far reported β-mannosidases such as β-mannosidase and β-mannanase judging from their substrate specificity. Endo-β-mannosidase can not hydrolyze *p*-nitrophenyl β-mannoside, while β-mannosidase and β-mannanase can (Table 2.6). Endo-β-mannosidase can barely hydrolyze β1,4-mannohexaose, which is a good substrate for β-mannanase (Table 2.6).[84]

F. Primary Structure of Endo-β-mannosidase

Endo-β-mannosidase was purified as an enzyme composed of three polypeptides followed by analysis of their N-terminal and internal amino acid sequences. Interestingly, all the amino acid sequences obtained from the three polypeptides were homologous to an *Arabidopsis* gene (At1g09010) encoding a putative glycoside hydrolase. Subsequently, a cDNA corresponding to this gene was cloned from *Arabidopsis* cDNAs. Although At1g09010 was registered as a gene encoding a protein with 650 amino acid residues, the cloned cDNA contained the entire nucleotide sequence of At1g09010 and encoded 944 amino acid residues with a molecular weight of 108 k corresponding to the sum of molecular weight of the three polypeptides of the lily enzyme (Fig. 2.12).[82] This gene product was expressed in *Escherichia coli* as a single polypeptide with a molecular weight of 111 k. The partially purified recombinant enzyme had the activity to hydrolyze Manα1-6Manβ1-4GlcNAcβ1-4GlcNAc-PA to Manα1-6Man and GlcNAcβ1-4GlcNAc-PA. The substrate specificity and other characteristics of the *Arabidopsis* endo-β-mannosidase were very similar to those of the lily enzyme, showing that this *Arabidopsis* cDNA encoded endo-β-mannosidase. A lily endo-β-mannosidase gene was also cloned from lily leaf cDNAs.[83] This cDNA had 2859 bp ORF encoding 952 amino acid residues. All the amino acid sequences of

Table 2.6 The substrate specificity of the purified lily endo-β-mannosidase

Substrate	Hydrolysis rate (%)
Manα1 ↘ 6 Manβ1-4GlcNAcβ1-4GlcNAc-PA	100
Manα1 ↗ 3 Manα1 ↘ 6 Manβ1-4GlcNAcβ1-4GlcNAc-PA	48
Manα1 ↘ 6 Manα1 ↗ 3 Manα1 ↘ 6 Manβ1-4GlcNAcβ1-4GlcNAc-PA	42
Manβ1-4GlcNAcβ1-4GlcNAc-PA	4.0
Manβ1-4GlcNAcβ1-4GlcNAc-PA Manα1 ↗ 3	<0.01
Manα1 ↘ 6 Manβ1-4GlcNAcβ1-4GlcNAc-PA 2 | Xylβ1	<0.01
Manα1 ↘ 6 Manβ1-4GlcNAcβ1-4GlcNAc-PA Manα1 ↗ 3	<0.01
Manα1 ↘ 6 Manα1 ↗ 3 Manα1 ↘ 6 Manβ1-4GlcNAcβ1-4GlcNAc-PA Manα1 ↗ 3	<0.01
Manα1 ↘ 6 Manβ1-4GlcNAcβ1-4GlcNAc-peptide	300
Manβ1-4GlcNAcβ1-4GlcNAc-peptide	9.8
Manα1 ↘ 6 Manβ1-4GlcNAcβ1-4GlcNAc-peptide Manα1 ↗ 3	<0.01
Manβ1-4Manβ1-4Manβ1-4Manβ1-4Manβ1-4Man-PA	<0.01
*p*NP-α-Man	<0.01
*p*NP-β-Man	<0.01

Relative hydrolysis rates to that of Manα1-6Manβ1-4GlcNAcβ1-4GlcNAc-PA are shown. [Reprinted with permission from Sasaki, A. *et al.* (2005) *J. Biochem.* **137**, 90]

```
A.thaliana     -----------MAEIGKTVLDFGWIAARSTEVDVNGVQLTTTNPPAISS--ESRWMEAAVPGTVLGTLVKNKAIPDPFYG  67
G.hirsutum     ----------MAEIGQKTLLDSGWLAARSTDVQLTGTQLTTTYPPTSP---TSPWMEAVVPGTVLATLVENKVVGDPFYG  67
O.sativa       MAAAAAAAAAAAAAEVGKRVLDTGWLAARSTEVALTGEQLTTTDPPPADPEPTAPWMHAAVPGTVLGTLLKNKLIPDPFYG  80
C.fimi         ------------------------MITQDLYDGWTLTAVSGPVPAELAG----VRVRARVPGTSHTALLDEGLIPDPYLD  52
T.neapolitana  --------------------------MRRIDLNGRWQIRDSKGEFSLVG---------LVPGVVQADLVREGLLPHPYVG  45
B.taurus       ----------------MLLRLLLLLAPCGAGFATKVVSISLRGNWKIHSGNGSLQLPAAVPGCVHSALFNKRIIKDPYYR  64
A.aculeatus    -----------MRALPTTATTLLGVLFFPSASRSQYVRDLGTEQWTLSSATLNRTVPAQFPSQVHMDLLREGIIDEPYND  69

A.thaliana     LENEAITDIADSGRDYYTFWFFTKFQCQRLLNQYVHLNFRAINYSAQVFVNGHKTELPKGMFRRHTLDVTDILHPE-SNL 146
G.hirsutum     LENETILDIADSGREYYTFWFFTKFQCKLSGAQHLDLNFRAINYSAEVYLNGHKRVLPKGMFRRHSLEVTDILNPDGSNL 147
O.sativa       LNNESIIDIAKSGRGHYTFWFFTTFQCAPAGHQHVSLNFRGINYSAEVYLNGHKEVLPKGMFRRHTLDITDVLRPDGKNL 160
C.fimi         RNE----DVLAWMRRTDWAYERELVLDPAAADERVDLVFGGIDTVGTVTFDGHELGRTANQHRSYRFDVRALLRPDTQRL 128
T.neapolitana  TNEDLFKEIEDREWIYEREFN---FREDLLDEDRVDLVFEGIDTLADVYLNDVYLGSTEDMFLEYRFDIKGVLKEK-NHL 121
B.taurus       FNNLDYRWIALDNWTYIKKFK---LHSDMSTWSKVNLVFEGIDTVAVVLLNSVPIGKTDNMFRRYSFDITHTVKAVNIIE 141
A.aculeatus    LNDFNLRWIADANWTYTSGKIEG----LGEDYESTWLVFDGLDTFASISFCGQFVGATDNQFRQYMFDVSSILKACPEEP 145

A.thaliana     LALIVHPPDHPGTIPP-EGGQGGDHEIG---------KDVAAQYVQGWDWICPIRDRNTGIWDEVSISVTGPVRIIDPHL 216
G.hirsutum     LAVLVHPPDHPGSIPP-VGGQGGDHEIG---------KDVATQYVEGWDWIAPVRDRNTGIWDEVSISVTGPVKIIDPHL 217
O.sativa       LAVLVHPPVHPGAIPP-QGGQGGDHEIG---------KDVATQYVEGWDWMCPIRDRNTGIWDEVSIPVTGPVRIMDPHL 230
C.fimi         RVDLRAAIVHAEAERERLGHRPLAYPQP---------FNMVRKMACSFGWDWGPDLQTAGLWKPVRVERWRTARLAS--- 196
T.neapolitana  RVFIKSPVKVPKTLEQNYGVLGGPEDSI---------RGYIRKAQYSYGWDWGARIVTSGIWRPVYIETYRKARLQDS-- 190
B.taurus       VRFQSPVVYANQRSERHTAYWVPPNCPPPVQDGECH-VNFIRKMQCSFGWDWGPSFPTQGIWKDVRIEAYNVCHLNYFMF 220
A.aculeatus    TLGIQFGSAPNIVDAIAQDPSSPTWPEGVQITYEYPNRWFMRKEQSDFGWDWGPAFAPAGPWKPGYVVQLKQAAPVYVRN 225

A.thaliana     VSTFFDDY-KRAYLHVTAELENKSTWNTECSVNIQITAELENGVCLVEHLQTENVLIPAQGRIQH---TFKPLYFYK-PE 291
G.hirsutum     VSSFFDRY-TRVYLHATTELENRSSWVAECSLNIQVTTELEGSVCLMEHLKTQHVSIPPRARIQY---TFPQLFFYK-PN 292
O.sativa       VSTFYDDF-KRSYLHCTLQLENRSSWLSDCKLKIQVSTELEGNICLVEHLQSYEISVPPNSVLEY---TIPPLFFYK-PN 305
C.fimi         -----------VRTHVTVDPDGTGRVRVLVDLERSGLPGGDAPVTLRARVLSADVAVTVPGTATS---AVVELEVPR-AP 261
T.neapolitana  ---------------TAYLVDLKGKDAVVKVNGFVYGEGDLSVDVFVNGEKMGSYQVMEKNGEK---FFEGVFSLKNVK 251
B.taurus       TP---------IYDNYMKTWNLKIESSFDVVSSKLVSGEAIVAIPELNIQQTNNIELQHGERTVE---LFVKIDKAIIVE 288
A.aculeatus    TDLDIYRLGQINYLPPDQTQPWVVNASLDYLGSLPENPSMAIEVKDLQSGEILASRPLTNITVTEGSVTGVTVLEGVDPK 305

A.thaliana     LWWPNGMGKQNLYDILITVVVNEFGESDSWMQPFGFRKIESVIDSVTGG-----------RLFKINGEPIFIRGGNWILS 360
G.hirsutum     LWWPNGMGKQSLYNVSITVDVKGHGESDSWGQLFGFRKIESHIDSATGG-----------RLFKVNGQPIFIRGGNWILS 361
O.sativa       LWWPNGMGKQSLYNVEIGVDANGFGESDSSNHHFGFRKIESTIDGSTGG-----------RIFKVNGEPVFIRGGNWILS 374
C.fimi         LWWPVGHGPQPLSDLTVTLATRDDEPLDSWSRRIGFRTVEVDTTPDEDGT--P-------FTFRVNGRPVFVKGANWIPD 332
T.neapolitana  LWYPWNVGEPYLYDFTFILKESE-KEVYREEKKIGLRKVRILQEPDGEGK--T-------FIFEINGEKVFAKGANWIPA 321
B.taurus       TWWPHGHGNQTGYNMSVIFELDG-GLRFEKSAKVYFRTVELVEEPIQNSPGLS-------FYFKINGLPIFLKGSNWIPA 360
A.aculeatus    LWWPQGLGDQNLYNVTISVTDGGNQSVAEVTKRTGFRTIFLNQRNITDAQLAQGIAPGANWHFEVNGHEFYAKGSNLIPP 385

A.thaliana     DGLLR-LSKERYRTDIKFHADMNMNMIRCWGGGLAERPEFYHFCDIYGLLVWQEFWITGDVDGRGVPVSNPNGPLDHELF 439
G.hirsutum     DCLLL-LSKERYKTDIKFHADMNLNMIRCWGGGLAERPEFYHYCDVYGLLVWQEFWITGDVDGRGVPVSNPNGPLDHDLF 440
O.sativa       DGLLR-LTRKRYMTDIKFHADMNFNMLRCWGGGLAERPDFYHFCDIYGLMVWQEFWITGDVDGRGIPISNPNGPLDHDLF 453
C.fimi         DHLLTRITRERLAHRLDQAVEANLNLLRVWGGGIYESEDFYDLCDERGLLVWQDFLLACAAYPEEQPIWD--------EL 404
T.neapolitana  DNILTWLKTEDYEKLVKMAKEANMNMLRVWGGGIYESEDFYKLCDELGIMVWQDFMYACLEYPDHLPWFR-------KLA 394
B.taurus       DSFQDRVTSAMLRLLLQSVVDANMNALRVWGGGVYEQDEFYELCDELGIMIWQDFMFACALYPTDKDFMDS--------V 432
A.aculeatus    DCFWTRVTEDTMTRLFDAVVAGNQNMLRVWSSGAYLHDYIYDLADEKGILLCSEFQFSDALYPTDDAFLEN--------V 457

                                  *
A.thaliana     LLCARDTVKLLRNHPSLALWVGGNEQVPPKDINEALKQDLRLHSYFET---------QLLSDKDSDPSVYLDGTRVYIQG 510
G.hirsutum     MLCARDTVKLLRNHPSLALWVGGNEQVPPADINTSLKNDLKLHPFFESQSENITS-VEGLSTAYKDPSQYLDGTRVYIQG 519
O.sativa       LLCARDTVKLLRNHASLALWVGGNEQVPPVDINKALKNDLKLHPMFVSNHTTKSP-GKDISEDPTDPSKYLDGTRVYVQG 532
C.fimi         EAEARENVARLTPHASLVLWNGGNENLWG---FMDWGWPQELEGRTWGYRLATEL-LKGVVAELDPTRPYADGSPYSPGF 480
T.neapolitana  NDEARKIVRKLRYHPSIVLWCGNNENNWG--FDEWGNMSRKVDGINLGNRLYLFD-FPRICAEEDPATPYWPSSPYGGE- 470
B.taurus       REEVTHQVRRLKSHPSIITWSGNNENEAALMMGWYDTKPGYLQTYIKDYVTLYVKNIRTIVLEGDQTRPFITSSPTNGAK 512
A.aculeatus    AAEVVYNVRRVNHHPSLALWAGGNEIESLMLLLVEAADPESYPFYVGEYEKMYISLFLPLVYENTRSISYSPSSTTEGYL 537

                                                       *
A.thaliana     SMWD--GFADGKGNFTDGPYEIQY---------PEDFFKDTYYKYGFNPEVGSVGMPVAETIRATMPPEGWTIPLFKKGL 579
G.hirsutum     SMWD--GFANGKGGFTDGPYEIQN---------PEDPFKDNFYKYGFNPEVGSVGIPVAATIRATMPREGWQIPLFKKLP 588
O.sativa       SMWD--GFANGKGDFTGGPYEIQY---------PESFFKDSFYKYGFNPEVGSVGVPVAATIRATMPSEGWSIPIFKKRI 601
C.fimi         ALDD--VHPNDPDHGTHHEWEVWN---------RVDYSAYRDDVPRFCSEFGFQGPPTWSTLTRAVRAD---------- 538
T.neapolitana  -------KANSEKEGDRHVWNVWSG--------WMNYDHYEKDTGKFISEFGFQGAPHMKTIEFFSKPQ---------- 524
B.taurus       TIAEGWLSPNPYDLNYGDVHFYDYV--------SDCWNWRTFPKARFVSEYGYQSWPSFSTLEKVSSEE---------- 573
A.aculeatus    DIDLSAPVPMAERYSNTTEGEYYGDTDHYNYDASIAFDYGTYPVGRFANEFGFHSMPSLQTWQQALTDP---------- 606

A.thaliana     DGFIKEVPNRMWDYHKYIPYSNPGKVHDQILMYGTP------ENLDDFCLKAQLVNYIQYRALFEGWSS------QMWTK 647
G.hirsutum     NGYTEEVPNPIWQYHKYLPYSKPGKVHDQIELYGTP------EDLDDFCLKAQLVNYIQYRALLEGWTS------RMWSK 656
O.sativa       DGYINEVPNPIWDYHKYIPYSKPGKVHDQIELYGHP------SDLDDFCEKAQLVNYVQYRALLEGWTS------FMWTK 669
C.fimi         DGGPLTKDDPTFLLHQKAEDGNGKLDRGLAPHLGVP------AGFVDWHWATQLNQARAVAFAIEHYR-------SWWPR 605
T.neapolitana  ---ERDPFHPVMLKHNKQVEGQERLIRFIYGNFGRC------RDFESFVYLSQLNQAEAIKFGVEHWR-------SRKYK 588
B.taurus       ---DWSYRSSFALHRQHLINGNNEMLHQIELHFKLPNSTDQLRRFKDTLYLTQVMQAQCVKTETEFYRRSRSEIVNGKGH 650
A.aculeatus    --ADLTFNSSVVMLRNHHYPAGGLMTDNYHNTVARHGRNDPGRAGLLPDAQHSVRPRGQLQRLVPRDPALPGGPLQVTNP 684

A.thaliana     YTGVLIWKNQNPWTGLRGQFYDHLLDQTASFYGCRSAAEPVHVQLNLASYFVEVV--NTTSKELSDVAIEASVWDLDGNC 725
G.hirsutum     YTGVLIWKTQNPWTGLRGQFYDHLLDQTAGFFGCRCAAEPIHVQLNLATYFIEVV--NTTAEELSNVAIEASVWDLEGAC 734
O.sativa       FTGVLIWKTQNPWTGLRGQFYDHLLDQTAGFYGCRCAAEPIHVQLNLDSYFIEVV--NTTADELRDVAVEISAWDLDGAS 747
C.fimi         TAGAIVWQLNDCWPVTSWAAIDGDERVKPLWHALRRAYAPRLLTVQPR----------DGRDELAVVNDTGGLWQGTVRL 675
T.neapolitana  TAGTLFWQLNDSWPVFSWSAVDYFKRPKALYYYARRFFADVLPVVKKV----------DGKVGLFVVSDLRELKKASVRF 658
B.taurus       TMGALYWQLNDIWQAPSWSSLEYGGKWKMLHYFARHFFAPLLPVGFED-------------KDMLFIYGASDLHSDQQMM 717
A.aculeatus    VLPAGQRAARTPARVPVLAARGHLAGALVGGDRVRRPLEGPHYVARDIYKPVIVSPFWNYTTGALDIYVTSDLWTAAAGS 764

A.thaliana     PYYKVFKIVSAPPKKVVKISEFKYPKAANPKHVYFLLLKLYTVSDKAVISRNFYWLHLPGKNYTLLEPYRKKQIPLKITC 805
G.hirsutum     PYYKVFDKLSLPPKKVVSISEMKYPKSKNPKPVFFLLLKLYHVSNYSIVSRNFYWLHVSGGDYKLLEPYRNKRIPLKITS 814
O.sativa       PYYRVTEKIAVPPKKVQQVTEMSYPKTKNPKPVYFLLLKLFKLSDNQVLSRNFYWLHLPGKDYKLLEQYRQKQIPLKINS 827
C.fimi         SRRTLDGATLAEVELGLAVGAWSVG--------LFALPDEVAAPDDAAGEVLVVLDGDVRTVHTWAQDVDLRLDPDPVSA 747
T.neapolitana  AAYRVSGDLVFEKFYDVTLPADSVT-----------LVDTVSTSDDLVFFVEVRVEGRSFKNYRILKKWREMDVPD---- 723
B.taurus       LTVRVHTWSSLELVCSESTNPFVIKAGESVLLYTKPVPELLKGCPGCTRQSCVVSFYLSTDGELLSPINYHFLSSLKN-- 795
A.aculeatus    VTLTWRDLSGKPIASNGGLPTKPLPFHVGALNSTRLYRMNMKQQPLPRHEDAILALELTATGSLPNTDEEVTFTHEQWFT 844

A.thaliana     NAVMVGSRYELEVNVHNTSRANLAKNVVQE-------------------------DEKRDLGLLQKLFSRCVVSADSNRG 860
G.hirsutum     KTFIKGSSYEVEMKVLNKSKKPDPKTLTYKNNFAVRNDDSDFDMTSLKPIPDTRTDLKQPTGLFQRLYRQFSRESD---G 891
O.sativa       KISISGSGYEVRMSIENRSKKPENANVSTMNLADANGSDRTG----------EEAIQDGHSSGLWGKIRRGLIITRSDDN 897
C.fimi         TVSPLQQDGYRVDVTARTFARS------------------------------------------------VTLHVDRLDP 778
T.neapolitana  ------------------------------------------------------------------------------PG 725
B.taurus       -------------------------------------------------------------------------------A 796
A.aculeatus    PAFP--------------------------------------------------------------------------KD 850

A.thaliana     LKVVEMKGSDSGVAFFLRFSVHNAETEKQ-----DTRILPVHYSDNYFSLVPGESMSFKISFAAPTGMKKSPRVMLQGWN 935
G.hirsutum     LRVAEINGSDGGVAFFLHFSVHGAKLEHEE--GEDSRILPVHYSDNYFSLVPGEEMSIKISFKVPPGVS--PRVTLRGWN 967
O.sativa       VRTVEVKGADSGVSFFLHFSVHTSEPSSSQDVYKDTRILPVHYSDNYFSLVPGEKMAIDISFEAPQGST--PRVILKGWN 975
C.fimi         DATVDDALVDVPAGETFSFHVRTSARFDAA----ALTRSPVLRTANDVVVPRGATAPAGAERDLQQSR------------ 842
T.neapolitana  ISVIEKENEIEITAERPAFGIKILSEELPE-----DDFLFLEPGEKIILKKPGGFFDVKSLFDYLER------------- 787
B.taurus       KGLHKANITATISQQGDTFVFDLKTSAVAPFVWLDVGSIPGRFSDNGFLMTEKTRTVFFYP------------------- 857
A.aculeatus    LDLVNLRVRVEYDAPLGKFAVEATAGVALYTWLEHPEGVVGYFEENSFVVVPGQKKVVGFVVQADETDGEWVHDVTVRSL 930
```

Fig. 2.12 Continue

```
A.thaliana      YPDRFSVFG                                       944
G.hirsutum      YHHGVHTVL                                       976
O.sativa        YHLDHAVTL                                       984
C.fimi          ---------                                       842
T.neapolitana   ---------                                       787
B.taurus        ---------                                       857
A.aculeatus     WDLNEGE--                                       937
```

Fig. 2.12 Amino acid sequence alignment of endo-β-mannosidases, their plant homologues and β-mannosidases. A. thaliana, *Arabidopsis thaliana* endo-β-mannosidase; G. hirsutum, *Gossypium hirsutum* (plant) putative glycoside hydrolase; O. sative, *Oryza sativa* (plant) putative glycosyl hydrolase; B. taurus, *Bos taurus* (mammal) β-mannosidase; A. aculeatus, *Aspergillus aculeatus* (fungus) β-mannosidase; C. fimi, *Cellulomonas fimi* (bacteria) β-mannosidase; T. neapolitana, *Thermotoga neapolitana* (thermophilic bacteria) β-mannosidase. Sequences were aligned using the CLUSTALW program. Gaps are marked by dashes. The conserved amino acid residues in all the β-mannosidases aligned are in bold. The proposed catalytic glutamic acid residues (Glu464 and Glu559 in AtEBM numbering) are marked by asterisks.
[Reprinted with permission from Ishimizu, T. *et al.* (2004) *J. Biol. Chem.* **279**, 38560]

the lily enzyme analyzed were identical to the putative amino acid sequence derived from the nucleotide sequence of the cloned cDNA, showing that the three polypeptides, components of endo-β-mannosidase, were encoded by a single gene and generated by proteolysis after translation. Amino acid sequence identity between lily and *Arabidopsis* endo-β-mannosidases is 67 %.

Highly homologous sequences (66 to 70 % identity in amino acid sequences) to those of endo-β-mannosidases were found in a public database (Fig. 2.12).[82, 83] They are *Oryza sativa* (rice) and *Gossypium hirsutum* (cotton) cDNAs, but their functions have not yet been analyzed. An endo-β-mannosidase-like sequence is found as a single copy gene in the genome of *Arabidopsis thaliana* and *Oryza sativa*. Many ESTs derived from various plant species are also found to encode endo-β-mannosidase-like sequences. Amino acid sequences of endo-β-mannosidases are slightly homologous to β-mannosidases from mammals[85-88] (participating in degradation of *N*-linked sugar chains), fungus[89, 90] and bacteria[91-93] (participating in hydrolysis of galactomannan) with 13~20 % identity (Fig. 2.12). All the homologous genes from species other than plant were β-mannosidases. Species sequencing their genome from other than plant species do not possess an endo-β-mannosidase-like gene. In addition, endo-β-mannosidase activity has not been detected in any species tested other than plant species. The above findings indicate that endo-β-mannosidase is a plant-specific endoglycosidase.[82, 83]

Endo-β-mannosidase is classified in the glycoside hydrolase family GH-A[94] to which β-mannosidase belongs. Enzymes belonging to this family hydrolyze the glycosidic bond with retention of the stereochemistry at the cleavage point and are proposed to form the $(\alpha/\beta)_8$ barrel fold structure.[95] Several dozen amino acid residues have been conserved in the sequence alignment of endo-β-mannosidases and β-mannosidases (Fig. 2.12).[82, 83] Of these, two glutamic acid residues, Glu-464 and Glu-549 (numbered *Arabidopsis* endo-β-mannosidases), are candidates for the general acid-base catalyst and the nucleophile, respectively, based on the results of *Cellulomonas fimi* β-mannosidase obtained by chemical modification experiments and site-directed mutagenesis (Fig. 2.12).[96, 97] In fact, alanine replacement of Glu-464 or Glu-549 eliminated enzyme activity.[82] The amino acid sequence of endo-β-mannanase,[98] which participates in mannan degradation, is slightly homologous only in the region around these two glutamic acid residues of endo-β-mannosidases.

The phylogenetic tree of endo-β-mannosidases and β-mannosidases was constructed (Fig. 2.13).[82, 83] Endo-β-mannosidases and their plant homologues form a cluster and are clearly distinguished from the β-mannosidases.

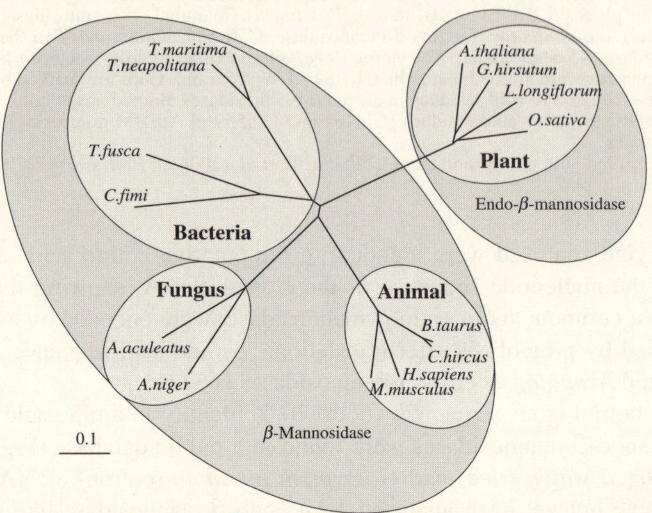

Fig. 2.13 Phylogenetic tree of endo-β-mannosidases, their plant homologues and β-mannosidases. Sequences used are from *Capra hircus*, *Homo sapiens*, *Mus musculus*, *Aspergillus niger*, *Thermobifida fusca*, *Thermotoga maritima*, in addition to the sequences used in Fig. 2.12. The bar represents the number of nucleotide substitutions.
[Reprinted with permission from Ishimizu, T. *et al.* (2004) *J. Biol. Chem.* **279**, 38561]

G. Possible Role of Endo-β-mannosidase in the Plant Cell

Endo-β-mannosidase was discovered[8] based on speculation about the processing pathway of *N*-linked sugar chains to generate an *N,N′*-diacetylchitobiose structure on S-RNase.[9] The question arises: is this endoglycosidase really involved in producing an *N,N′*-diacetylchitobiose structure during the processing and/or degradation of *N*-linked sugar chains? This has not yet been clarified, but the role of endo-β-mannosidase is speculated from its substrate specificity. This enzyme hydrolyzes the Manβ1-4GlcNAc linkage of oligomannose-type *N*-linked sugar chains, but does not hydrolyze sugar chains containing the Manα1-3Manβ structure. Interestingly, this substrate specificity of endo-β-mannosidase is complementary to that of plant α-mannosidases, such as jack bean α-mannosidase, which hydrolyzes Manα1-3Manβ linkage faster than Manα1-6Manβ.[99-101] That is, oligomannose-type *N*-linked sugar chains of which Manα1-3Manβ linkage has been hydrolyzed with an α-mannosidase become good substrates for endo-β-mannosidase. α-Mannosidase and endo-β-mannosidase may act cooperatively on the trimannosyl core *N*-linked sugar chain to produce the *N,N′*-diacetylchitobiose structure on a glycoprotein such as S-RNase[81] (Fig. 2.14). This speculation agrees well with the observation that a crude enzyme hydrolyzed Manα1-6(Manα1-

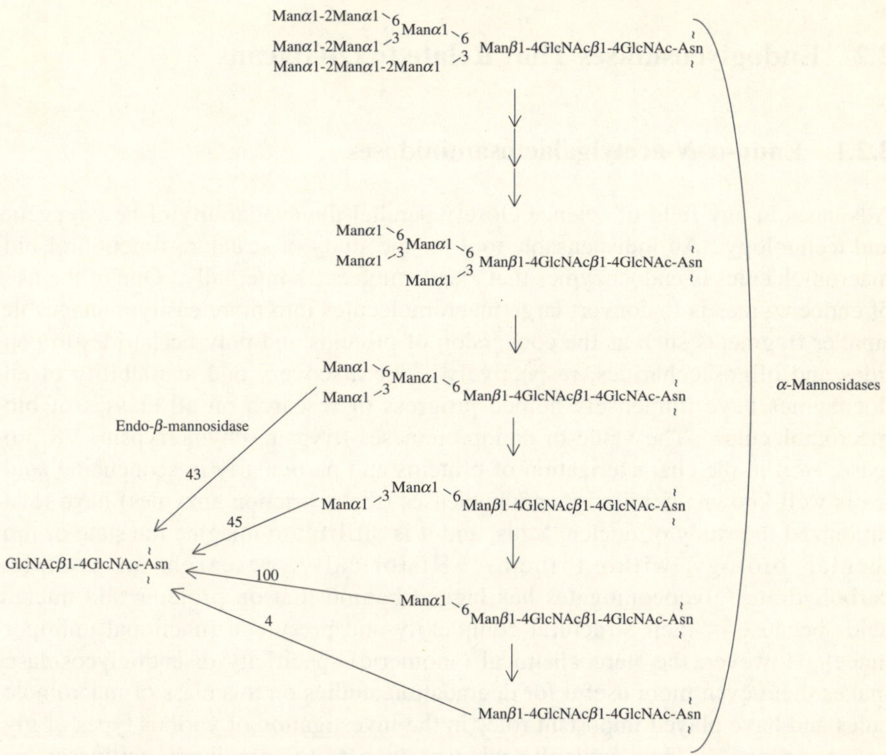

Fig. 2.14 Proposed pathway for degradation and/or processing of oligomannose-type *N*-linked sugar chains to an *N,N'*-diacetylchitobiose structure in plants. The numbers near the arrows indicate relative hydrolysis rate by endo-*β*-mannosidase. The rate against Man*α*1-6Man*β*1-4GlcNAc*β*1-4GlcNAc-PA is taken as 100.

3)Man*β*1-4GlcNAc*β*1-4GlcNAc-PA biphasically (Fig. 2.10a). During the secretion of S-RNase, which is a secretory protein bearing an *N,N'*-diacetylchitobiose as a major *N*-linked sugar chains, it must pass through the organelle where *α*-mannosidase and endo-*β*-mannosidase are localized. On the other hand, mammals not possessing an endo-*β*-mannosidase-like gene have a *β*-mannosidase and an *α*-1,6-mannosidase specific for Man*α*1-6Man*β* in lysosome.[102] *β*-Mannosidase and *α*-1,6-mannosidase may cooperatively be involved in the degradation of Man*α*1-6(Man*α*1-3)Man*β*1-4GlcNAc*β*1-4GlcNAc to GlcNAc*β*1-4GlcNAc in mammals. Localization of endo-*β*-mannosidase in the plant cell should be determined to prove the proposed pathway. In addition, endo-*β*-mannosidase-deficient mutant plants will provide information on the functions of this endoglycosidase in plant cell.

2.2 Endoglycosidases That Relate to *O*-Glycans

2.2.1 Endo-α-*N*-acetylgalactosaminidases

Advances in any field of science closely parallel the availability of new reagents and technology. An indispensable tool for the study of structure-function of biomacromolecules is endoenzymes that cleave molecules internally. One of the uses of endoenzymes is to convert large macromolecules into more easily manageable, smaller fragments such as the conversion of proteins and polysaccharides to peptides and oligosaccharides, respectively. The discovery and availability of endoenzymes have immensely helped progress of research on all classes of biomacromolecules. The value of endoproteinases (trypsin, chymotrypsin, V8 protease, *etc.*) in the characterization of proteins and particularly in sequencing studies is well known. Similarly, endonucleases (the restriction enzymes) have revolutionized the study of nucleic acids, and it is difficult to imagine the state of molecular biology without them. Historically, research on complex carbohydrates/glycoconjugates has lagged behind that on proteins and nucleic acids because of their structural complexity and presumed functional unimportance! However, the stereochemical (anomeric) specificity of endoglycosidases makes them even more useful for degradation studies on this class of macromolecules and have played important roles in the investigation of various types of glycoconjugates.[103] Thus, endoglycosidases such as the amylases, chitinases and lysozymes greatly benefited the early studies on the elucidation of structures of starch, glycogen, chitin and the peptidoglycans of bacterial cell wall. In the case of the very complex animal structural glycoconjugates, glycosaminoglycans and proteoglycans, the increase in our knowledge of their structures has paralleled the discovery of endoglycosidases acting on them. Thus, hyaluronidase first described by Karl Meyer and coworkers,[104] was invaluable in establishing the structures of the repeating disaccharide units of not only hyaluronan but also the chondroitin sulfates. Further, chondroitinases, heparitinase/heparinase and keratanases have played important roles in elucidating the detailed structures of the respective glycosaminoglycans. The more recent discoveries by Masahiko Endo and coworkers of endoglycosidases that act on the linkage region of proteoglycans and the exciting applications of these enzymes are topics of this monograph. Similarly, the research on the *N*-linked glycoproteins greatly benefited by the discovery of endo-β-*N*-acetylglucosaminidase D (Endo-D) by Muramatsu,[1] followed by that of many others with related activity. For example, the differing specificities of Endo-D and Endo-H played a major role in elucidating the biosynthetic and processing pathway of the *N*-linked glycoproteins. In contrast, the progress in the field of *O*-linked glycoproteins and more specifically of epithelial mucins has been slow partly due to the lack of endoglycosidases that act on these macromolecules. The discovery of enzyme(s) similar to *N*-glycanases that can release the intact *O*-linked saccharides from mucin glycoproteins would be a major contribution to the field.

By definition, endo-α-*N*-acetylgalactosaminidases (Endo-α-GalNAc-ases) are enzymes that cleave an internal α-*N*-acetylgalactosaminyl linkage in a molecule. However, in the field of glycoconjugates the term Endo-α-GalNAc-ase is also used to describe enzymes that are capable of releasing intact saccharides, di- or larger, by hydrolysis of the linkage between α-*N*-acetylgalactosamine and serine or threonine residues. In reality, the only well-characterized Endo-α-GalNAc-ase known to-date is one that has strict specificity for releasing the disaccharide Galβ1-3GalNAcα- from natural or synthetic substrates.

Iwase and Hotta have previously reviewed the topic of Endo-α-GalNAc-ases.[105] In this chapter, first the structure and function of the glycoproteins containing the α-*N*-acetylgalactosaminyl serine or threonine linkages are briefly discussed. This is followed by a review of the purification, properties and applications of the currently known Endo-α-GalNAc-ases. Finally, a perspective of the possible future direction of research on this subject is presented.

A. Glycoproteins Containing GalNAcα-Serine/Threonine Linkages

Glycoproteins in which oligosaccharides are linked via GalNAc to the hydroxyl group of serine and threonine are referred to as *O*-linked glycoproteins or *O*-glycosylated proteins. These glycoconjugates are also called mucin-type glycoproteins, since this is the primary mode of linkage in mucins. There are a great variety of Ser/Thr-linked saccharides based on the different core structures. A general structure for the saccharides linked to Ser/Thr could be depicted as follows:

Peripheral	**Core**	**Linkage region**
Sialic acid, fucose, sulfate, GalNAc	Gal, GalNAc, GlcNAc	GalNAc

The eight different core structures found so far in *O*-linked glycoproteins are listed in Table 2.7. The numbering approximately reflects the frequency of occurrence since types 1 to 4 cores are widely distributed and types 5 to 8 have been found so far only in a few glycoproteins. It is clear from the depicted core structures that Ser/Thr-linked saccharides can be unbranched or branched. The *O*-linked saccharide chains are terminated typically, but not exclusively, by the addition of sialic acids, fucose and sulfate residues. Sialic acid and fucose are always linked by an α anomeric linkage. In certain mucin-type glycoproteins the saccharide chains are terminated by α-linked Gal or GalNAc, which confer blood group A or B status, respectively. When sulfation occurs it is on Gal, GlcNAc or GalNAc residues, either terminal or internal. The combination of different core structures, chain elongation, formation of branch chains and variation of terminations give rise to a bewildering variety of saccharides in the *O*-linked sialoglycoproteins. The number of oligosaccharides per molecule of protein range from a few to a few hundreds occurring as dense clusters along the protein backbone. In fact, there is no reliable estimate of the exact number of oligosaccharide units for any epithelial mucin

Table 2.7 The core structures of saccharides *O*-glycosidically linked to serine or threonine

Core Type	Structure
Core 1	Galβ1-3GalNAcα-
Core 2	Galβ1-3GalNAcα- 6 | GlcNAcβ1
Core 3	GlcNAcβ1-3GalNAcα-
Core 4	GlcNAcβ1-3GalNAcα- 6 | GlcNAcβ1
Core 5	GalNAcα1-3GalNAcα1-
Core 6	GalNAcα- 6 | GlcNAcβ1
Core 7	GalNAcα- 6 | GalNAcα1
Core 8	Galα1-3GalNAcα-

molecule, which usually has a high content (50 to 80% by weight) of carbohydrate compared to other sialoglycoproteins. There are two clearly distinguishable sub-classes of *O*-linked sialoglycoproteins.

Class one *O*-linked sialoglycoproteins primarily represent the cell surface, membrane-associated molecules such as erythrocyte membrane glycophorins, human white blood cell-associated leukosialin (CD43, sialophorin), platelet membrane-associated CD42b (glycocalcin), mouse macrosialin (CD68), synovial fluid lubricin, human milk fat granule membrane glycoproteins, *etc*. The expression of this sub-class of membrane-associated cell surface mucin glycoproteins is markedly increased in malignantly transformed cells.[106] Because of their important role in the pathobiology of cancer there has been increased interest in this class of sialoglycoproteins. The *O*-linked saccharides are also widely implicated as ligands for selectins, which are involved in thrombosis, inflammation, allergy, autoimmunity and cancer metastasis. For example, the endothelial cell-associated mucin type glycoproteins such as GlyCAM-1, CD34 and MAdCAM-1 have all been reported to be ligands for L-selectin, and the leukocyte membrane mucin-type glycoprotein PSGL-1 has been shown to be a ligand for P-selectin and E-selectin.

Class two *O*-linked sialoglycoproteins comprise the mucins, which are the major components of epithelial mucus secretions. Secretory mucins generally have poorly defined but apparently very high molecular weights (of the order of millions) and in solution yield highly viscous solutions or gels. The effective functioning of mucin is dependent on its ability i) to maintain the appropriate viscoelastic properties of the mucus secretion and ii) to provide saccharides as decoys for the adherence of pathogens and thereby inhibit infection of the epithelial cells. Abnormality in mucins is implicated in the pathobiology of several human diseases. These include cystic fibrosis, chronic bronchitis, Crohn's disease, duodenal ulceration, colonic adenocarcinomas, infertility problems and inflammatory ulcerative colitis. Despite the importance of the epithelial mucins in human health and disease, information on their structure, function, biosynthesis and regulation of expression is fragmentary. Major reasons for lack of progress are the structural complexity of these macromolecules and unavailability of tools such as the endoglycosidases that deglycosylate mucin glycoproteins without affecting protein backbone integrity.

B. Endoglycosidases That Act on the *O*-Linked Glycoproteins

1) Endo-α-*N*-acetylgalactosaminidase from *Diplococcus* (*Streptococcus*) *pneumoniae* (Endo-α-GalNAc-ase-D)

The only endoenzyme capable of acting on this class of glycoconjugates available to date was discovered more than 25 years ago. While investigating the mucin-type glycopeptides isolated from mouse melanoma we unexpectedly discovered an endo-α-*N*-acetylgalactosaminidase (Endo-α-GalNAc-ase-D) activity in the culture filtrates of *Diplococcus pneumoniae*.[107)] Our report was soon followed by publications by Endo and Kobata,[108)] and Glasgow *et al.*,[109)] who had independently discovered the same activity. We purified the enzyme and initially investigated its substrate specificity for natural substrates.[110)] The purification of this enzyme is straightforward except for the removal of sialidase activity, which strongly associates and coelutes with the Endo-α-GalNAc-ase in standard procedures of purification including DEAE-Sephadex chromatography. The two activities could be separated either by affinity chromatography, isoelectric focusing, or repetitive ion-exchange chromatography with varying gradients. Subsequent studies on the activity of the enzyme on synthetic, chromogenic substrates, done in collaboration with Khushi Matta, helped to establish its strict substrate specificity for the core 1 disaccharide, Galβ1-3GalNAc.[111)] Some 20 years later, the substrate specificity of the Endo-α-GalNAc-ase from *D. pneumoniae* was re-examined by Brooks and Savage.[112)] These investigators used desialylated bovine submaxillary mucin containing six of the known core structures (Table 2.7) and sialylated antifreeze glycoprotein as substrates and confirmed the strict specificity of the enzyme for the non-sialylated Galβ1-3GalNAc. Essentially, this enzyme is able to release Galβ1-3GalNAc from a variety of synthetic and natural substrates including macromolecules such as asialoepiglycanin,[113)] and asialoepitectin.[114)] The pH optimum of Endo-α-GalNAc-ase-D is in the range of 6.0 to 7.6 depending on the buffer used. Thus Endo-α-GalNAc-ase-D is well suited for treating intact cells while maintaining viability. Like most glycosidases, this enzyme exhibits some specificity towards the aglycon. Thus, with synthetic substrates the enzyme did not hydrolyze the methyl α-glycoside and showed 70% activity towards the phenyl α-glycoside compared to glycosides of *p*-nitrophenyl- and *o*-nitrophenyl α-glycosides of Galβ1-3GalNAc.[111)] During the studies on the deglycosylation of epitectin (MUC1 glycoprotein) an unidentified product with molecular weight higher than Galβ1-

3GalNAc was detected.[114] The product was characterized as Galβ1-3GalNAcα1-1(3)glycerol formed by the transfer of the disaccharide to glycerol, which was present in the enzyme preparation as a protectant against denaturation.[115] Therefore, when using this enzyme it is essential to remove glycerol, which may have been added to commercial samples to stabilize its activity. Another, precaution to prevent transglycosylation is to avoid using buffers containing potential acceptors such as Tris or ethanolamine. We also documented the ability of the enzyme to transfer Galβ1-3GalNAc to other acceptors including *p*-nitrophenol, threonine, serine and hexoses.[115] The use of transglycosylation activity of the Endo-α-GalNAc-ases for synthesis of neoglycoconjugates is discussed in Chapter 4.

Since 1977, this enzyme has been marketed by several companies and is widely used by glycobiologists. Unfortunately, the very strict specificity of this enzyme limits its general utility. We would like to point out that the enzyme is sometimes sold as *O*-glycanase, which implies broad substrate specificity for *O*-linked saccharides. This misleading name, probably chosen as a marketing strategy, has resulted in confusion about its substrate specificity among investigators using the commercial enzyme.[112]

2) Endo-α-*N*-acetylgalactosaminidase from *Clostridium perfringens*
As early as 1972, the presence of an Endo-α-GalNAc-ase activity in culture filtrates of *Clostridium perfringens* was reported.[116] The authors referred to the activity as α-*N*-acetylgalactosaminyl oligosaccharase implying its ability to release oligosaccharides from *O*-linked glycoproteins. Subsequently, it was reported that the purified enzyme released oligosaccharides,[117] and the enzyme preparation was marketed as an oligosaccharase (Bethesda Research Laboratories, MD). However, subsequent studies by other investigators on the purified enzyme established that the Endo-α-GalNAc-ase produced by *Clostridium perfringens* has the same specificity as that of the *D. pneumoniae* enzyme.[118]

3) Endo-α-*N*-acetylgalactosaminidase from *Alcaligenes* sp. (Endo-α-GalNAc-ase-A)
Tochikura and coworkers isolated a microorganism from soil capable of producing Endo-α-GalNAc-ase. The organism was identified as *Alcaligenes* sp., a nonpathogenic aerobe.[119] The enzyme could be induced by pig gastric mucin and was readily purified from the proteins precipitated by ammonium sulfate of the culture filtrate by sequential chromatography on DEAE-Sephadex and Sephadex G-200.[120] This enzyme has an acidic pH optimum of 4.5, which is not suitable for studies on viable cells. Galactose and EDTA, both of which significantly inhibit Endo-α-GalNAc-ase-D, do not affect the Endo-α-GalNAc-ase-A activity. Another difference between the two enzymes is in their ability to hydrolyze the Galβ1-3GalNAc linked to single serine or threonine residues. The *Alcaligenes* enzyme uses these glycopeptides, isolated from the urine of patients with Kanzaki disease, [121, 122] efficiently with a higher activity on the serine than on the threonine derivative. In contrast, the *Diploccocus* enzyme acts poorly on these substrates confirming that the nature of the aglycon influences the overall activity of these

endoglycosidases.[123)]

4) Endo-α-*N*-acetylgalactosaminidase from *Bacillus* sp. (Endo-α-GalNAc-ase-B)
An Endo-α-GalNAc-ase with broad aglycon activity has been purified from another bacterium (*Bacillus* sp.) isolated from soil. Endo-α-GalNAc-ase-B exhibits significant transglycosylation activity and this property has been utilized for the synthesis of neoglycolipids such as Galβ1-3GalNAcα-Hexyl.[124)]

5) Endo-α-*N*-acetylgalactosaminidase from *Streptomyces* sp. OH-11242 (Endo-α-GalNAc-ase-S)
Hotta and coworkers reported an Endo-α-GalNAc-ase with a different substrate specificity compared to that of the endoenzyme discussed above. This enzyme was produced by a strain of *Streptomyces* sp. (OH-11242) isolated from soil and cultured in media containing pig gastric mucin as the only carbon source. The purified enzyme was able to release the tetrasaccharide Galβ1-3(Galβ1-4GlcNAcβ1-6)GalNAc with the core 2 structure in addition to Galβ1-3GalNAc from asialofetuin.[125)] There is no information on whether the enzyme can act on sialylated *O*-linked saccharides. Endo-α-GalNAc-ase-S was found to preferentially release oligosaccharides linked to serine rather than threonine. However, the findings on this enzyme are yet to be independently substantiated and to date neither the enzyme nor the microorganism producing it are available to other investigators.

6) Endoglycoprotease from *Pasteurella haemolytica*
Even though it is not an Endo-α-GalNAc-ase, this enzyme is included here because it has interesting substrate specificity. It is an endoprotease with high specificity for *O*-glycoproteins which was discovered in *Pasteurella haemolytica* by Mellors and coworkers. It was shown to cleave substrates such as glycophorin A, leukosialin and fetuin.[126)] In collaboration with Alan Mellors, we demonstrated the ability of this enzyme to completely degrade epitectin, a malignancy associated mucin type glycoprotein, in solution as well as on the cell surface.[127)] Thus, although not an endoglycosidase it has limited use for structure-function studies on cell surface mucin glycoproteins.

7) Endo-α-*N*-acetylgalactosaminidases from other sources
Ever since the discovery of Endo-α-GalNAc-ase-D we have been interested in similar enzyme activities but with broader substrate specificity. We have conducted a very limited search for such enzymes in bacteria that thrive in a mucus environment. We selected and tested bacteria that showed increased growth in defined medium containing 0.25 % pig gastric mucin as the sole carbon source. The secreted proteins in the culture filtrate and the proteins extracted from the pelleted cells were collected by ammonium sulfate precipitation and assayed for Endo-α-GalNAc-ase activities by incubating with a variety of mucin glycopeptides as substrates. Generally, the culture filtrates showed significant glycosidase activity compared to the very low activity in the extracts of the bacteria. The incubation mixture was first analyzed for release of saccharides terminating with *N*-acetyl-

hexosamine (GalNAc at the reducing end) by a colorimetric assay. If positive, the reaction mixture was chromatographed on a calibrated column of Bio-Gel P-2 and the fractions analyzed for the presence of oligosaccharides larger than Gal-GalNAc. The bacteria we tested initially include *Helicobacter pylori* associated with peptic ulcer and gastric cancer, *Pseudomonas aeruginosa* associated with lung infection of cystic fibrosis patients and *Streptococcus mutans* associated with dental plaque. Of these *H. pylori* was found to produce an Endo-α-GalNAc-ase. The activity was partially purified and shown to release only Galβ1-3GalNAc from various mucins.[128)] Because of the same strict substrate specificity as Endo-α-GalNAcase-D it was not further investigated. Recently, we have tested additional bacteria, particularly those associated with the gastrointestinal tract of humans as further discussed in the next section. The same Gal-GalNAc-releasing endo-α-*N*-acetylgalactosaminidase activity was detected in *Ruminococcus torques* and *B. fragilis* (Muttusamy and Bhavanandan, unpublished results).

From the above discussion it is clear that all the reported Endo-α-GalNAc-ases (with the possible exception of the *Streptomyces* OH-11242 enzyme), have a very restricted specificity for the glycan moiety of the substrate. No extension or modification of the disaccharide, Galβ1-3GalNAc, structure appears to be tolerated by this class of Endo-α-GalNAc-ases. However, the nature of the aglycon of the substrate seems to influence the catalytic activity of these enzymes differently, as illustrated by the comparison of the aglycon preferences of Endo-α-GalNAc-ase-D and Endo-α-GalNAc-ase-A.[123)] The ability of a variety of bacteria to release the Galβ1-3GalNAc suggests that these organisms lack a β-galactosidase capable of cleaving Galβ1-3 linkages from intact glycoproteins. This observation also suggests the importance of Galβ1-3GalNAc for these organisms. One possibility is that this disaccharide may be functioning as an essential factor for their growth.

C. Applications of Endo-α-*N*-acetylgalactosaminidases

The importance of Endo-α-GalNAc-ases for structure-function studies on mucins and other *O*-linked glycoproteins has been documented.[129)] Chemical methods of deglycosylation such as treatment with anhydrous hydrogen fluoride or trifluoromethane sulfonic acid are not only destructive to the saccharides but are also inefficient for the complete removal of *O*-linked saccharides. Further, these treatments invariably result in degradation of the protein backbone. At present it is not feasible to obtain the oligosaccharides of mucins with the reducing end intact. The only satisfactory chemical method for cleavage of saccharides linked to serine and threonine is alkaline borohydride treatment that leads to the reduction of the terminal GalNAc of the released saccharides. Anhydrous hydrazine treatment has been reported as a method of releasing intact *O*-linked saccharides.[130)] However, in extensive trials we have been unsuccessful in isolating intact saccharides from mucins by hydrazinolysis. Typically, anhydrous hydrazine treatment of bovine submaxillary mucin, epitectin, pig gastric mucin and fetuin resulted in a complex mixture of degraded saccharides from which it was impossible to fractionate any undegraded oligosaccharide that might have been present (Bhavanandan, unpublished results). Thus, a major use of the Endo-α-GalNAc-ases with broad substrate specificity will be to prepare the intact oligosaccharides and the intact core proteins under mild physiological conditions.

Mucin oligosaccharides with the reducing end intact will be valuable for the investigation of their specific function such as receptors for pathogens. Isolation of intact oligosaccharides will facilitate the identification of the multivalent saccharide structures in tracheobronchial mucins that interact with the bacterial pathogens that infect airways of humans. For example, chronic *Pseudomonas aeruginosa* infection is an important prognostic factor in cystic fibrosis since a highly significant correlation exists between the presence of the bacteria in sputum, impaired pulmonary function and mortality. Our studies revealed that cystic fibrosis mucin is much more effective than normal lung mucin in inhibiting the interaction of *P. aeruginosa* with asialo GM_1, the putative cell surface receptor.[131] Therefore, in extensions of these studies it will be important to identify the mucin saccharides which are involved in the inhibition of this pathogen's interaction with host cell receptors. Such inhibition studies will require the preparation of reactive saccharides suitable for synthesizing neoglycoconjugates and clustered glycosides (carbohydrate-based dendrimers). Further, the very small amounts of human mucins and cancer cell surface mucin glycoproteins available for research strongly dictate the need for endo-enzymatic approaches for obtaining structurally defined Ser/Thr-linked saccharides from natural sources for studies on their anti-infective potentials. Despite the strict substrate specificity of the available Endo-α-GalNAc-ases they have had a number of applications in research.[105]

Finally, the use of transglycosylation to prepare neoglycoconjugates is highly advantageous since this approach does not require nucleotide donors, the reaction is stereospecific, and in the case of endoglycosidases, large blocks of preformed oligosaccharides can be transferred in one step. Thus, an Endo-α-GalNAc-ase with broad substrate specificity and exhibiting transglycosylation activity will have wide application in chemoenzymatic synthesis of the glycosides and glycopeptides containing *O*-linked saccharides.

D. Search for Endo-α-*N*-acetylgalactosaminidases with Broader Substrate Specificity (A True *O*-Glycanase)

The Endo-α-GalNAc-ases reported so far have limited utility in research because of their very restricted substrate specificity. This is in marked contrast to the situation where endoglycosidases are available for releasing almost all of the *N*-linked saccharide types on proteins. Since there are a great variety of *O*-linked saccharides based on the different core structures, Endo-α-GalNAc-ases with different and broader substrate specificities must exist.

Several observations suggest the possible presence of mucin-degrading endoglycosidases in nature. One line of circumstantial evidence for the presence of Endo-α-GalNAc-ases with broader substrate specificity is the excretion of saccharides, which are derived from mucins or *O*-linked glycoproteins, in the urine of rat[132] and human.[133] This suggests the strong possibility of the presence of Endo-α-GalNAc-ases in animals, as pointed out by Lecat *et al.*[121] Another reason why animal sources must be seriously considered is that a number of the presently known endoglycosidases are from animals. These include hyaluronidase, heparti-nase, endo-β-*N*-acetylglucosaminidase, endo-β-galactosidase from human urine, peptide-N^4-(*N*-acetyl-β-D-glucosaminyl)asparagine amidase, endoglycoceramidase and the endoglycosidases acting on the linkage region of proteoglycans. Further, lysosomes of animal cells have a full complement of glycosidases capable of degrading glycoconjugates for disposal or recycling. The severe storage diseases (e.g., Pompe's, Hurlers', Tay-Sachs, I-Cell disease and mannosidosis) that arise due to deficiencies in these lysosomal glycosidases are well known. Specifically, the patients with mucolipidosis I and Kanzaki disease are reported to excrete in their urine sialyl saccharides linked to single serine residues or short peptides.[121, 122] The enzymatic defect in these disorders is not fully understood but probably involves a deficiency in an enzyme required for catabolism of *O*-linked saccharide. There is a good precedence for the possibility of the presence of Endo-α-GalNAc-ases in the lysosomes of animal tissues. Specifically, the detection of urinary glycosaminoglycans with xylose, galactose and uronic acids at their reducing ends and the subsequent discovery in rabbit liver of endoenzymes cleaving the xylosidic and galactosidic linkages in the linkage region of proteoglycans.[134] Thus, liver and other animal organs, particularly the spleen since this is the site of degradation of large amounts of glycophorin, must be carefully investigated for the presence of Endo-α-GalNAc-ases.

An alternate possibility of the source of the urine saccharides mentioned above is that they are the results of mucin degradation by microorganisms in the colon, followed by absorption into the circulation. Substantial quantities of intestinal mucins are known to undergo degradation by the indigenous enteric bacteria as evidenced by the excretion of large amounts of mucin in the feces of germ-free rats compared to their conventional counterparts.[135, 136] Further, a number of microbes are associated with the gastrointestinal tract and their population is estimated to be greater than that of the somatic and germ cells in humans.[137] Therefore, even if only a very small subpopulation of the intestinal microorganisms of animals is capable of degrading mucins they may be a major source of glycosidases acting on mucins. Several researchers have investigated the glycoconjugate-degrading activities of the ecosystem of the mammalian colon. A few intestinal microorganisms including the two anaerobes *B. fragilis*, and *R. torques*, reveal varying levels of exoglycosidase activities.[138, 139] As mentioned above, we have tested these two organisms for Endo-α-GalNAc-ases but found only the Galβ1-3 GalNAc-releasing activity. A microorganism for special consideration is *Bacteroides thetaiotaomicron* since it is known to metabolize host-derived mucins and glycosaminoglycans in addition to the ability to degrade a variety of dietary plant polysaccharides.[140] The genomic sequence of this dominant microorganism of the indigenous human gut flora has been recently published.[140] Of the 4779 predicted proteins in its proteome, 172 have been recognized as glycoside hydrolases, whereas 1149 (24%) did not have appreciable homology to entries in databases. These 1149 proteins could very well include the type of novel mucin-degrading endoglycosidases for which we are searching.

A major problem in the screening for endoglycosidases, particularly in crude enzyme preparations, is the selection and use of appropriate substrates. Since any potential source of endoglycosidases is also likely to have substantial exoglycosidase activities, the typical natural and synthetic test substrates are likely to be degraded (destroyed) by the combined action of the exoglycosidases. Similarly, any product of endoglycosidase would be modified or destroyed by the exoglycosidases minimizing the chances of detecting the released oligosaccharides. In exceptional cases it is possible that the products of endoglycosidases would be detected in crude preparations due to the absence of a crucial exoglycosidase(s), differences in pH optimum or preferential cleavage. The absence of exogalactosidase capable of hydrolyzing Galβ1-3 linkage seems to be the case with the above-discussed core 1 disaccharide-specific, Endo-α-GalNAc-ase that has been detected in a number of bacteria. There is no simple solution to the problem of competing exoglycosidases during the general screening for endoglycosidases. In order to reach definitive conclusions regarding the presence or absence of endoglycosidases in a preparation it is necessary to test the potential source against many types of substrates and under different assay conditions.

The following are two potential approaches. One approach is to minimize the competition between exo- and endo-glycosidases by modification of the nonreducing end of the potential substrates to render them resistant to sequential degradation by exoglycosidases. The majority of oligosaccharides in mucin glycoproteins terminate in sialic acid. Therefore, one strategy is to modify the nonreducing end sialyl residues of appropriate native glycoconjugate substrates. We developed a method to modify sialyl residues in sialoglycoproteins with the primary goal of introducing various β- and γ-emitter radiolabels by a variation of the procedure to introduce tritium label into terminal sialyl residues. In the first step, the sialoglycoprotein was subjected to mild periodate oxidation in order to generate exocyclic aldehyde on the sialyl residues. In the second step, ^{14}C-labeled amino compounds such as, ethanolamine or glycine were coupled to the aldehyde by reductive amination using sodium cyanoborohydride.[141)] The modified sialoglycoproteins were found to have increased half-life of circulation in animals, enhanced immunogenecity, and were resistant to various sialidases.[141)] This last property suggests the potential use of the sialyl modified mucin type glycoproteins or mucins as substrates in the search for endoglycosidases. A significant proportion of oligosaccharides in mucins also terminate in galactose and this proportion can be increased by desialylation. The nonreducing galactose residues of asialomucins and asialoglycoproteins could then be similarly modified by oxidation with galactose oxidase followed by coupling of amino compounds to the generated aldehyde group. The most likely reason that renders the modified nonreducing end insensitive to exoglycosidases is the introduction of a bulky group. Thus, one must be aware that these modifications may also render the entire saccharide resistant to endoglycosidase, particularly if the modified terminal residue is crucial for recognition by the enzyme. A method to modify the terminal sialic acid without introducing a bulky group is the reduction of the -COOH group. The nonreducing terminal *N*-acetylneuraminic acid residues of α_1-acid glycoprotein glycopeptides have been modified by esterification with carbodiimide followed by reduction with sodium borohydride. The sialic acids modified in this manner were found to be resistant to hydrolysis by sialidase.[142] Similarly, the terminal galactose residues can be made resistant to exo-β-galactosidases by conversion to galacturonic acid. This can be done by first treating asialoglycoproteins with galactose oxidase to generate an aldehyde group on carbon 6 of galactose. The aldehyde is then converted to a carboxyl by the oxidation with hypoiodite.

The second approach not involving nonreducing end modification is the use of inhibitors for exoglycosidases. This approach has been successfully used to discover the rabbit liver endoglycosidases acting on the linkage region of proteoglycans where 0.05 M saccharo-1,4-lactone was used to inhibit exo-β-glucuronidase and thereby the degradation of the glycosaminoglycan chain. Thus, sialidase inhibitors such as *N*-(4-nitrophenyl)-oxamic acid, Neu5Ac2en and neuramin, and galactosidase inhibitors such as galactonolactone and galactostatin (5-amino-5-deoxygalactose) could be included in the incubation mixture.

2.2.2 An Unusual GlcNAcα1-4Gal-releasing Endo-β-galactosidase

Commercially available sialidases prepared from *Clostridium perfringens* ATCC10543 were found to be contaminated with an unusual endo-β-galactosidase (Endo-β-Gal-ase) capable of releasing a disaccharide, GlcNAcα1-4Gal, from *O*-glycans specifically expressed in the gastric gland mucous cell-type mucin.[143, 144] This chapter describes the isolation, characterization and molecular cloning of this enzyme named Endo-β-Gal$_{GnGa}$.

A. Enzyme Assay

Porcine gastric mucin (PGM) that had been dialyzed exhaustively against distilled water was used as a substrate for assaying the Endo-β-Gal$_{GnGa}$ activity. The 10 μL reaction mixture containing 80 μg of PGM and an appropriate amount of enzyme in 20 mM sodium acetate buffer, pH 6.0 (Buffer A), was incubated at 37°C for a predetermined time. The reaction was stopped by adding 2 μL of glacial acetic acid. An aliquot of the reaction mixture was applied onto a silica gel TLC plate and developed twice with 1-butanol/ acetic acid/ water (2/1/1, v/v/v), sprayed with the diphenylamine-aniline-phosphoric acid reagent, and heated at 110°C for 20 min to visualize glycoconjugates.[145] The quantitative analysis of each glycoconjugate band on a TLC plate was carried out using an image scanner and the NIH Image 1.61 program. One unit of enzyme activity was defined as the amount that releases 1 μmol of GlcNAcα1-4Gal from PGM per min at 37°C under the conditions described.

B. Purification of Endo-β-Gal$_{GnGa}$ by Sephacryl S-200 Affinity Chromatography

Clostridium perfringens ATCC10543 was cultured under anaerobic conditions at 37°C for 28 h in 3.5% Todd-Hewitt broth supplemented with 0.18% K_2HPO_4, 0.25% NaCl, 0.15% glucose and 0.005% cysteine. The culture supernatant was brought to 80% saturation with solid ammonium sulfate and left standing overnight at 4°C. The precipitate was collected by centrifugation, dissolved in Buffer A, and dialyzed against the same buffer. This crude enzyme preparation was applied onto a Sephacryl S-200 HR column (Amersham Pharmacia Biotech) previously equilibrated with Buffer A. The enzyme was retained by this column probably because of its affinity for α-linked D-glucose polymer. The column was washed with Buffer A followed by washing with Buffer A containing 0.5 M NaCl to remove the unadsorbed proteins. The enzyme retained by the column was eluted with 50 mM methyl α-D-glucopyranoside in Buffer A containing 0.5 M NaCl. Under this condition, Endo-β-Gal_{GnGa} was released from the column. The active fractions were concentrated by ultrafiltration using Centricon YM-10 (Millipore) and dialyzed against Buffer A to remove methyl α-D-glucopyranoside and NaCl. By this simple procedure, Endo-β-Gal_{GnGa} was purified 3,000-fold from the culture supernatant of *C. perfringens* with an overall recovery of about 50%. The purified enzyme migrated as a single protein band of about 46 kDa in SDS-PAGE.[143)]

C. Liberation of GlcNAcα1-4Gal from PGM by Endo-β-Gal_{GnGa}

The crude enzyme prepared from the culture supernatant of *C. perfringens* released several monosaccharides and oligosaccharides from PGM (Fig. 2.15A, lane 4). The purified Endo-β-Gal_{GnGa}, on the other hand, released only one oligosaccharide from PGM (Fig. 2.15A, lane 5). Using a Dionex HPAEC-PAD analyzer, this oligosaccharide was found to contain *N*-acetylglucosamine and galactose in an equimolar ratio. The structure of this oligosaccharide was subsequently established to be GlcNAcα1-4Gal by NMR spectroscopy. A part of the ^{1}H-NMR spectrum showing the anomeric proton region is presented in Fig. 2.15B. Other details have been described in Ashida *et al.*[143)]

D. Substrate Specificity of Endo-β-Gal_{GnGa}

To characterize the specificity of Endo-β-Gal_{GnGa}, TLC and MS were used to analyze the enzymatic hydrolysis of GlcNAcα1-4Galβ1-4GlcNAcβ1-6(GlcNAcα1-4Galβ1-3)GalNAc-ol (hexasaccharide-alditol) prepared from PGM. As expected, this hexasaccharide-alditol was refractory to jack bean β-*N*-acetylhexosaminidase (β-Hex) (Fig. 2.16A, lane 3). However, Endo-β-Gal_{GnGa} converted the hexasaccharide-alditol to an oligosaccharide and a disaccharide with TLC mobility identical to that of the authentic GlcNAcα1-4Gal (Fig. 2.16A, lane 5). Incubation of the hexasaccharide-alditol with both Endo-β-Gal_{GnGa} and β-Hex resulted in the conversion of the slow moving oligosaccharide to a fast moving oligosaccharide with concomitant release of *N*-acetylglucosamine (Fig. 2.16A, lane 6). These results indicate that the enzyme released the GlcNAcα1-4Gal linked to the core 2 branched GlcNAc through a β1-4 linkage. This conclusion was supported further by ESI (electrospray ionization)-MS analyses in the positive ion mode (Fig. 2.16B). The structures and the monoisotopic molecular masses of the substrate and products of these reactions are shown in Fig. 2.16C. Fig. 2.16B(1) shows the ESI-MS of the hexasaccharide-alditol. The ion peaks of $[M+H]^+$, $[M+Na]^+$, $[M+2H]^{2+}$, $[M+Na+H]^{2+}$, $[M+K+H]^{2+}$ and $[M+2Na]^{2+}$ appear at *m/z*

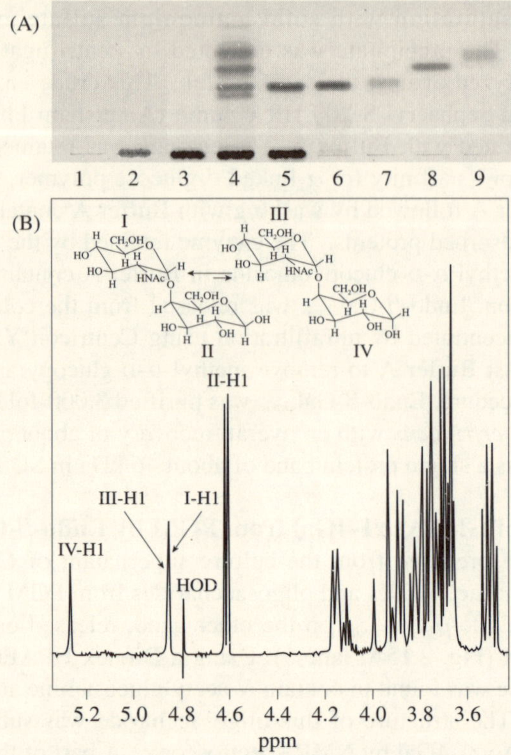

Fig. 2.15 TLC and ^{1}H-NMR analyses of the disaccharide GlcNAcα1-4Gal released from PGM by Endo-β-Gal$_{GnGa}$. (A) TLC analysis. Lane 1, Endo-β-Gal$_{GnGa}$; lane 2, crude clostridial enzyme; lane 3, PGM; lane 4, PGM + crude clostridial enzyme; lane 5, PGM + Endo-β-Gal$_{GnGa}$; lane 6, GlcNAcα1-4Gal; lane 7, Galβ1-4GlcNAc; lane 8, D-galactose; lane 9, *N*-acetylglucosamine. (B) Part of the 500-MHz ^{1}H-NMR spectrum of GlcNAcα1-4Gal. [Reprinted with permission from Ashida, H. *et al.* (2001) *J. Biol. Chem.* **276**, 28228]

1157.4, 1179.1, 579.3, 590.3, 598.2 and 601.2, respectively. Fig. 2.16B(2) shows the ESI-MS of the products after the substrate has been digested with Endo-β-Gal$_{GnGa}$. As expected, two products appeared at *m/z* 384.2 (protonated form of GlcNAcα1-4Gal, sodium adduct at *m/z* 406.2) and at *m/z* 792.1 (protonated form of the counterpart of the digested substrate, sodium adduct at *m/z* 814.1). The ESI-MS spectrum of the products, generated by incubating the substrate with Endo-β-Gal$_{GnGa}$ and β-Hex, is shown in Fig. 2.16B(3). Notably, two new compounds appeared at *m/z* 222.2 (protonated form of *N*-acetylglucosamine, sodium adduct at *m/z* 244.2) and *m/z* 589.2 (protonated form of GlcNAcα1-4Galβ1-3GalNAc-ol, sodium adduct at *m/z* 611.1) with the disappearance of the product peaks (*m/z* 792.1 and 814.1) in Fig. 2.16B(2). The GlcNAcα1-4Gal peaks (*m/z* 384.2 and 406.2) remain. The above results prove conclusively that Endo-β-Gal$_{GnGa}$ has released the GlcNAcα1-4Gal moiety from the GlcNAcα1-4Galβ1-4GlcNAc-branch in the core 2 *O*-glycan in preference to the GlcNAcα1-4Galβ1-3GalNAc-ol in the core 1 *O*-glycan of the hexasaccharide-alditol.

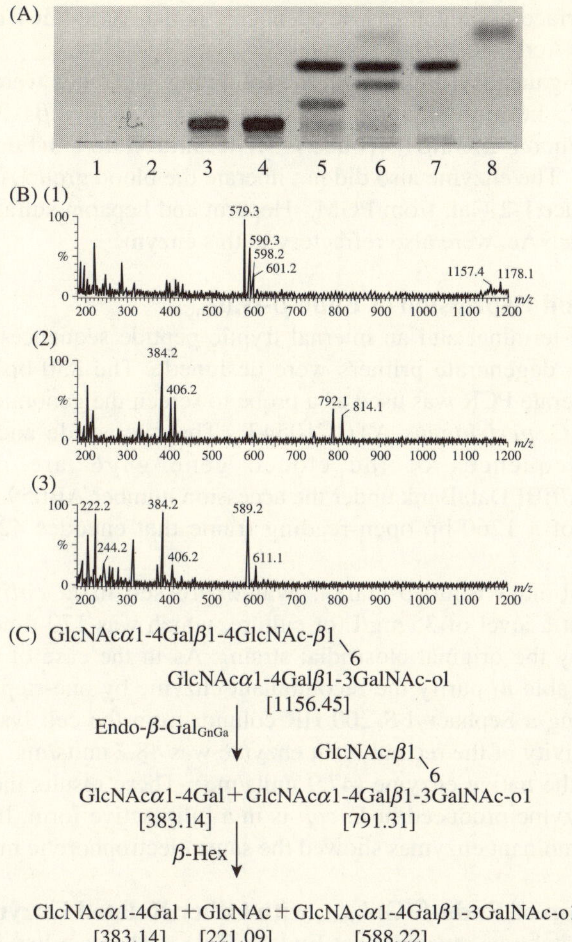

Fig. 2.16 Hydrolysis of the hexasaccharide-alditol (Hex-ol) by Endo-β-Gal$_{\mathrm{GnGa}}$ and β-*N*-acetylhexosaminidase (β-Hex). (A) TLC analysis. Lane 1, β-Hex; lane 2, Endo-β-Gal$_{\mathrm{GnGa}}$; lane 3, Hex-ol + β-Hex; lane 4, Hex-ol; lane 5, Hex-ol + Endo-β-Gal$_{\mathrm{GnGa}}$; lane 6, Hex-ol + Endo-β-Gal$_{\mathrm{GnGa}}$ + β-Hex; lane 7, GlcNAcα1-4Gal; lane 8, *N*-acetylglucosamine. (B) ESI-MS analyses. (1), Hex-ol; (2), Hex-ol + Endo-β-Gal$_{\mathrm{GnGa}}$; (3), Hex-ol + Endo-β-Gal$_{\mathrm{GnGa}}$ + β-Hex. (C) Hydrolysis reactions of Hex-ol catalyzed by Endo-β-Gal$_{\mathrm{GnGa}}$ and β-Hex. The numbers in brackets represent the monoisotopic molecular mass for each component.
[Reprinted with permission from Ashida, H. *et al.* (2001) *J. Biol. Chem.* **276**, 28231]

Although the GlcNAcα1-4Galβ1-3GalNAc-ol branch was resistant to Endo-β-Gal$_{\mathrm{GnGa}}$, the aglycon, GalNAc-ol, was not in a pyranose form. When PGM was incubated with both Endo-β-Gal$_{\mathrm{GnGa}}$ and β-Hex, it became Tn antigen-positive,[143] indicating the exposure of GalNAcα1-Ser/Thr. This result strongly suggests that Endo-β-Gal$_{\mathrm{GnGa}}$ hydrolyzes the endo-β-galactosyl linkage not only in the GlcNAcα1-4Galβ1-4GlcNAc sequence of the core 2 branched *O*-glycan but also in the GlcNAcα1-4Galβ1-3GalNAcα1-Ser/Thr sequence of the core 1 structure.

Endo-β-Gal$_{\mathrm{GnGa}}$ was also able to remove the disaccharide GlcNAcα1-4Gal from the cell surface of human gastric adenocarcinoma AGS-4GnT cells [146] stably expressing GlcNAcα1-4Galβ1-R residues.[143]

The endo-β-galactosyl linkages in the following substrates were refractory to Endo-β-Gal$_{GnGa}$: keratan sulfate, Galα1-3Galβ1-4GlcNAcβ1-3Galβ1-4Glc, GalNAcα1-3(Fucα1-2)Galβ1-4(Fucα1-3)Glc, and Galα1-3(Fucα1-2)Galβ1-4(Fucα1-3)Glc. The enzyme also did not liberate the blood group A trisaccharide, GalNAcα1-3(Fucα1-2)Gal, from PGM. Heparin and heparan sulfate, which contain α-linked GlcNAc, were also refractory to this enzyme.

E. Cloning and Expression of Endo-β-Gal$_{GnGa}$

Based on the N-terminal and an internal tryptic peptide sequences of the native Endo-β-Gal$_{GnGa}$, degenerate primers were designed. The 850-bp fragment obtained by degenerate PCR was used as a probe to screen the genomic DNA library prepared from *C. perfringens* ATCC10543. The nucleotide and the deduced amino acid sequences of the cloned gene *gngC* are deposited in DDBJ/GenBank/EBI DataBank under the accession number AB059351. The gene *gngC* consists of a 1260-bp open reading frame that encodes 420 amino acid residues.

The recombinant Endo-β-Gal$_{GnGa}$ was expressed in *E. coli* DH5α using pUC18 vector at a level of 35 mg/L of culture, which was 170 times higher than that produced by the original clostridial strain. As in the case of the native enzyme, we were able to purify the recombinant enzyme by one-step affinity chromatography using a Sephacryl S-200 HR column from the cell lysate of *E. coli.* The specific activity of the recombinant enzyme was 48.2 units/mg, which is identical to that of the native enzyme (47.9 units/mg). These results indicate that the recombinant enzyme produced in *E. coli* is in a fully active form. In addition, the native and recombinant enzymes showed the same electrophoretic mobility.

F. Comparison of Endo-β-Gal$_{GnGa}$ with Other Related Enzymes

Searching the databases revealed that Endo-β-Gal$_{GnGa}$ shares a low but significant similarity with Endo-β-Gal-ase from *Flavobacterium keratolyticus*[147] and Endo-β-Gal-ase-C from *C. perfringens* ATCC10873 [148] (16% and 21% amino acid identity, respectively). While the molecular size of Endo-β-Gal$_{GnGa}$ is similar to that of *Flavobacterium* Endo-β-Gal-ase (420 vs 422 amino acids), Endo-β-Gal-ase-C is much larger in size (845 amino acids). Endo-β-Gal$_{GnGa}$ also shares some similarity (~18% amino acid identity) with bacterial endo-β-glucanases of the glycoside hydrolase family 16,[149] such as 1,3-1,4-β-glucan 4-glucanohydrolase (EC 3.2.1.73), 1,3-β-glucan 3-glucanohydrolase (EC 3.2.1.39) and 1,3(1,3;1,4)-β-glucan 3(4)-glucanohydrolase (EC 3.2.1.6). These endo-β-glucanases were found to possess a common motif, EXDX(X)E, and the two Glu residues in this motif have been shown to be the active site catalytic residues of 1,3-1,4-β-glucan 4-glucanohydrolase from *Bacillus*.[150, 151] This motif was also found in the fungal cellulases (EC 3.2.1.4) and cellobiohydrolases (EC 3.2.1.91) of the glycoside hydrolase family 7, formerly called cellulase family C. Interestingly, as shown in Fig. 2.17, the EXDX(X)E motif was found to be conserved in Endo-β-Gal$_{GnGa}$ as well as in *C.*

Enzyme / Organism	Position		Motif	
Endo-β-galactosidase				
C. perfringens (Endo-β-Gal$_{GnGa}$)	162	NSKQTG	EIDILE	TFFSKKDTWRIAAY
C. perfringens (Endo-β-Gal-ase-C)	510	DETGHD	EIDVLE	YLGQDPWGAWTTNH
F. kelatolyticus	165	GWPSCG	EIDSME	HVNNESVMYHTIHN
1,3-1,4-β-Glucan 4-glucanohydrolase				
Bacillus macreans	122	HGTQWD	EIDI-E	FLGKDTTKVQFNYY
Bacillus licheniformis	128	DGTPWD	EIDI-E	FLGKDTTKVQFNYY
Bacillus amyloliquefaciens	124	EGTPWD	EIDI-E	FLGKDTTKVQFNYY
Clostridium thermocellum	130	DNNPWD	EIDI-E	FLGKDTTKVQFNWY
1,4-β-Glucan 4-glucanohydrolase (Cellulase)				
Trichoderma reesei	212	QGFCCN	EMDILE	GNSRANALTPHSCT
Fusarium oxysporum	209	QGVCCN	ELDIWE	ANSRATHIAPHPCS

Fig. 2.17 Sequence alignment of the putative active sites of endo-β-galactosidases with other related enzymes. The EXDX(X)E motif is boxed. Accession numbers are as follows: Endo-β-Gal$_{GnGa}$ (AB059351); Endo-β-Gal-ase-C from *C. perfringens* (AB038772); Endo-β-Gal-ase from *Flavobacterium keratolyticus* (AF083896); 1,3-1,4-β-glucan 4-glucanohydrolases from *Bacillus macreans* (X55959), *B. licheniformis* (X57279), *B. amyloliquefaciens* (M15674), and *C. thermocellum* (X58392); 1,4-β-glucan 4-glucanohydrolases (cellulases) from *Trichoderma reesei* (M15665) and *Fusarium oxysporum* (L29378).
[Reprinted with permission from Ashida, H. *et al.* (2002) *Biochemistry* **41**, 2392, Copyright (2002) American Chemical Society]

perfringens Endo-β-Gal-ase-C and *F. keratolyticus* Endo-β-Gal-ase.

G. Effect of Site-directed Mutagenesis on the Amino Acid Residues in the Putative Catalytic Site

To determine whether the EIDILE sequence (Glu-168 to Glu-173) of Endo-β-Gal$_{GnGa}$ could be a catalytic motif, we constructed three specific mutants each containing a single amino acid substitution, E168Q, D170N and E173Q. Although the expression levels of these three mutant proteins in *E. coli* were found to be comparable to that of the wild type, the enzyme activities toward the hydrolysis of PGM and hexasaccharide-alditol were drastically reduced by the point mutation. The mutant D170N showed a very low level of specific activity (0.08 units/mg), and E168Q showed only trace activity (0.004 units/mg). On the other hand, E173Q did not show any detectable activity. These results suggest that Glu-173 and Glu-168 are essential for the enzyme activity and that the side chains of these two Glu residues are involved in the catalytic activity of Endo-β-Gal$_{GnGa}$. The replacement of Asp-170 with Asn also severely attenuated the enzyme activity, suggesting that this acidic amino acid residue is also important for the Endo-β-Gal$_{GnGa}$ activity.

H. Catalytic Mechanism of Endo-β-Gal$_{GnGa}$ Monitored by NMR Spectroscopy

The glycoside hydrolases have been divided into retaining and inverting enzymes. Each enzyme in these two classes exhibits a unique stereochemical course of reaction resulting in the release of the glycon with either retention or inversion of the original anomeric configuration at the cleavage site. To determine the catalytic mechanism of Endo-β-Gal$_{GnGa}$, NMR spectroscopy was used to monitor the hydrolysis of the hexasaccharide-alditol. Fig. 2.18 presents the time course of the hydrolysis showing the appearance of the resonance signals for the anomeric protons (H1s) of the released products, the disaccharide GlcNAcα1-4Gal and the tetrasaccharide-alditol GlcNAcβ1-6(GlcNAcα1-4Galβ1-3)GalNAc-ol, as well as the disappearance of several key anomeric protons of the substrate. The spectra acquired during the first 22.5 min after the addition of the enzyme showed the appearance of the cleavage product, GlcNAcα1-4Gal, with the Gal being almost exclusively in the β-configuration (Fig. 2.18B). The spectra acquired between 30 and 105.5 min (Fig. 2.18C) showed the conversion of β-Gal in GlcNAcα1-4Gal to α-Gal due to the mutarotation. These results clearly indicate that the mechanism of Endo-β-Gal$_{GnGa}$ involves stereochemical retention of the anomeric configuration of the product.

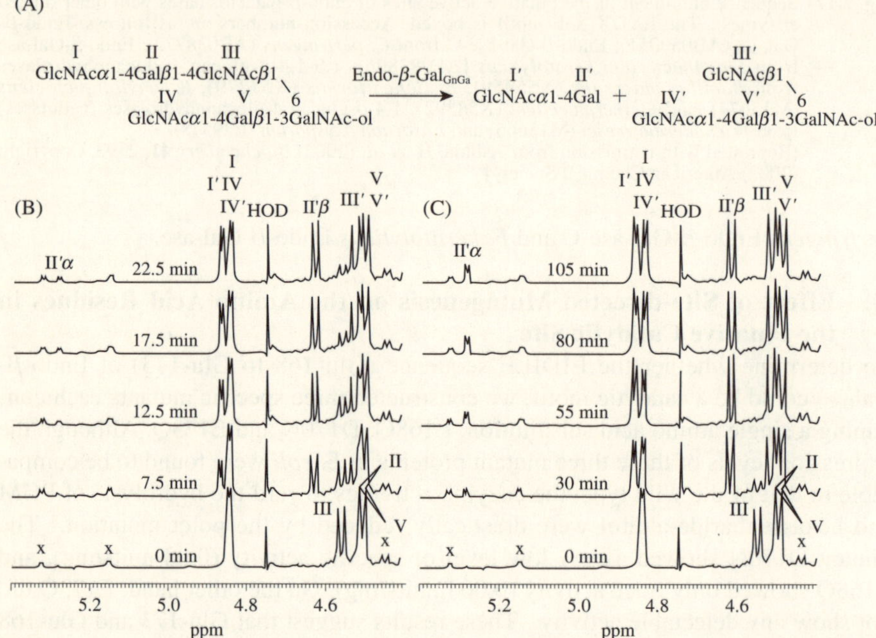

Fig. 2.18 ^{1}H-NMR spectra of the anomeric proton region showing the stereochemical course of hydrolysis of the hexasaccharide-alditol by Endo-β-Gal$_{GnGa}$. (A) Structural representation of the overall reaction. (B) The spectra acquired during the first 22.5 min after the addition of the enzyme, showing the disappearance of I, II and III, and the appearance of I', II'β, and III'. (C) The spectra acquired between 30 and 105 min showing the conversion of II'β to II'α due to mutarotation. The small resonances at ~5.32, 5.15 and 4.43 ppm (marked x) originate from the minor contaminants present in the substrate.
[Reprinted with permission from Ashida, H. *et al.* (2002) *Biochemistry* **41**, 2393, Copyright (2002) American Chemical Society]

I. Crystal Structure of Endo-β-Gal$_{GnGa}$

The crystal structure of Endo-β-Gal$_{GnGa}$ was determined using data extending to 1.82 Å resolution.[152, 153)] The model was deposited at the Protein DataBank with ID 1UPS. The peptide is folded into two domains that are flexibly connected. The larger N-terminal domain extending to residue 279 resembles a sandwich of antiparallel β-sheets also observed in other glycosidases. It exhibits significant structural similarity to κ-carrageenase from *Pseudoalteromonas carrageenovora* (PDB ID 1DYP)[154)] and also to 1,3-1,4-β-glucan 4-glucanohydrolase from *Bacillus* (PDB ID 1BYH).[151)] The additional small C-terminal domain beginning at residue 290 is a cylindrical barrel composed of six double-stranded β-sheets found in a different class of carbohydrate-binding proteins, such as basic fibroblast growth factor (PDB ID 1BFG). Basic fibroblast growth factor has affinity for heparin, an α-linked GlcNAc containing polysaccharide.[155)] Thus, the C-terminus of Endo-β-Gal$_{GnGa}$ may function as the substrate recognition domain.

J. Classification of Endo-β-galactosidases

The microbial endo-β-galactosidases can be classified into the following four groups by their substrate specificities: (I) Endo-β-Gal-ase that cleaves the endo-β-galactosyl linkages in polylactosaminoglycans (see Section 2.4.2); (II) Endo-β-Gal-ase that releases blood group A and B trisaccharides from blood group A and B substances[156)]; (III) Endo-β-Gal-ase-C that releases the Galα1-3Gal from the xenoantigen Galα1-3Galβ1-R[148)]; and (IV) Endo-β-Gal$_{GnGa}$ that specifically releases GlcNAcα1-4Gal from *O*-glycan on gastric mucin.

The substrates for enzymes belonging to groups I, II and III may exist in *O*-glycans as well as in other glycans. Recently, the present authors have cloned the group II Endo-β-Gal from *C. perfringens* ATCC10543, which released blood group A and B trisaccharides from *O*-glycans of PGM and human ovarian cyst glycoprotein.[157)]

2.3 Endoglycosidases That Relate to Proteoglycans

Details of the catabolic mechanism of proteoglycans, which are components of extracellular matrix, remain unexplained. Chondroitin sulfate proteoglycan (ChS-PG) and hyaluronan (HA) show similar distribution patterns in living organisms, and both can be degraded by hyaluronidases, an endo-β-*N*-acetylhexosaminidase (EC 3.2.1.35). While ChS is easily excreted into urine, HA is not. Although there is more chondroitin 4-sulfate (Ch4S) than chondroitin 6-sulfate (Ch6S) in blood, this ratio is reversed in urine. In addition, ChS shows various molecular weight distributions.[158–160)] Thus it is reasonable to conclude that the catabolic mechanism of glycosaminoglycans (GAGs) is quite complicated and that the mechanism may involve a number of enzymes that have not yet been identified. The present author has been interested in this mechanism for many years and has so far discovered several kinds of novel endoglycosidases that act on the linkage region between a core protein and a GAG chain in proteoglycan (Fig. 2.19). In order to develop glycotechnological studies on GAGs, the discovery of many glycosidases is eagerly anticipated, especially the endoglycosidases for releasing oligosaccharide chains. This chapter outlines the background of the discovery of several new enzymes and the problems involved in the preparation of substrates for new enzymes.

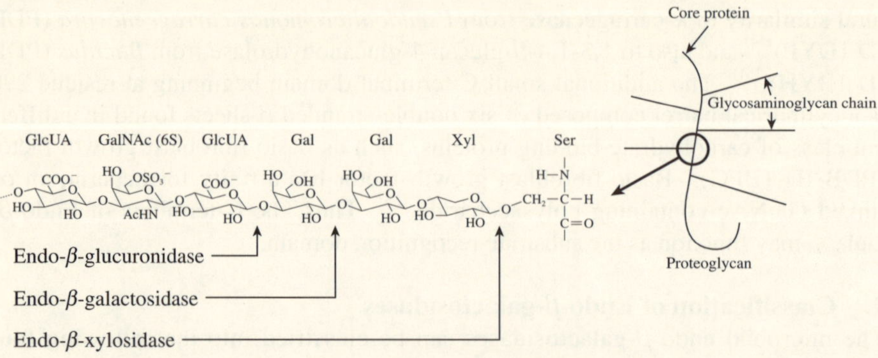

Fig. 2.19 Endoglycosidases acting on the linkage region of proteoglycan. GlcUA, glucuronic acid; GalNAc, *N*-acetylgalactosamine; Gal, galactose; Xyl, xylose; Ser, serine.

The endoglycosidases discovered by the author and his coworkers, endo-β-xylosidase, endo-β-galactosidase and endo-β-glucuronidase, are described below.

Since GAGs in urine are the final metabolic products of GAGs in the living body, the analysis of their chemical structure, especially the sugar residues at their reducing ends, provides information on endoglycosidases, enzymes that hydrolyze an internal bond of the sugar chain, thereby taking part in the degradation of GAGs. Although our studies suggested through the chemical analysis of GAGs in urine that a part of the reducing ends of GAGs is released into urine,[161)] we had not yet isolated an endoglycosidase. To know the catabolic mechanism of intravital proteoglycan, urinary ChSs, which are the final metabolic products of the proteoglycan, were purified and their sugar residues identified at their reducing ends. We then dialyzed 200 L of urine from normal healthy subjects and obtained the GAG fraction that precipitated with cetylpyridinium chloride from the nondialyzable fraction. The sugar residues at the reducing ends of GAGs in this fraction were investigated.[160, 162)] As a result, it was shown that the sugar at the reducing end of ChS in urine is xylose, galactose, or glucuronic acid. It was suggested that three enzymes that act on the ChS chain: endo-β-xylosidase, endo-β-galactosidase, and endo-β-glucuronidase (Fig. 2.19). The former two enzymes act on the linkage structure between the peptide and sugar chain in ChS-PG and the latter acts on a glucuronide bond, the internal structure of a long sugar chain (Fig. 2.19). The presence of endo-α-iduronidase that acts on a long dermatan sulfate (DS) chain was also suggested by a similar experiment.

To confirm the presence of these enzymes in animal tissue, ChS-PG as a substrate was incubated with a 30-50% ammonium sulfate precipitate of rabbit liver homogenate as a partially purified enzyme. Then the ChS chains isolated from proteoglycan were collected and the sugars at the reducing ends were analyzed.[163, 164)] The sugars were identified as glucuronic acid, xylose or galactose at the reducing end, suggesting the presence of the endoglycosidases described above.

2.3.1 Endo-β-xylosidase

Endo-β-xylosidase acts on Xyl-Ser of the GlcUAβ1-3Galβ1-3Galβ1-4Xylβ1-*O*-Ser, (GlcUA-Gal-Gal-Xyl-Ser) structure, which is a linkage region between a core protein and a GAG chain on the proteoglycan (Fig. 2.19).

A. Preparation of Fluorogenic GAG as a Substrate for Endo-β-xylosidase and Determination of Its Activity

Chondroitin sulfate peptidoglycan (ChS-peptide), obtained by digesting the ChS-PG with a protease, is the original substrate for endo-β-xylosidase. To measure the activity of this enzyme, it is essential to detect specifically xylose, which is the sugar at the reducing end of the oligosaccharide produced after reaction with this enzyme. There had been, however, no successful experimental methods for determination of the xylose. Xylosyl-(4-methylumbelliferone) (Xyl-MU) has been used to inhibit the biosynthesis of proteoglycan. It was demonstrated that incubation of chondrocytes with Xyl-MU resulted in elongation of GAGs from Xyl-MU and their secretion into culture conditioned medium,[165] suggesting that Xyl-MU replaces Xyl-Ser as a primer for ChS. The GAG-MU has a common structure of Gal-Gal-Xyl, which is a linkage structure between a peptide and a sugar chain in proteoglycan, at the nonreducing end of the GAG moiety, and the xylose is bound to MU. This report provided useful information on how to prepare a substrate for endo-β-xylosidase. Then, it was found that GAG-MU is available as a substrate to measure endo-β-xylosidase activity. It is easy to assess the activity by measuring fluorescent intensity of liberating MU since GAG-MU liberates MU if the Xyl-MU bond on it was hydrolyzed (Fig. 2.20).[166, 167] Therefore GAG-MU was prepared from cultured human skin fibroblasts. Briefly, human skin fibroblasts were cultured in the presence of 0.5 mM of Xyl-MU, and the culture supernatant was collected and dialyzed with water. Then the nondialyzable fractions including synthesized GAG-MU were subjected to gel filtration. To remove protein and HA,

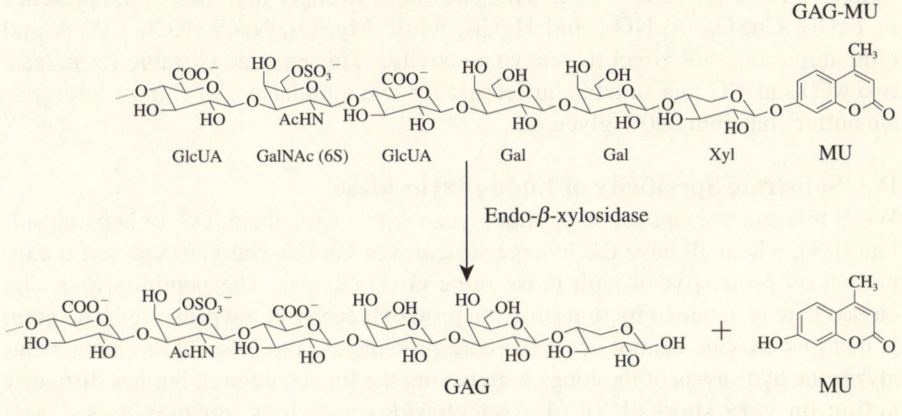

GAG-MU

GlcUA GalNAc (6S) GlcUA Gal Gal Xyl MU

Endo-β-xylosidase

GAG + MU

Fig. 2.20 GAG-MU as a substrate of endo-β-xylosidase. MU, 4-methylumbelliferone; GAG, glycosaminoglycan; GlcUA, glucuronic acid; GalNAc, *N*-acetylgalactosamine; Gal, galactose; Xyl, xylose; Ser, serine.

fractions including GAG-MU were pooled and digested with protease followed by *Streptomyces* hyaluronidase, hyaluronan lyase (EC 4.2.2.1), which is a very specific to hyaluronan. GAG-MU was then purified by ion-exchange column chromatography after cetylpyridinium chloride precipitation. These isolated GAGs-MU have GAG chains composed of a mixture of 70% ChS and 30% DS with average molecular weight of around 10,000. Enzyme assay is performed as follows. One μM of GAG-MU as a substrate is incubated with an enzyme source in 0.1 M sodium acetate buffer, pH 4.0, containing 25 mM glucono-δ-lactone for 2 h at 4°C. After the reaction is stopped and the fluorescence developed by adding 0.2 M glycine-NaOH buffer, pH 10.4, the fluorescence is measured with an excitation wavelength of 350 nm and an emission wavelength of 448 nm. Glucono-δ-lactone is an inhibitor of exo-β-xylosidase. One unit is defined as the amount of enzyme that liberates 1 nmol of MU from GAG-MU per minute.

B. Purification of Endo-β-xylosidase

This enzyme was isolated from the midgut gland of *Patinopecten yessoensis* by monitoring its activity using the assay system described above.[168] Proteins were extracted from acetone-treated powder of the midgut gland of *Patinopecten yessoensis* by stirring in 10 mM Tris-HCl buffer, pH 7.0, containing protease inhibitors. After the protein extract was treated with protamine sulfate, ammonium sulfate precipitation was done. The 50-70% saturated ammonium sulfate precipitate fraction was collected and successively applied onto Sephacryl S-200, and DEAE-Sephacel. Finally, this enzyme was purified and no activity was found for the following enzymes: exo-glycosidases, endo-β-glucuronidase, hyaluronidase, sulfatase, phosphatase and protease.

C. Properties of Endo-β-xylosidase

The molecular weight of endo-β-xylosidase is 78,000, the optimal pH around 4.0, and the isoelectric point about 7.0. The K_m value for GAG-MU (Mr 10,000) was estimated to be 1.7×10^{-2} mM. This enzyme is strongly inhibited in the presence of $FeCl_3$, $CuSO_4$, $AgNO_3$, and $HgCl_2$, while $MgSO_4$, NaCl, KCl, EDTA and dithiothreitol do not affect the enzyme activity. This enzyme is stable for at least two weeks at 4°C in a suitable buffer and for two months at –20°C in an appropriate buffer containing 20% glycerol.

D. Substrate Specificity of Endo-β-xylosidase

While this enzyme can act on peptidoglycan with a ChS chain, DS, or heparan sulfate (HS), which all have the linkage structure of GlcUA-Gal-Gal-Xyl-Ser, it cannot act on proteoglycan with those same GAG chains. The peptidoglycan discussed here is obtained by digesting the proteoglycan with the protease. The point is that this enzyme cannot act if the core proteins are too long. The enzyme catalyzes the hydrolysis of the long chain having the linkage region, but has difficulty acting on very short chain oligosaccharides and does not cleave Xyl-MU (Table 2.8). Therefore this enzyme is extremely valuable for releasing intact GAGs from peptidoglycan. Moreover, this enzyme was also used for the prepara-

Table 2.8 Comparative activities of endo-β-xylosidase

MU derivatives	Relative activity[a] (%)
- (IduUA/ GlcUA-GalNAc(OSO_3^-))$_n$-GlcUA-Gal-Gal-Xyl-MU	100
ΔGlcUA-GalNAc-GlcUA-Gal-Gal-Xyl-MU	30
ΔGlcUA-Gal-Gal-Xyl-MU	30
Gal-Gal-Xyl-MU	12
Gal-Xyl-MU	7
Xyl-MU	<1

GlcUA, glucuronic acid; GalNAc, *N*-acetylgalactosamine; Gal, galactose; Xyl, xylose; IduUA, iduronic acid; MU, 4-methylumbelliferone; Δ, unsaturated.

[a] The relative activity when -(IduUA/ GlcUA-GalNAc(OSO_3^-))$_n$-GlcUA-Gal-Gal-Xyl-MU was used as a substrate was assumed to be 100%.

[Reprinted with permission from Takagaki, K. *et al.* (1990) *J. Biol. Chem.* **265**, 859]

tion of a substrate for endo-β-galactosidase, described later.

2.3.2 Endo-β-galactosidase

A. Preparation of Fluorogenic GAG as a Substrate for Endo-β-galactosidase and Determination of Its Activity

To prepare the substrate of this enzyme, we conceived the idea that ChS chains from proteoglycan should be labeled with 2-aminopyridine (2-pyridylamine, PA), fluorescence material at the reducing end, applying the method of Hase *et al.*[78)] However, disjunction of GAG chains from proteoglycan with alkaline makes the structure at Gal-Gal-Xyl at the reducing end uneven due to peeling reaction. Moreover, reduction using sodium borohydride changes the sugar at the reducing end of GAG chains to sugar alcohol. Therefore the sugar cannot be labeled with fluorescence. For this reason, endo-β-xylosidase, which had previously been purified was used for the preparation of ChS having a xylose at the reducing end. The xylose was labeled with PA. Preparation of chondroitin sulfate-PA (ChS-PA) as a substrate for endo-β-galactosidase succeeded in this way[169, 170)] (Fig. 2.21). Digestion with endo-β-xylosidase made it possible to also label DS and HS with PA. The enzyme activity of endo-β-galactosidase is monitored as follows. One μM of ChS-PA as substrate is incubated with an enzyme source in 0.1 M sodium acetate buffer, pH 4.0, containing 10 mM D-galactal and 10 mM saccharo-1,4-lactone for 12 h at 37°C. The reaction is stopped by boiling and the reaction products are analyzed by HPLC. D-Galactal and saccharo-1,4-lactone are inhibitors for exo-β-galactosidase and exo-β-glucuronidase, respectively. The amount of the hydrolyzed product was estimated from the fluorescence intensity using Gal-Xyl-PA as the standard under the wavelengths of excitation 320 nm and emission 400 nm. One unit was defined as the amount of the enzyme that liberated 1 nmol of Gal-Xyl-PA (Fig. 2.21) from the ChS-PA/min.

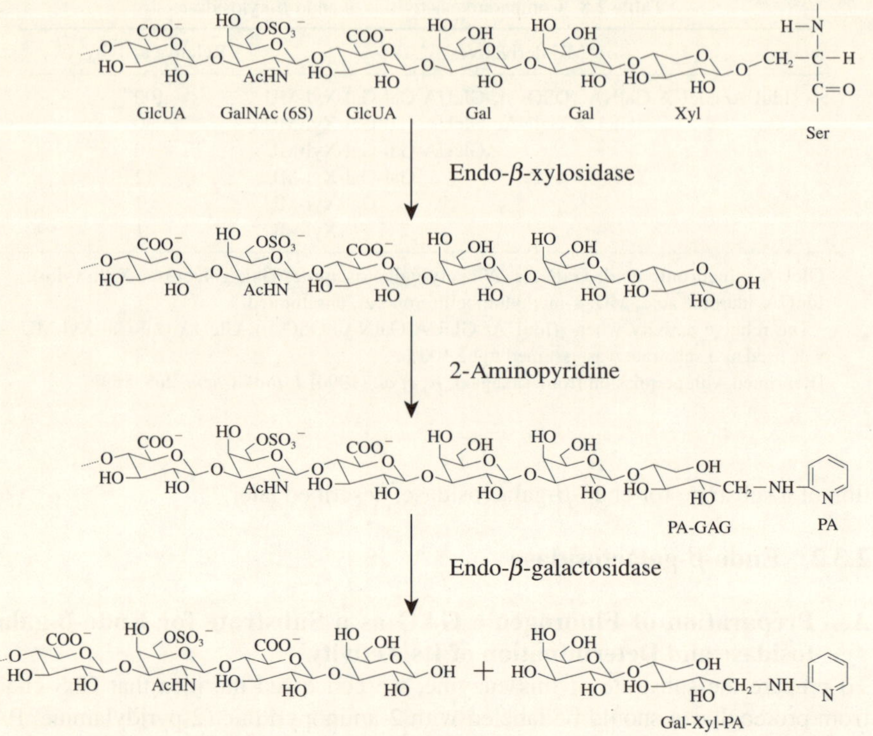

Fig. 2.21 Preparation of GAG-PA as a substrate of endo-*β*-galactosidase. GAG-PA (pyridylaminated GAG) was prepared by endo-*β*-xylosidase digestion followed by pyridylamination, and digested with the endo-*β*-galactosidase.

B. Purification of Endo-*β*-galactosidase

This enzyme was isolated from the midgut gland of *Patinopecten yessoensis* using the above-mentioned assay system to monitor its activity.[171)] Proteins were extracted from acetone-treated powder of the midgut gland of the *Patinopecten yessoensis* by stirring in 10 mM Tris-HCl, pH 7.0, containing protease inhibitors. After the protein extract was treated with protamine sulfate, ammonium sulfate precipitation was done. The 50-70% saturated ammonium sulfate precipitate fraction was collected and successively applied onto Sephacryl S-200, DEAE-Toyopearl 650. The enzyme was further purified by HPLC using a TSK gel Phenyl-5PW RP column.

C. Properties of Endo-*β*-galactosidase

The molecular weight of endo-*β*-galactosidase is 160,000 and the optimal pH around 4.0. This enzyme is strongly inhibited in the presence of $CuSO_4$ and $HgCl_2$, while $MgSO_4$, NaCl, KCl and dithiothreitol do not affect the enzyme activity.

D. Substrate Specificity of Endo-*β*-galactosidase

Reaction products of endo-*β*-galactosidase are long glycans that have a galactose at the reducing end. It should be determined which of the two galactoses in Gal-Gal-Xyl of the linkage structure between a core peptide and a sugar chain in proteoglycan will be hydrolyzed by this enzyme. By the activity measurement described above, it became clear whether this enzyme hydrolyzes Gal-Gal or Gal-Xyl in Gal-Gal-Xyl of the linkage region between a core peptide and a GAG chain. It was found that the enzyme acts only on the Gal-Gal bond of the Gal-Gal-Xyl structure in GAG (Fig. 2.19), since Gal-Xyl-PA, and not Xyl-PA, was detected as a hydrolyzed product. Several endo-β-galactosidases have been isolated from different sources.[172-175] This endo-β-galactosidase from *Patinopecten yessoensis* is different from either of the previously known endo-β-galactosidases that act on glycoproteins or glycolipids. This enzyme acts on ChS, DS and HS but does not act on KS. It does not act on proteoglycans having a large core protein but does act on peptidoglycans having a small core protein. The enzyme activity is unaffected by the chain length of GAG. The minimum size required for hydrolysis by this enzyme is a hexasaccharide in the linkage region between the core protein and GAG chain.

2.3.3 Endo-β-glucuronidase

The reaction product of endo-β-glucuronidase is Chs chains having a glucuronic acid at the reducing end. Although many endo-type glucuronidases such as hyaluronidase from leeches that acts on HA,[176] an enzyme acting on heparin[177] and heparanase acting on HS[178–181] have been discovered, this is the only enzyme that acts on ChS. It acts on the glucuronide bond in the linkage region between a core protein and a GAG chain on proteoglycans.

A. Preparation of a Substrate for Endo-β-glucuronidase and Determination of Endo-β-glucuronidase Activity

ChS-peptide, which is obtained by thoroughly digesting ChS-PG with protease, is an excellent substrate for this enzyme,[182] and ChSs on the market can also be used as substrates. Some ChSs, however already possess glucuronic acid at their reducing ends, presenting with high background in the assay system for endo-β-glucuronidase activity. In this case, ChS is used as a substrate after it is converted to sugar alcohol by reducing with sodium borohydrate, *etc*. ChS with at least a Gal-Gal disaccharide unit of the linkage region is sufficient for the substrate of this enzyme (Fig. 2.19). It has been difficult to measure selectively the glucuronic acid formed at the reducing ends of substrate after incubation with this enzyme. One problem in the assay systems of glycosidases based on their reduction activity, such as the Park-Johnson method, is that the measured value is much higher than that of the actual reduction activity due to the peeling reaction of sugar chains, a side reaction caused by heating under alkaline conditions (Table 2.9). This causes monosaccharides to be liberated from the reducing end of the sugar chain sequentially. Therefore, this kind of assay system based on reduction activity is unsuitable to assess the activity of endoglycosidases. However, the Milner-Avigad method, which measures reducing activity under acidic conditions, can measure

Table 2.9 The rate of coloring of various sugars by the Milner-Avigad method

Sugar	Milner-Avigad method (%)[a]	Park-Johnson method (%)[a]
Glucuronic acid	100	100
N-Acetylglucosamine	0	127
N-Acetylgalactosamine	0	120
Xylose	0	128
Galactose	0	118
Gulonic acid	0	0
N-Acetylglucosaminitol	0	0
N-Acetylgalactosaminitol	0	0
GlcUAβ1-3GalNAc	0	187
GlcUAβ1-3GalNAcβ1-4GlcUA	30	205
(4-GlcUAβ1-3GalNAc-)$_2$	2	329
(4-GlcUAβ1-3GalNAc-)$_2$-GlcUA	32	278
(4-GlcUAβ1-3GalNAc-)$_3$	4	404

[a] The value shows the relative rate when the reduction power of glucuronic acid is 100%.

GlcUA, glucuronic acid; GalNAc, *N*-acetylgalactosamine.

the activity of glucuronic acid, an acidic sugar at the reducing end, without causing a peeling reaction (Table 2.9).[183] This method is extremely reliable for accessing endo-β-glucuronidase activity because its specificity is much higher than that of other methods, although the color intensity of the glucuronic acid at the reducing ends in long GAG chains is lower than that of glucuronic acid as a monosaccharide.[184] Saccharo-1,4-lactone is added to inhibit the exo-β-glucuronidase activity in the reaction mixture.[182]

B. Purification of Endo-β-glucuronidase

Endo-β-glucuronidase was purified from rabbit liver using the above method to monitor its activity.[182] The supernatant of the 10% homogenate of rabbit liver in 5 mM sodium phosphate buffer, pH 7.4, containing protease inhibitors was treated with protamine sulfate. The resulting precipitate fraction, 50-70% saturated ammonium sulfate precipitation, was collected and successively applied onto DEAE-cellulose, Sephacryl S-300 and heparin-Sepharose CL-6B. The enzyme was further purified by polyacrylamide gel electrophoresis.

C. Properties of Endo-β-glucuronidase

The molecular weight of endo-β-glucuronidase is 35,000, the optimal pH around 4.0 and the isoelectric point about 5.4. The K_m value for ChS (average M_r was estimated as 40,000) was estimated to be 7×10^{-3} M. This enzyme is strongly inhibited in the presence of $HgCl_2$. Saccharo-1,4-lactone, a specific inhibitor of exo-β-glucuronidase does not inhibit endo-β-glucuronidase activity.

D. Substrate Specificity of Endo-β-glucuronidase

This enzyme acts on both Ch4S and Ch6S to the same extent. It does not act on the ChS-PG but act on ChS-peptide, which has a small core protein (Table 2.10). It does not act on any internal glucuronide bond other than that in the linkage region between a core protein and a GAG chain. This enzyme does not act on proteoglycan but on peptidoglycan obtained by digesting proteoglycan by protease. It can act specifically on the glucuronide bond, GlcUAβ1-3Gal-, in the linkage between a ChS chain and peptide, but cannot act on the other glucronide bond within a GAG chain. The enzyme does not act on a short oligosaccharide although it acts on a long ChS chain.

Table 2.10 Substrate specificity of endo-β-glucuronidase

Substrate	Relative activity[a] (%)
ChS-peptide	100
ChS-PG	3
Ch6S oligosaccharides, (4GlcUAβ1-3GalNAc(6-OSO_3^-)1-$)_2$	0
Ch6S oligosaccharides, (4GlcUAβ1-3GalNAc(6-OSO_3^-)1-$)_3$	0
Hyaluronan	0

[a] The relative activity when ChS-PG was used as a substrate was assumed to be 100%.
ChS, chondroitin sulfate; ChS-PG, chondroitin sulfate proteoglycan; GlcUA, glucuronic acid; GalNAc, *N*-acetylgalactosamine.

It is presumed that there are some glycosidases having different specificity in view of the structures of urinary ChSs as terminal products of ChSs. In order to identify these new enzymes, it is necessary to establish an assay system that can prove their existence and assess their activities, including how to prepare their substrates. As noted above, no artificial substrate in existence is adequate to measure activity of endoglycosidase. Discovery of these endoglycosidases leads not only to breakthroughs in the metabolic mechanism of the proteoglycan and structural analysis and but also to the preparation of new substrates and discovery of new enzymes (Fig. 2.19). The fact that endo-β-xylosidase made a substrate of endo-β-galactosidase is a good example.

Other investigators have found endoglycosidases that act on GAG, for instance, endo-β-glucuronidase that acts on heparin[177)] and heparan sulfate,[177–181)] and hyaluronidase that acts on HA as an endo-β-*N*-acethylhexosaminidase. It is believed that the discovery of a number of endo- or exoglycosidases of differing specificities, including the above enzymes, will facilitate the development of glycotechnological research of GAGs.

2.4 Endoglycosidases That Relate to Glycosphingolipids

2.4.1 Endoglycoceramidase

Recently, glycosphingolipids have emerged as bioactive modulators/transducers for various cell functions.[185)] Glycosphingolipids are also well-known receptors for various microbes and their toxins.[186)] To analyze the structures and functions of glycosphingolipids, the development of glycosphingolipid-specific enzymes are required. All glycosphingolipid-degrading enzymes reported so far can act not only on glycosphingolipids but also on oligosaccharides and glycoproteins. In this context, real glycosphingolipid-specific enzymes had not been reported until the discovery of endoglycoceramidase (EGCase, also known as ceramide glycanase) described in this chapter. EGCase (ceramide glycanase) cleaves the glycosidic linkage between oligosaccharides and ceramides of various glycosphingolipids.[187, 188)] The enzyme recognizes both the sugar and lipid portions of substrates and is therefore specific to glycolipids.[189–191)] This chapter describes the discovery, distribution, specificities, molecular cloning, three-dimensional modeling structure and applications of EGCase. Very recently, a biological role of animal EGCase was revealed using a hydra, *Hydra magnipapillata*, as a model animal.[192)] This chapter also describes briefly the unique catabolic pathway of glycosphingolipids involving EGCase in hydra.

A. Discovery

The discovery of DNA restriction enzymes has led to an amazing new technology in the life sciences, *i.e.*, gene technology. There is no doubt that many endoglycosidases have contributed to the clarification of structures and functions of various glycoconjugates. However, in the early nineties there were few reports about endo-type glycosidases acting on glycolipids.

Endo-β-galactosidase from *Escherichia freundii*[193)] and *Flavobacterium keratolyticus*[173)] seems to be the first endo-type enzyme acting on glycosphingolipids. The enzyme cleaves the internal β-galactosidic linkage of *N*-acetyllactosamine (galactosyl-*N*-acetylglucosamine) repeating units in keratan sulfate, glycoproteins, milk oligosaccharides and glycosphingolipids. The enzyme was then widely used for the detection and structural analysis of *N*-acetyllactosamine-containing glycoconjugates (see Section 2.4.2). To analyze the structure and functions of glycosphingolipids, new endoglycosidases acting on glycosphingolipids were further required in those days.

Based on the successful experience of the finding of *Flavobacterium* endo-β-galactosidase in Kitamikado's laboratory at Kyushu University, the author (M. I), who was working in Yamagata's laboratory at Mitsubishi Kasei Institute of Life Sciences, began to search for new glycosphingolipid-degrading enzymes from microorganisms in natural habitats. He employed an "enrichment culture technique" which is often used for the screening of the microbes capable of degrading specif-

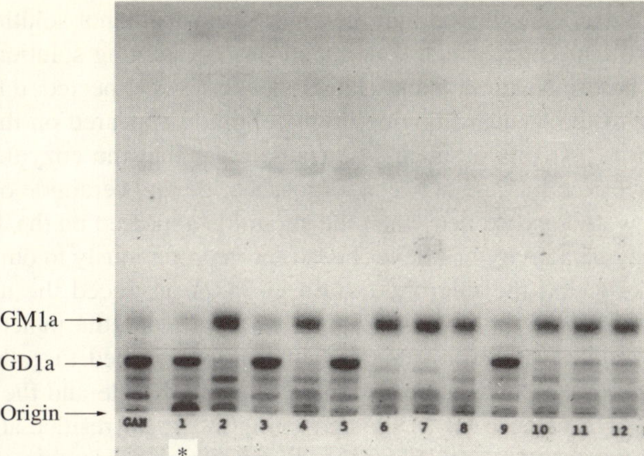

Fig. 2.22 TLC showing the action of EGCase from strain G-74-2. Screening was conducted to search for microbes capable of producing endoglycosidases acting on glycosphingolipids. Many bacteria were isolated from natural habitats using synthetic medium containing gangliosides as the sole source of carbon and examined for their ability to decompose glycosphingolipids. The bacterial strains were cultivated with bovine brain gangliosides and the supernatants were analyzed by TLC which was developed with chloroform/methanol/0.2%$CaCl_2$ (120/80/18, v/v/v) and visualized with orcinol-H_2SO_4 reagent. This TLC showed EGCase activity for the first time during screening. Lane 1 shows the hydrolysis of gangliosides by strain G-74-2. In lane 1, oligosaccharides released from gangliosides possibly by the action of enzyme were retained on the origin of the TLC under the conditions used. GAN, bovine brain ganglioside mixture.

ic substances. Using a synthetic medium containing bovine brain gangliosides as the sole source of carbon, more than 1,000 strains were isolated from land soils, river waters, sea waters, etc. These isolates were separately inoculated into a peptone-yeast extract (PY) medium containing bovine brain gangliosides then incubated at 30°C for several days. Then, bacterial cells were removed by centrifugation and the degradation products of gangliosides in the culture supernatant were analyzed by TLC. Fig. 2.22 is a TLC which represents one example of many trials, showing that several isolates produced sialidase capable of degrading gangliosides to generate GM1a (lanes 2, 4, 6, 7, 8, 10, 11, 12). We noticed that in lane 1 the origin of the TLC (where the sample was applied) was strongly stained with orcinol-H_2SO_4, although the pattern of gangliosides on the TLC was little affected and no monosaccharides were produced (Fig. 2.22). Peculiarly, however, the reducing power in this culture supernatant significantly increased, suggesting that oligosaccharides had been generated. It was then speculated that (1) the strain could produce the enzyme which cleaves the internal glycosidic linkage of gangliosides, and (2) the sialic acid-containing oligosaccharides produced could be developed very slowly on the TLC in the developing solvent used in this experiment. Interestingly, the TLC showed no newly generated glycosphingolipids, suggesting that an orcinol-negative product (nonsugar product) had been generated by the action of enzyme. Was it ceramide? Promptly, we tried to detect free ceramide in the culture supernatant using the method just reported in *Analytical Biochemistry* by Nakamura and Handa at Tokyo Medical and Dental School.[194)]

The TLC plate was then dipped into the chloroform-methanol solution containing Coomassie Brilliant Blue, which was a well-known staining solution for proteins but not lipids before Nakamura and Handa's paper. As expected, duplicate bands corresponding to the standard bovine brain ceramide appeared on the TLC in the dipping solution. At this moment, it became clear that the enzyme was able to cleave the glycosidic linkage between oligosaccaride and ceramide of glycosphingolipids. We were very excited when the ceramide appeared on the TLC plate because such enzyme activity had never been reported previously to our knowledge.

It was likely that the microbe (strain G-74-2) produced the novel enzyme which could hydrolyze the glycosphingolipids to produce the intact oligosaccharides and ceramides. Based on this speculation, we rushed to gather more evidence, performing structural analysis of the oligosaccharide and the ceramide released from GM1a using FAB-MS, determining the sugar residue at the reducing end of the oligosaccharide, purifying the enzyme from the culture supernatant of strain G-74-2, determining substrate specificity of the enzyme, *etc.* Takashi Mikawa at Center Research Institute of Mitsubishi Kasei was asked to identify the strain G-74-2, and he identified the isolate *Rhodococcus equi.* All data collected clearly indicated that *Rhodococcus equi* strain G-74-2 produced the novel enzyme which hydrolyzes the β-glucosidic linkage between oligosaccharides and ceramides of various acidic and neutral glycosphingolipids. However, partially purified enzyme did not hydrolyze glucosylceramide (Glcβ1-1′Cer), galactosylceramide (Galβ1-1′Cer), neogalatriaosylceramide (Galβ1-6Galβ1-6Galβ1-1′Cer), sphingomyelin or various glycoproteins, indicating that the enzyme is specific to glycosphingolipids (but not sphingomyelin or glycoproteins) and hydrolyzes the β-glucosyl linkage (but not the β-galactosyl linkage) between ceramides and oligosaccharides (but not monosaccharides). The minimum structural requirement for hydrolysis of glycosphingolipids by EGCase seemed to be lactosylceramide (Galβ1-4Glcβ1-1′Cer). Based on the these results, we tentatively designated the enzyme "endo-type glycosylceramidase" that is endoglycoceramidase (EGCase) for short.[187] Fig. 2.23 shows the action of EGCase on GM1a ganglioside. From the experiment using $H_2{}^{18}O$, the enzyme was demonstrated to hydrolyze the β1,1′-glycosidic linkage between sugar chain and ceramide in glycosphingolipids. EC number 3.2.1.123 was given to the EGCase and the systematic name for the enzyme, "oligoglycosylglucosylceramide glycohydrolase", was recommended by the Nomenclature Committee of the International Union of Biochemistry and Molecular Biology in 1992. Yu-Teh Li and his associates in New Orleans reported similar enzyme activity in leeches, and they called it ceramide glycanase.[188]

B. Distribution

EGCase (ceramide glycanase) has been found in both prokaryotes (microorganisms) and eukaryotes (invertebrates) *i.e.*, actinomycetes (*Rhodococcus* sp.),[187, 195] bacteria (*Corynebacterium* sp.),[196] leeches (*Hirudo medicinalis*),[188] hard-shelled clam (*Merecenaria merecenaria*),[197] earthworm (*Lumbricus terrestaris*),[198] jellyfish (*Cyanea nozakii*),[199] and hydra (*Hydra magnipapillata*).[192] In addition, EGCase activity was widely distributed in invertebrates belonging to the phyla

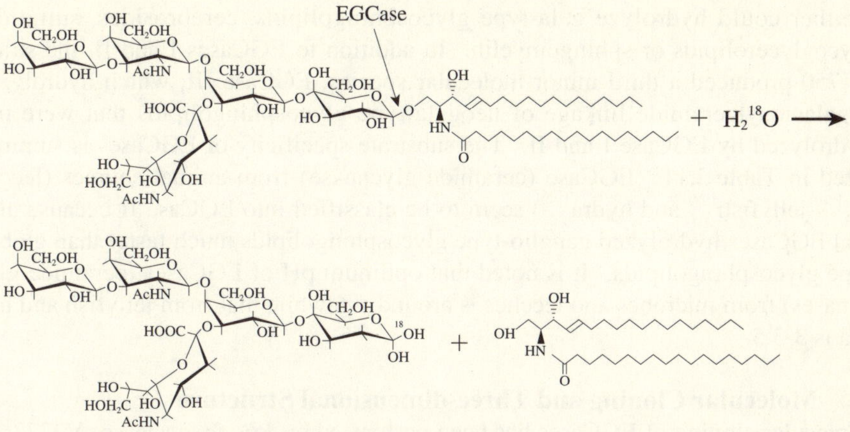

Fig. 2.23 Action of EGCase on GM1a. GM1a dissolved in $H_2{}^{18}O$ or H_2O was incubated with EGCase. After digestion, oligosaccharide and ceramide were separated by partition with a chloroform/methanol mixture then subjected to FAB-MS. The *m/z* of the oligosaccharide from the reaction in $H_2{}^{18}O$ increased by two daltons compared to that in H_2O, while the *m/z* of the ceramide from both reactions are the same, indicating that ^{18}OH was transferred to the reducing end of the oligosaccharide. This experiment clearly shows that the enzyme hydrolyzes the glycosidic linkage between the oligosaccharide and the ceramide of GM1a.

Cnidaria, *Mollusca* and *Annelida*.[192] However, EGCase is likely to be missing in vertebrates including mammals. The presence of EGCase in plants has not been proved. The gene encoding EGCase seems to have been lost during evolution like cellulase. In this context, it is interesting to note that the catalytic domain of EGCase is very similar to that of cellulase (endo-β-1,4-glucanase).[200-202] It was long controversial whether invertebrates *per se* synthesize EGCase. Alternatively, it was suspected that symbionts such as bacteria produce EGCase. Molecular cloning of the EGCase gene from jellyfish[199] and hydra[192] has settled the dispute, *i.e.*, (1) both animal EGCase genes, which show about 20% identity to the gene of *Rhodococcus* EGCase II, contained the polyA signals which are characteristics of eukaryote genes, (2) EGCases from jellyfish and hydra were soluble *N*-glycosylated proteins which have never been observed in prokaryotes, (3) whole mount *in situ* hybridization and immunocytochemistry revealed that EGCase is distributed specifically in the endodermal layer throughout the body of hydra *per se*.

C. Specificity and Properties

Three molecular species of EGCase were found in the culture supernatant of *Rhodococcus* sp. M-750, the mutant strain generated from G-74-2, and two of three isoenzymes were completely purified.[189] The apparent molecular mass, determined by SDS-PAGE, was 55.9 kDa for EGCase I and 58.9 kDa for EGCase II, and their pIs were 5.3 and 4.5, respectively. Both were capable of hydrolyzing the β-glucosidic linkage between the sugar chain and ceramide in ganglio-type, lacto-type and globo-type glycosphingolipids. Globo-type glycosphingolipids were strongly resistant to hydrolysis by EGCase II in comparison with EGCase I.

Neither could hydrolyze gala-type glycosphingolipids, cerebrosides, sulfatide, glycoglycerolipids or sphingomyelin. In addition to EGCases I and II, the strain M-750 produced a third minor molecular species, EGCase III, which hydrolyzes β-galactosylceramide linkage of neogala-type glycosphingolipids that were not hydrolyzed by EGCase I and II. The substrate specificity of EGCases is summarized in Table 2.11. EGCase (ceramide glycanase) from animal sources (leeches,[190] jellyfish[199] and hydra[192]) seem to be classified into EGCase II because animal EGCases hydrolyzed ganglio-type glycosphingolipids much faster than globo-type glycosphingolipids. It is noted that optimum pH of EGCases (ceramide glycanases) from microbes and leeches is around 4-6 while that from jellyfish and hydra is 3-3.5.

D. Molecular Cloning and Three-dimensional Structure

Molecular cloning of EGCases has been performed on *Rhodococcus* sp. M-777,[200] *Rhodococcus* sp. C9,[195] jellyfish[199] and hydra.[192] The open reading frame of *Rhodococcus* EGCase II showed 20% and 25% identity to that of the jellyfish and hydra enzymes, respectively, at the amino acid level. Eight amino acid residues, which are essential for the catalytic activity of GH5 endo-β-1,4-glucanases (cellulases),[202] were all conserved in these three EGCases; Glu-233 and Glu-351 are thought to be an acid/base catalyst and a nucleophile, respectively, and His-135 to be important for interaction with glucose (Fig. 2.24). Point mutagenesis in one of the eight conserved residues in the *Rhodococcus* EGCase II caused a marked loss of the activity, indicating that these conserved amino acids are all essential for EGCase activity. Three-dimensional modeling of the *Rhodococcus* EGCase II predicted that the enzyme is a $(\beta/\alpha)_8$ barrel, so-called "TIM-barrel" structure and the eight residues form an active site like endo-β-1,4-glucanases (cellulases) (Fig. 2.25), although the identity of both enzymes is less than 20%. This result suggests that EGCase is a retaining glycohydrolase, the reaction of which occurs via a double-displacement mechanism like endo-β-1,4-glucanases. However, the substrate specificity of EGCase is completely different from that of endo-β-1,4-glucanase, *i.e.*, the former hydrolyzes glycosphingolipids but not cellulose whereas the later hydrolyzes cellulose but not glycosphingolipids. This discrepancy may be explained by the three-dimensional modeling of both enzymes in which the substrate-binding clefts of both enzymes are somewhat different (data not shown). Collectively, it is strongly suggested that EGCases and endo-β-1,4-glucanases evolved from the same ancestral gene.

E. Activator Proteins

A detergent was required for the hydrolysis of glycosphingolipids *in vitro* by EGCase. *Rhodococcus* sp. M-777, the mutant strain from G-74-2, was found to produce activator proteins which stimulated the hydrolysis of glycosphingolipids by EGCases in the absence of detergent.[203] The molecular mass and pI of the activator (designate activator II) were 69.2 kDa and 4.0, respectively. The activator II enhanced the hydrolysis of various glycosphingolipids by EGCase II in the absence of detergents in a concentration-dependent manner. Interestingly, activator

Table 2.11 Specificity of endoglycoceramidase

Substrate			EGCase		
Class	Name	Structure	I*	II*	III***
			Hydrolysis (%)		
Ganglio series	GT1b	NeuAcα2-3Galβ1-3GalNAcβ1-4 (NeuAcα2-8NeuAcα2-3) Galβ1-4Glcβ1-1′Cer	16	14	ND
	GD1a	NeuAcα2-3Galβ1-3GalNAcβ1-4 (NeuAcα2-3) Galβ1-4Glcβ1-1′Cer	16	16	ND
	GM1a	Galβ1-3GalNAcβ1-4 (NeuAcα2-3) Galβ1-4Glcβ1-1′Cer	48	38	>5
	GM2	GalNAcβ1-4(NeuAcα2-3)Galβ1-4Glcβ1-1′Cer	25	28	ND
	GM3	NeuAcα2-3Galβ1-4Glcβ1-1′Cer	18	85	ND
	asialo GM1	Galβ1-3GalNAcβ1-4Galβ1-4Glcβ1-1′Cer	72	65	>5
Globo series	Gb5Cer	GalNAcα1-3GalNAcβ1-3Galα1-4Galβ1-4Glcβ1-1′Cer	5 (88**)	>1 (9**)	ND
	Gb4Cer	GalNAcβ1-3Galα1-4Galβ1-4Glcβ1-1′Cer	2 (87**)	>1 (5**)	0
	Gb3Cer	Galα1-4Galβ1-4Glcβ1-1′Cer	18 (100**)	>1 (12**)	ND
Lacto series	Sialosylparagloboside	NeuAcα2-3Galβ1-4GlcNAcβ1-3Galβ1-4Glcβ1-1′Cer	52	100	ND
	Lactosylceramide	Galβ1-4Glcβ1-1′Cer	51	86	0
Gala series	Trigalactosylceramide	Galβ1-6Galβ1-6Galβ1-1′Cer	0	0	74
	Galactosylceramide	Galβ1-1′Cer	0	0	ND
	Sulfatide	HSO_3-Galβ1-1′Cer	0	0	0

*Condition I, 0.3 milliunits of the enzyme was incubated with 50 nmol of the appropriate substrate at 37°C for 2 h. ** Condition II, 1.5 milliunits of the enzyme were incubated with 50 nmol of the appropriate substrate at 37°C for 16 h. ***Condition III, 0.5 milliunits of the enzyme was incubated with 50 nmol of appropriate substrate at 37°C for 8 h. ND, not determined.

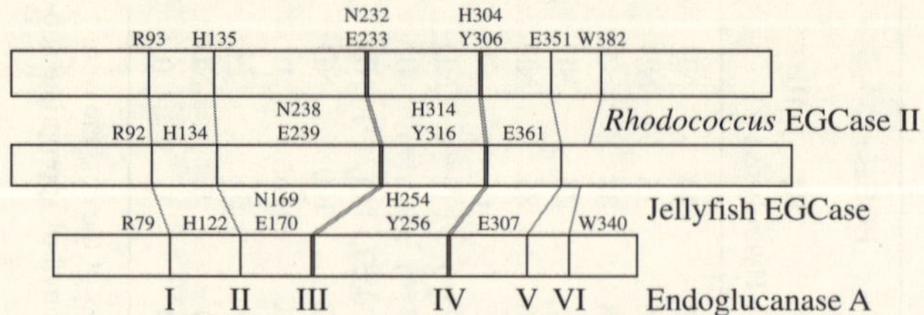

Fig. 2.24 Schematic illustration of the eight conserved amino acids in EGCase and family 5 endo-β-1,4-glucanase (family 5 cellulase).

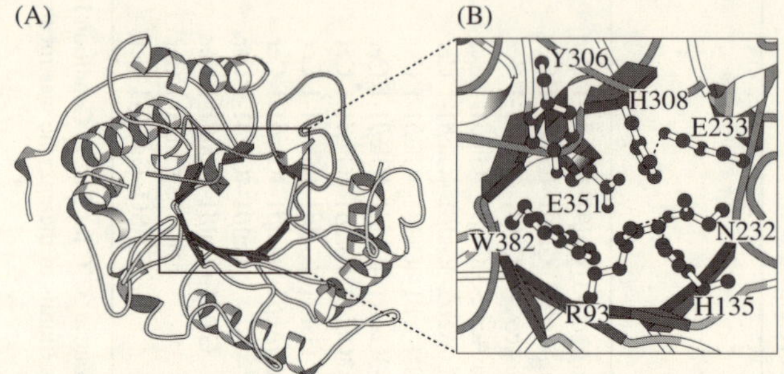

Fig. 2.25 Structural model of *Rhodococcus* sp. M-777 EGCase II. (A) The three-dimensional model of the EGCase was constructed by homology modeling using the catalytic domain of *Clostridium* endoglucanase A as a template using the program MOLSCRIPT.[204)] (B) Close-up view of the active site of the EGCase model.

II stimulated the activity of EGCase II much more strongly than EGCase I. In addition to activator II, *Rhodococcus* sp. M-777 produced a minor molecular species of activator protein which specifically stimulated the hydrolysis of glycosphingolipids by EGCase I (designated activator I). It was demonstrated that EGCase II can hydrolyze the cell-surface glycosphingolipids with the assist of activator II instead of detergents without impairing cell viability, *e.g.*, no hemolysis was observed when about 50% of cell-surface glycosphingolipids were hydrolyzed by the enzyme in the presence of activator.[205, 206)] Thus, activator proteins enable researchers to perform the analysis of biological functions of cell-surface glycosphingolipids by EGCases. The mechanism for the stimulation of EGCases by the activator proteins remains to be clarified. These activator proteins have not been found in animals possessing EGCases (ceramide glycanases). It is noteworthy that in mammals all the hydrolysis reactions of glycosphingolipids *in vivo* by exoglycosidases require specific activator proteins which can be replaced by detergents *in vitro*.

F. Biological Functions

EGCase (ceramide glycanase) is distributed widely in microbes and invertebrates but not vertebrates. The biological functions of EGCase remain to be elucidated. A previous study has revealed that many opportunistic microbes can produce EGCase and a relationship between pathogenesis and EGCase is suspected.[207)]

Recently, EGCase gene was found in the genome of *Propionibacterium acnes*, a commensal of human skin.[208] On the other hand, a new acidic EGCase was cloned from hydra and the function of this animal EGCase was clarified.[192] Hydra are the most primitive group of animals with a defined body and widely used for the study of pattern formation, cell differentiation and morphogenesis. Using whole mount *in situ* hybridization and western blotting with specific anti-hydra EGCase antibody, the enzyme was found to be present in mainly digestive cells of endodermal layers and transiently released into the gastric cavity during the feeding process. The hydra EGCase was able to hydrolyze glycosphingolipids of dietary brine shrimp to produce oligosaccharides and ceramides. Radioisotope-labeled GM1a was directly injected into the gastric cavity and the metabolic products monitored. It was found that the radioisotope substrates (labeled at sugar or lipid moieties) were catabolized directly to ceramide and oligosaccharide. The former was further converted to fatty acid and fatty acid-derived metabolites (glycerophospholipids and triacylglycerols) while the latter was converted to shorter-chain oligosaccharides, possibly GM2- and GM3-oligosaccharides. It was, however, noted that GM1a *per se* was resistant to hydrolysis by hydra exoglycosidases. These results indicate that EGCase is involved in the catabolic processing of dietary glycosphingolipids in hydra. This catabolic pathway is completely different from that of mammals, which are likely to be missing the EGCase. The catalytic pathway involving EGCase seems to be unique but quite ubiquitous in invertebrates except insects since EGCase activity is widely distributed in *Cnidaria*, *Mollusca*, *Annelida*, *Porifera*, *Echinodermata*, *etc*.

G. Applications

EGCases (ceramide glycanases) from *Rhodococcus* sp.[209, 210], leech[211] and hard-shell clam[197] have been used successfully for the structural analysis of glycosphingolipids. In this monograph, analysis of sugar chain structures using EGCase is described in Chapter 7. Leech,[212] coryneform bacteria[213] and jellyfish[214] enzymes were found to catalyze transglycosylation activity. This unique activity was used for the preparation of various neoglycoconjugates and fluorescence-labeled glycosphingolipids described in Chapter 6. *Rhodococcus* EGCase II assisted by an activator protein has been proved to be a good tool for cleaving the cell-surface glycosphingolipids and analyzing the functions of endogenous glycosphingolipids.[204] For example, hydrolysis of cell surface glycosphingolipids by EGCase II reduced EGF-receptor phosphorylation of human epidermoid carcinoma cell line A431[215] and inhibited synchronous oscillations of intracellular Ca^{2+} in cultured cortical neurons.[216] These results suggest the significance of glycosphingolipids in EGF-mediated signaling and synapse functions. Furthermore, homeostatic regulation of cell-surface glycosphingolipids coupled to glucosylceramide synthase (GlcT-1) was discovered by a method using EGCase II and its activator.[207, 217] Recently, this method was used for the clarification of the functions of glycosphingolipids involved in TAG-1 mediated signaling in lipid microdomains.[218]

2.4.2 Endo-β-galactosidase

Endo-β-galactosidase is a generic term for enzymes that act on internal β-galactosyl linkages within the sugar chains of various glycoconjugates. This section deals with endo-β-galactosidases capable of hydrolyzing glycoconjugates possessing the *N*-acetyllactosamine structure. This type of enzyme was first discovered as a keratan sulfate-degrading enzyme,[219, 220] later demonstrated to hydrolyze the glycosphingolipids, glycoproteins and oligosaccharides containing *N*-acetyllactosamine unit.[172, 193] The endo-β-galactosidase from *Escherichia freundii* is mainly described, focusing on the history of the discovery of the enzyme, specificity for glycosphingolipids and oligosaccharides, and its contribution in finding the lactosaminoglycan family.

A. Discovery

In 1970 Manabu Kitamikado of Kyushu University, Japan, discovered an enzyme that degrades keratan sulfate during his search for new glycosaminoglycan-degrading enzymes.[219] Keratan sulfate, mainly distributed in cornea and cartilage of vertebrates, is a glycosaminoglycan composed of lactosaminoglycan repeating units linking alternately galactose and *N*-acetylglucosamine through β-glycosyl linkages while the greater part of galactose and part of *N*-acetylglucosamine are sulfated. This enzyme was secreted from *E. freundii*, human intestinal bacterium, cultured in a medium containing keratan sulfate.[219] Allowing the purified enzyme to act on keratan sulfate, di-, tetrasaccharides and oligosaccharides of larger molecular size were generated. Critical analysis of the oligosaccharides revealed that all the oligosaccharides generated from keratan sulfate by the action of *E. freundii* enzyme had galactose at their reducing end.[220] The main product was demonstrated to be a disaccharide in which the *N*-acetylglucosamine residue at the nonreducing end was sulfated (HO_3S-6GlcNAcβ1-3Gal). Tetrasaccharide had three sulfate groups, *i.e.*, internal galactose and two *N*-acetylglucosamines were sulfated.

From these results, it was concluded that the enzyme hydrolyzes the internal β1,4 galactosidic linkage in keratan sulfate sugar chains, *i.e.*, the enzyme should be classified as an endo-β-galactosidase.[220] It is likely that the enzyme could not cleave the internal galactosidic linkage if the sugar is substituted by a sulfate group, thus producing oligosaccharides larger than disaccharides from keratan sulfate.

Similar enzyme activity was also reported by Hirano and Meyer in 1973,[221] but no detailed information was given thereafter. In 1975, Nakazawa and Suzuki reported another type of keratan-sulfate degrading enzyme from a soil bacterium, *Pseudomonas* sp.[222] This enzyme, called keratanase, hydrolyzes the internal β-galactosidic linkage within keratan sulfate sugar chains like the *E. freundii* enzyme. Thus, both enzymes have been confused and used as similar enzymes for a long time. However, the specificity of these two enzymes is clearly different.[223]

Fig. 2.26 represents the sugar chains having the structure of R-GlcNAcβ1-3Galβ1-4-R′ at four locations. Only structure (1) is subject to hydrolysis by the *Pseudomonas* enzyme, but the *E. freundii* enzyme can cleave linkages (1) and (2).

(1) (2)
OSO_3^-
R-GlcNAcβ1-3Galβ1 - 4GlcNAcβ1-3Galβ1 - 4-R′

(3) (4)
OSO_3^- OSO_3^- OSO_3^-
R-GlcNAcβ1-3Galβ1 - 4GlcNAcβ1-3Galβ1 - 4-R′

Fig. 2.26 Cleaving points of endo-β-galactosidases from *E. freundii* and *Pseudomonas* sp. *E. freundii*-type endo-β-galactosidase (the enzyme from *E. freundii*, *F. keratolyticus* or *B. fragilis*) hydrolyzes linkages (1) and (2) while *Pseudomonas*-type enzyme (keratanase I) hydrolyzes (1) but not (2). Linkages (3) and (4) are not hydrolyzed by either type of enzyme.

Neither enzyme can cleave (3) or (4), since the internal galactose is sulfated. In other words, *E. freundii* enzyme acts on the *N*-acetyllactosamine structure not only in keratan sulfate but also glycoproteins and glycosphingolipids while the *Pseudomonas* enzyme only hydrolyzes the sulfated *N*-acetyllactosamine structure in keratan sulfate, even though these enzymes are classified as keratan-sulfate-degrading enzymes. *E. freundii*-type enzyme has also been found in *Flavobacterium keratolyticus*[173)] and *Bacteroides fragilis*.[224)] *F. keratolyticus* enzyme was produced by the bacterium without induction of keratan sulfate[173)] making preparation of the enzyme much easier than *E. freundii* enzyme, which was produced by the bacterium only in the presence of keratan sulfate.[219)] *E. freundii* and *B. fragilis* enzymes are now commercially available. Recently, molecular cloning of the *F. keratolyticus* enzyme has been achieved.[147)]

B. Actions on Glycosphingolipids and Oligosaccharides

E. freundii-type endo-β-galactosidase can hydrolyze the internal β-galactosidic linkage of not only keratan sulfate but also glycosphingolipids, glycoproteins and oligosaccharides.[172, 193)] Table 2.12 shows the specificity of *E. freundii* and *F. keratolyticus* endo-β-galactosidases toward glycosphingolipids and oligosaccharides.[173, 193)] Both enzymes hydrolyze sialyl neolactotetraosylceramide (NeuAcα2-3Galβ1-4GlcNAcβ1-3Galβ1-4Glcβ1-1′Cer) and neolactotetraosylceramide (Galβ1-4GlcNAcβ1-3Galβ1-4Glcβ1-1′Cer) very efficiently. Substitution of fucose at the C-4 of the penultimate *N*-acetylglucosamine does not affect the susceptibility of *E. freundii* enzyme while fucosylation of terminal galactose greatly reduced it. Neither enzyme hydrolyzes the internal β-galactosidic linkage of globo-type or ganglio-type glycosphingolipids because they do not have an *N*-acetyllactosamine unit. Hydrolysis of glycosphingolipids by the enzymes was performed under conditions including detergents; for example, the reaction mixture should include taurodeoxycholate at a final concentration of 1 mg/mL. Detergents are likely to dissolve (or disperse) glycosphingolipid substrates in the reaction mixture. Interestingly, removal of the ceramide (Cer) moiety from glycosphingolipids reduced the susceptibility of both enzymes toward substrates, *i.e.*, lacto-*N*-tetraose (Galβ1-4GlcNAcβ1-3Galβ1-4Glc) is hydrolyzed much more slowly than neolactotetraosylceramide (Galβ1-4GlcNAcβ1-3Galβ1-4Glcβ1-1′Cer). Fucosylation of oligosaccharides resulted in a decrease of susceptibility of the substrate to the en-

Table 2.12 Specificity of endo-β-galactosidases from *E. freundii* and *F. keratolyticus* for glycosphingolipids and oligosaccharides

Substrate	Hydrolysis (%)	
	E. freundii	*F. keratolyticus*
NeuAcα2-3Galβ1-4GlcNAcβ1-3Galβ1-4Glcβ1-1′Cer	100	100
Galβ1-4GlcNAcβ1-3Galβ1-4Glcβ1-1′Cer	100	100
Fucα1-2Galβ1-3GlcNAcβ1-3Galβ1-4Glcβ1-1′Cer	20	ND
Galβ1-3[Fucα1-4]GlcNAcβ1-3Galβ1-4Glcβ1-1′Cer	100	ND
Fucα1-2Galβ1-3[Fucα1-4]GlcNAcβ1-3Galβ1-4Glcβ1-1′Cer	0	0
Galβ1-3GalNAcβ1-4[NeuAcα2-3]Galβ1-4Glcβ1-1′Cer	0	0
GalNAcβ1-3Galα1-4Galβ1-4Glcβ1-1′Cer	0	0
Galβ1-4GlcNAcβ1-3Galβ1-4Glc	5	50
Galβ1-3GlcNAcβ1-3Galβ1-4Glc	5	50
Fucα1-2Galβ1-3GlcNAcβ1-3Galβ1-4Glc	0	5

The incubation mixture contains 10 nmol of the substrate and 50 units of the enzyme in 100 μL of 0.01M sodium acetate buffer, pH 5.8. The incubation was carried out at 37°C for 17 h. For the hydrolysis of glycosphingolipids, 100 μg of sodium taurodeoxycholate was also included. One unit of the enzyme was defined as the amount of enzyme required to liberate 1 μg of reducing sugar equivalent to that of galactose per hour using keratan sulfate as the substrate. ND, not determined.

zyme. Interestingly, *F. keratolyticus* enzyme seemed to hydrolyze oligosaccharides much faster than *E. freundii* enzyme (Table 2.12).

C. Specificity

E. freundii-type endo-β-galactosidase was proved to be an endo-β-galactosidase with wide specificity which catalyzes the following reaction: R-GlcNAcβ1-3Galβ1-4-R′ + H_2O → R-GlcNAcβ1-3Gal + R′. The most important specificity of the enzyme is the sugar moiety adjacent to galactose. The adjacent sugar must be an *N*-acetylglucosamine. If the *N*-acetylglucosamine is replaced by other sugars, the hydrolysis of internal galactosidic linkage will not occur. This strict specificity of the enzyme allowed the researchers to identify the existence of the repeated structure of *N*-acetyllactosamine in various glycoconjugates. By using the specificity of the enzyme, glycoconjugates with repeating units of *N*-acetyllactosamine, called "lactosaminoglycans" hereafter, have been reported in various cells and tissues. For example, they were found in mouse early embryos,[225)] teratocarcinoma cells,[226)] cancer-related antigens,[227)] human erythrocytes[228)] and granulocytes.[229)] Shur and Hall reported that lactosaminoglycans could be sperm receptors involved in the process of mouse fertilization.[230)] In their hypothesis, a galactosyl transferase on the surface of sperms recognizes the *N*-acetylglucosamine residues of the lactosaminoglycan structure of the zona pellucida and this process could be involved in the first step of fertilization. Very recently, Eda *et al.* reported that sialyl lactosaminoglycans on CD43 are the marker for phagocytic recognition when they are gathered on the surface of Jurkat cells during the apoptotic process caused by the Fas ligand.[231)] Collectively, lactosaminoglycan family, which was first discovered using the specificity of *E. freundii* endo-β-galactosidase, are distributed widely in vertebrates and plays an important biological role.

D. Further Comments

As described above, endo-β-galactosidase of *E. freundii* was first discovered as a keratan sulfate-degrading enzyme,[219, 220] and later demonstrated to hydrolyze the internal β-galactosidic linkage of *N*-acetyllactosamine units in glycosphingolipids, glycoproteins and oligosaccharides.[172, 193] It was shown that sulfation of *N*-acetylglucosamine in the *N*-acetyllactosamine unit is not necessary for the *E. freundii*-type enzymes such as *F. keratolyticus*[173] and *B. fragilis*[224] enzymes. In contrast, the sulfation is essential for hydrolysis of the substrate by the *Pseudomonas* enzyme.[223] Thus, the two enzymes are clearly distinguishable, *i.e.*, the former shows wide specificity to the internal β-galactosidic linkage in lactosaminoglycans, which are glycoconjugates containing *N*-acetyllactosamine units, and the latter shows strict specificity to keratan sulfate.[223] In other words, *E. freundii*-type enzymes are designated "lactosaminoglycan-degrading endo-β-galactosidase" and *Pseudomonas*-type enzymes as "keratan-sulfate degrading endo-β-galactosidase"or "keratanase". Recently, another type of keratanase, called keratanase II, was found in *Bacillus circulans*; this enzyme hydrolyzes the internal β-*N*-acetylglucosamine linkage in keratan sulfate,[232] and clearly distinguished from endo-β-galactosidase. It is recommended that the *Pseudomonas* endo-β-galactosidase be designated "keratanase I," and that keratanase II be classified as an endo-β-*N*-acetylglucosaminidase.

Since the discovery of endo-β-galactosidase from *E. ferundii* in 1970,[219] lactosaminoglycans were reported one after another as a result of the use of the specificity of the enzyme.[225–231] This demonstrates that the discovery of new enzymes can open doors to new research areas, much like the discovery of restriction enzymes for DNAs, which led to the explosive development of gene technology.

References

1. Muramatsu, T. (1971) *J. Biol. Chem.* **246**, 5535-5537
2. Tarentino, A.L., Plummer, T.H., and Maley, F. (1972) *J. Biol. Chem.* **247**, 2629-2631
3. Muramatsu, T., Koide, N., and Maeyama, K. (1978) *J. Biochem.* **83**, 363-370
4. Tarentino, A.L. and Maley, F. (1974) *J. Biol. Chem.* **249**, 811-817
5. Koide, N. and Muramatsu, T. (1974) *J. Biol. Chem.* **249**, 4897-4904
6. Tai, T., Yamashita, K., Ogata-Arakawa, M., Koide, N., Muramatsu, T., Iwashita, S., Inoue, Y., and Kobata, A. (1975) *J. Biol. Chem.* **250**, 8569-8575
7. Tai, T., Yamashita, K., and Kobata, A. (1977) *Biochem. Biophys. Res. Commun.* **78**, 434-441
8. Ito, S., Muramatsu, T., and Kobata, A. (1975) *Arch. Biochem. Biophys.* **171**, 78-86
9. Morinaga, T., Kitamikado, M., Iwase, H., Li, S.-C., and Li, Y.-T. (1983) *Biochim. Biophys. Acta* **749**, 211-213
10. Yamamoto, K., Tanaka, T., Fujimori, K., Kang, C.-S., Ebihara, H., Kanamori, J., Kadowaki, S., Tochikura, T., and Kumagai, H. (1998) *Biotechnol. Appl. Biochem.* **28**, 235-242
11. Yamamoto, K., Kadowaki, S., Takegawa, K., Kumagai, H., and Tochikura, T. (1986) *Agric. Biol. Chem.* **50**, 421-429
12. Roberts, G., Tarelli, E., Homer, K.A., Philpott-Howard, J., and Beighton, D. (2000) *J. Bacteriol.* **182**, 882-890
13. Collin, M. and Olsen, A. (2001) *EMBO J.* **20**, 3046-3055
14. Barreaud, J.P., Bourgerie, S., Julien, R., Guespin-Michel, J.F., and Karamanos, Y. (1995) *J. Bacteriol.* **177**, 916-920

15. Takegawa, K., Nakoshi, M., Iwahara, S., Yamamoto, K., and Tochikura, T. (1989) *Appl. Environ. Microbiol.* **55**, 3107-3112
16. Hitomi, J., Murakami, Y., Saitoh, F., Shigematsu, N., and Yamaguchi, H. (1985) *J. Biochem.* **98**, 527-533
17. Elder, J.H. and Alexander, S. (1982) *Proc. Natl. Acad. Sci. USA* **79**, 4540-4544
18. Plummer, T.H., Elder, J.H., Alexander, S., Phelan, A.W., and Tarentino, A.L. (1984) *J. Biol. Chem.* **259**, 10700-10704
19. Tarentino, A.L., Gomez, C.M., and Plummer, T.H. (1985) *Biochemistry* **24**, 4665-4671
20. Tarentino, A.L. and Plummer, T.H. (1987) *Methods Enzymol.* **138**, 770-778
21. Lee, K.-O., Gesundheit, N., Chen, H.-C., and Weintraub, B.D. (1986) *Biochem. Biophys. Res. Commun.* **138**, 230-237
22. Moonen, P., Mermod, J.-J., Ernst, J.F., Hirshi, M., and DeLamarter, J.F. (1987) *Proc. Natl. Acad. Sci. USA* **84**, 4428-4431
23. Kuhn, J.E., Eing, B.R., Brossmer, R., Munk, K., and Braun, R.W. (1988) *J. Gen.Virol.* **69**, 2847-2858
24. Trimble, R.B. and Tarentino, A.L. (1991) *J. Biol. Chem.* **266**, 1646-1651
25. Plummer, T.H. and Tarentino, A.L. (1991) *Glycobiology* **1**, 257-263
26. Kadowaki, S., Yamamoto, K., Fujisaki, M., Kumagai, H., and Tochikura, T. (1988) *Agric. Biol. Chem.* **52**, 2387-2389
27. Nishigaki, M., Muramatsu, T., and Kobata, A. (1974) *Biochem. Biophys. Res. Commun.* **59**, 638-645
28. Norden, C.E., Lundblad, A., Svensson, S., Ockerman, P.-A., and Autio, S. (1974) *Biochemistry* **13**, 871-874
29. Pierce, R.J., Spik, G., and Montreuil, J. (1979) *Biochem. J.* **180**, 673-676
30. Tachibana, Y., Yamashita, K., Kawaguchi, M., Arashima, S., and Kobata, A. (1981) *J. Biochem.* **90**, 1291-1296
31. Ito, K., Okada, Y., Ishida, K., and Minamiura, N. (1993) *J. Biol. Chem.* **268**, 16074-16081
32. Kato, T., Hatanaka, K., Mega, T., and Hase, S. (1997) *J. Biochem.* **122**, 1167-1173
33. Kato, T., Fujita, K., Takeuchi, M., Kobayashi, K., Natsuka, S., Ikura, K., Kumagai, H., and Yamamoto, K. (2002) *Glycobiology* **12**, 581-587
34. Suzuki, T., Yano, K., Sugimoto, S., Kitajima, K., Lennarz, W.J., Inoue, S., Inoue, Y., and Emori, Y. (2002) *Proc. Natl. Acad. Sci. USA* **99**, 9691-9696
35. Tachibana, Y., Yamashita, K., and Kobata, A. (1982) *Arch. Biochem. Biophys.* **214**, 199-210
36. Moore, S.E.H. (1999) *Trends Cell Biol.* **9**, 441-446
37. Moore, S.E.H. and Spiro, R.G. (1994) *J. Biol. Chem.* **269**, 12715-12721
38. Parodi, A.J. (2000) *Annu. Rev. Biochem.* **69**, 69-93
39. Saint-Pol, A., Bauvy, C., Codogno, P., and Moore, S.E.H. (1997) *J. Cell Biol.* **136**, 45-59
40. Li, S.-C., Arakawa, M., Hirabayashi, Y., and Li, Y.-T. (1981) *Biochim. Biophys. Acta* **660**, 278-283
41. Yet, M.G. and Wold, F. (1988) *J. Biol. Chem.* **263**, 118-122
42. Kimura, Y., Iwata, K., Sumi, Y., and Takagi, S. (1996) *Biosci. Biotechnol. Biochem.* **60**, 228-232
43. Berger, S., Menudier, A., Julien, R., and Karamanos, Y. (1995) *Biochimie* **77**, 751-760
44. Vuylsteker, C., Cuvellier, G., Berger, S., Faugeron, C., and Karamanos, Y. (2000) *J. Exp. Bot.* **346**, 839-845
45. Kimura, Y., Matsuo, S., and Takagi, S. (1998) *Biosci. Biotechnol. Biochem.* **62**, 253-261
46. Kimura, Y., Tokuda, T., Ohno, A., Tanaka, H., and Ishiguro, Y. (1998) *Biochim. Biophys. Acta* **1381**, 27-36
47. Kimura, Y., Matsuo, S., Tsurusaki, S., Kimura, M., Hara-Nishimura, I., and Nishimura, M. (2002) *Biochim. Biophys. Acta* **1570**, 38-46
48. Kimura, Y. (2000) *Trends Glycosci. Glycotechnol.* **12**, 103-112
49. Robbins, P.W., Trimble, R.B., Wirth, D.F., Hering, C., Maley, F., Maley, G.F., Das, R., Gibson, B.W., Royal, N., and Biemann, K. (1984) *J. Biol. Chem.* **259**, 7577-7583
50. Tarentino, A.L., Quinones, G., Schrader, W.P., Changchien, L.-M., and Plummer, T.H. (1992) *J. Biol. Chem.* **267**, 3868-3872
51. Tarentino, A.L., Quinones, G., Changchien, L.-M., and Plummer, T.H. (1993) *J. Biol. Chem.* **268**, 9702-9708
52. Takegawa, K., Mikami, B., Iwahara, S., Morita, Y., Yamamoto, K., and Tochikura, T. (1991) *Eur. J. Biochem.* **202**, 175-180
53. Kuranda, M.J. and Robbins, P.W. (1991) *J. Biol. Chem.* **266**, 19758-19767
54. Takegawa, K., Yamabe, K., Fujita, K., Tabuchi, M., Mita, M., Izu, H., Watanabe, A., Asada, Y.,

Sano, M., Kondo, A., Kato, I., and Iwahara, S. (1997) *Arch. Biochem. Biophys.* **338**, 22-28
55. Muramatsu, H., Tachikui, H., Ushida, H., Song, X.-J., Qiu, Y., Yamamoto, S., and Muramatsu, T. (2001) *J. Biochem.* **129**, 923-928
56. Fujita, K., Kobayashi, K., Iwamatsu, A., Takeuchi, M., Kumagai, H., and Yamamoto, K. (2004) *Arch. Biochem. Biophys.* **432**, 41-49
57. Rao, V., Cui, T., Guan, C., and Van Roey, P. (1999) *Protein Sci.* **8**, 2338-2346
58. Waddling, C.A., Plummer, T.H., Tarentino, A.L., and Van Roey, P. (2000) *Biochemistry* **39**, 7878-7885
59. Kadowaki, S., Yamamoto, K., Fujisaki, M., Izumi, K., Tochikura, T., and Yokoyama, T. (1990) *Agric. Biol. Chem.* **54**, 97-106
60. Yamamoto, K., Kadowaki, S., Fujisaki, M., Kumagai, H., and Tochikura, T. (1994) *Biosci. Biotech. Biochem.* **58**, 72-77
61. Yamamoto, K., Kadowaki, S., Watanabe, J., and Kumagai, H. (1994) *Biochem. Biophys. Res. Commun.* **203**, 244-252
62. Haneda, K., Inazu, T., Yamamoto, K., Kumagai, H., Nakahara, Y., and Kobata, A. (1996) *Carbohyd. Res.* **292**, 61-70
63. Yamamoto, K. (2001) *J. Biosci. Bioeng.* **92**, 493-501
64. Tai, T., Yamashita, K., Ito, S., and Kobata, A. (1977) *J. Biol. Chem.* **252**, 6687-6694
65. Takegawa, K., Nakoshi, M., Yamamoto, K., Tochikura, T., and Iwahara, S. (1991) *J. Ferment. Technol.* **71**, 279-281
66. Koide, N. and Muramatsu, T. (1975) *Biochem. Biophys. Res. Commun.* **66**, 411-416
67. Takegawa, K., Yamaguchi, S., Kondo, A., Iwamoto, H., Nakoshi, M., Kato, I., and Iwahara, S. (1991) *Biochem. Int.* **24**, 849-855
68. Nilsson, K.G.I. (1988) *Trends Biotechnol.* **6**, 256-264
69. Cote, G.L. and Tao, B.Y. (1990) *Glycoconj. J.* **7**, 145-162
70. Trimble, R.B., Atkinson, P.H., Tarentino, A.L., Plummer, T.H., Jr., and Tomer, F. (1986) *J. Biol. Chem.* **261**, 12000-12005
71. Takegawa, K., Yamaguchi, S., Kondo, A., Kato, I., and Iwahara, S. (1991) *Biochem. Int.* **25**, 829-835
72. Fan, J.-Q., Takekawa, K., Iwahara, S., Kondo, A., Kato, I., Abeygunawardana, C., and Lee, Y.C. (1995) *J. Biol. Chem.* **270**, 17723-17729
73. Fan, J.-Q., Huynh, L.H., Reinhold, B.B., Reinhold, V.N., Takegawa, K., Iwahara, S., Kondo, A., Kato, I., and Lee, Y.C. (1996) *Glycoconj. J.* **13**, 643-652
74. Fujita, K., Miyamura, T., Sano, M., Kato, I., and Takegawa, K. (2002) *J. Biosci. Bioeng.* **93**, 614-617
75. Takahashi, N. (1977) *Biochem. Biophys. Res. Commun.* **76**, 1194-1201
76. Sugahara, K., Okumura, T., and Yamashina, I. (1972) *Biochim. Biophys. Acta* **268**, 488-496
77. Sasaki, A., Yamagishi, M., Mega, T., Norioka, S., Natsuka, S., and Hase, S. (1999) *J. Biochem.* **125**, 363-367
78. Hase, S., Ikenaka, T., and Matsusima, Y. (1978) *Biochem. Biophys. Res. Commun.* **85**, 257-263
79. Hase, S. (1994) *Methods Enzymol.* **230**, 225-237
80. Ishimizu, T., Norioka, S., Kanai, M., Clarke, A.E., and Sakiyama, F. (1996) *Eur. J. Biochem.* **242**, 627-635
81. Ishimizu, T., Mitsukami, Y., Shinkawa, S., Natsuka, S., Hase, S., Miyagi, M., Sakiyama, F., and Norioka, S. (1999) *Eur. J. Biochem.* **263**, 624-634
82. Ishimizu, T., Sasaki, A., Okutani, S., Maeda, M., Yamagishi, M., and Hase, S. (2004) *J. Biol. Chem.* **279**, 38555-38562
83. Sasaki, A., Ishimizu, T., and Hase, S. (2005) *J. Biochem.* **137**, 87-93
84. Villarroya, H. and Petek, F. (1976) *Biochim. Biophys. Acta* **438**, 200-211
85. Chen, H., Leipprandt, J.R., Traviss, C.E., Sopher, B.L., Jones, M.Z., Cavanagh, K.T., and Friderici, K.H. (1995) *J. Biol. Chem.* **270**, 3841-3848
86. Leipprandt, J.R., Kraemer, S.A., Haithcock, B.E., Chen, H., Dyme, J.L., Cavanagh, K.T., Friderici, K.H., and Jones, M.Z. (1996) *Genomics* **37**, 51-56
87. Alkhayat, A.H., Kraemer, S.A., Leipprandt, J.R., Macek, M., Kleijer, W.J., and Friderici, K.H. (1998) *Hum. Mol. Genet.* **7**, 75-83
88. Beccari, T., Bibi, L., Stinchi, S., Stirling, J.L., and Orlacchio, A. (2001) *Biosci. Rep.* **21**, 315-323
89. Takada, G., Kawaguchi, T., Kaga, T., Sumitani, J., and Arai, M. (1999) *Biosci. Biotechnol. Biochem.* **63**, 206-209
90. Ademark, P., de Vries, R.R., Hägglund, P., Stålbrand, H., and Visser, J. (2001) *Eur. J. Biochem.* **268**, 2982-2990

91. Stoll, D., Stålbrand, H., and Warren, A.J. (1999) *Appl. Environ. Microbiol.* **65**, 2598-2605
92. Béki, E., Nagy, I., Vanderleyden, J., Jäger, S., Kiss, L., Fülöp, L., Hornok, L., and Kukolya, J. (2003) *Appl. Environ. Microbiol.* **69**, 1944-1952
93. Parker, K.N., Chhabra, S.R., Lam, D., Callen, W., Duffaud, G.D., Snead, M.A., Short, J.M., Mathur, E.J., and Kelly, R.M. (2001) *Biotechnol. Bioeng.* **75**, 322-333
94. Henrissat, B. and Bairoch, A. (1996) *Biochem. J.* **316**, 695-696
95. Durand, P., Lehn, P., Callebaut, I., Fabrega, S., Henrissat, B., and Mornon, J.-P. (1997) *Glycobiology* **7**, 277-284
96. Stoll, D., He, S., Withers, S.G., and Warren, A.J. (2000) *Biochem. J.* **351**, 833-838
97. Zechel, D.J., Reid, S.P., Stoll, D., Nashiru, O., Warren, A.J., and Withers, S.G. (2003) *Biochemistry* **42**, 7195-7204
98. Bewley, J.D., Burton, R.A., Morohashi, Y., and Fincher, G.B. (1997) *Planta* **203**, 454-459
99. Berman, E. and Allerhand, A. (1981) *J. Biol. Chem.* **256**, 6657-6662
100. Maley, F. and Trimble, R.B. (1981) *J. Biol. Chem.* **256**, 1088-1090
101. Oku, H., Hase, S., and Ikenaka, T. (1991) *Anal. Biochem.* **185**, 331-334
102. Gasperi, R.D., Daniel, P.F., and Warren, C.D. (1992) *J. Biol. Chem.* **267**, 9706-9712
103. Kobata, A. (1979) *Anal. Biochem.* **100**, 1-14
104. Meyer, K., Dubos, R., and Smyth, E. M. (1937) *J. Biol. Chem.* **118**, 71-78
105. Iwase, H. and Hotta, K. (1992) in *CRC Handbook of Endoglycosidases and Glycoamidases* (Takahashi, N. and Muramatsu, T., eds.). Chapter 3, pp. 41-53, CRC Press, Florida
106. Bhavanandan, V.P. (1991) *Glycobiology* **1**, 493-503
107. Bhavanandan, V.P., Umemoto, J., and Davidson, E.A. (1976) *Biochem. Biophys. Res. Comm.* **70**, 738-745
108. Endo, Y. and Kobata, A. (1976) *J. Biochem.* **80**, 1-8
109. Glasgow, L.R., Paulson, J.C., and Hill, R.L. (1977) *J. Biol. Chem.* **252**, 8615-8623
110. Umemoto, J., Bhavanandan, V.P., and Davidson, E.A. (1977) *J. Biol. Chem.* **252**, 8609-8614
111. Umemoto, J., Matta, K.L., Barlow, J.J., and Bhavanandan, V.P. (1978) *Anal. Biochem.* **91**, 186-193
112. Brooks, M.M. and Savage A.V. (1997) *Glycoconj. J.* **14**, 183-190
113. Bhavanandan, V.P. and Codington, J.F. (1983) *Carbohydr. Res.* **118**, 81-89
114. Bardales, R., Bhavanandan, V.P., Wiseman, G., and Bramwell, M.E. (1989) *J. Biol. Chem.* **264**, 1980-1987
115. Bardales, R. and Bhavanandan, V.P. (1989) *J. Biol. Chem.* **264**, 19893-19897
116. Huang, C.C. and Aminoff, D. (1972) *J. Biol. Chem.* **247**, 6737-6742
117. Bell, W.C., Pomato, N., and Aminoff, D. (1978) *Carbohydr. Res.* **61**, 447-455
118. DiCioccio, R.A., Klock, P.J., Barlow, J.J., and Matta, K.L. (1980) *Carbohydr. Res.* **81**, 315-322
119. Fan, J.-Q., Kadowaki, S., Yamamoto, K., Kumagai, H., and Tochikura, T. (1988) *Agric. Biol. Chem.* **52**, 1715-1723
120. Fan, J.-Q., Yamamoto, K., Kumagai, H., and Tochikura, T. (1990) *Agric. Biol. Chem.* **54**, 233-234
121. Lecat, D., Lemonnier, M., Derappe, C., Lhermitte, M., van Halbeek, H., Dorland, L., and Vliegenthart, J.F.G. (1984) *Eur. J. Biochem.* **140**, 415-420
122. Hirabayashi, Y., Matsumoto, Y., Matsumoto, M., Toida, T., Iida, N., Matsubara, T., Kanzaki, T., Yokota, M., and Ishizuka, I. (1990) *J. Biol. Chem.* **265**, 1693-1701
123. Fan J.-Q., Yamamoto K., Matsumoto Y., Hirabayashi Y., Kumagai H., and Tochikura T. (1990) *Biochem. Biophys. Res. Commun.* **169**, 751-757
124. Ashida, H., Yamamoto, K., and Kumagai, H. (2001) *Carbohydr. Res.* **330**, 487-493
125. Ishii-Karakasa, I., Iwase, H., Hotta, K., Tanaka, Y., and Omura, S. (1992) *Biochem. J.* **288**, 475-482
126. Abdullah, K., Udoh, E.A., Shewan, P., and Mellors, A. (1992) *Infect. Immun.* **60**, 56-62
127. Hu, R., Mellors, A., and Bhavanandan, V.P. (1994) *Arch. Biochem. Biophys.* **310**, 300-309
128. Devaraj, H. and Bhavanandan, V.P. (1990) *FASEB J.* **4**, 2305
129. Yamamoto, K. (1994) *J. Biochem.* **116**, 229-235
130. Patel, T., Bruce, J., Merry, A., Bigge, C., Wormald, M., Jaques, A., and Parekh, R. (1993) *Biochem*istry **32**, 679-693
131. Devaraj, N., Sheykhanazari, M., Warren, W.S., and Bhavanandan, V.P. (1994) *Glycobiology* **4**, 307- 316
132. Maury, P. (1971) *Biochim. Biophys. Acta* **252**, 48-57
133. Parkkinen, J. and Finne, J. (1987) *Eur. J. Biochem.* **136**, 355-361
134. Endo, M., Takagaki, K., and Nakamura, T. (1992) in *CRC Handbook of Endoglycosidases and*

Glycoamidases (Takahashi, N. and Muramatsu, T., eds.), Chapter 5, pp. 105-132, CRC Press, Florida
135. Hoskins, L.C. and Zamcheck, N. (1968) *Gastroenterology* **54**, 210-217
136. Hoskins, L.C. (1978) *The Glycoconjugates* (Horowitz, M.I. and Pigman, W., eds.) Vol II, pp. 235-253
137. Hopper, L.V. and Gordon, J.I. (2001) *Glycobiology* **11**, 1R-10R
138. MacFarlane, G.T. and Gibson, G.R. (1991) *FEMS Microbiol. Lett.* **77**, 289-294
139. Hoskins, L.C., Boulding, E.T., and Larson, G. (1997) *J. Biol. Chem.* **272**, 7932-7939
140. Xu, J., Bjursell, M.K., Himrod, J., Deng, S., Carmichael, C., Chiang, H.C., Hooper, L.V., and Gordon, J.I. (2003) *Science* **299**, 2074-2076
141. Bhavanandan, V.P., Murray, M.C., and Davidson, E.A. (1988) *Glycoconj. J.* **5**, 467-498
142. Amemura, A., Shah, R.H., Bahl, O.P., and Merrick, J.M. (1978) *Carbohydr. Res.* **63**, 241-151
143. Ashida, H., Anderson, K., Nakayama, J., Maskos, K., Chou, C.-W., Cole, R.B., Li, S.-C., and Li, Y.-T. (2001) *J. Biol. Chem.* **276**, 28226-28232
144. Ashida, H., Maskos, K., Li, S.-C., and Li, Y.-T. (2002) *Biochemistry* **41**, 2388-2395
145. Anderson, K., Li, S.-C., and Li, Y.-T. (2000) *Anal. Biochem.* **287**, 337-339
146. Nakayama, J., Yeh, J.C., Misra, A.K., Ito, S., Katsuyama, T., and Fukuda, M. (1999) *Proc. Natl. Acad. Sci. USA* **96**, 8991-8996
147. Leng, L., Zhu, A., Zhang, Z., Hurst, R., and Goldstein, J. (1998) *Gene* **222**, 187-194
148. Ogawa, H., Muramatsu, H., Kobayashi, T., Morozumi, K., Yokoyama, I., Kurosawa, N., Nakao, A., and Muramatsu, T. (2000) *J. Biol. Chem.* **275**, 19368-19374
149. Henrissat, B. (1991) *Biochem. J.* **280**, 309-316
150. Juncosa, M., Pons, J., Dot, T., Querol, E., and Planas, A. (1994) *J. Biol. Chem.* **269**, 14530-14535
151. Keitel, T., Simon, O., Borriss, R., and Heinemann, U. (1993) *Proc. Natl. Acad. Sci. USA* **90**, 5287-5291
152. Deng, L., Liu, Z.-J., Ashida, H., Li, S.-C., Li, Y.-T., Rose, J., and Wang, B.-C. (2004) *Acta Crystallogr. D. Biol. Crystallogr.* **60**, 537-538
153. Tempel, W., Liu, Z.-J., Horanyi, P.S., Deng, L., Lee, D., Newton, M.G., Rose, J.P., Ashida, H., Li, S.-C., Li, Y.-T., and Wang, B.-C. (2005) *Proteins* **59**, 141-144
154. Michel, G., Chantalat, L., Duee, E., Barbeyron, T., Henrissat, B., Kloareg, B., and Dideberg, O. (2001) *Structure* **9**, 513-525
155. Faham, S., Hileman, R.E., Fromm, J.R., Linhardt, R.J., and Rees, D.C. (1996) *Science* **271**, 1116-1120
156. Takasaki, S. and Kobata, A. (1976) *J. Biol. Chem.* **251**, 3603-3609
157. Anderson, K., Ashida, H., Maskos, K., Dell, A., Li, S.-C., and Li, Y.-T. (2005) *J. Biol. Chem.* **280**, 7720-7728
158. Varadi, D.P., Cifonelli, J.A., and Dorfman, A. (1967) *Biochim. Biophys. Acta* **141**, 103-117
159. Wessler, E. (1971) *Biochem. J.* **122**, 373-384
160. Matsue, H. and Endo, M. (1987) *Biochim. Biophys. Acta* **923**, 470-477
161. Endo, M., Yamamoto, M., Munakata, H., Yamamoto, R., Namiki, O., and Yosizawa, Z. (1980) *Tohoku J. Exp. Med.* **131**, 355-361
162. Majima, M., Nakamura, T., Igarashi, S., Matsue, H., and Endo, M. (1984) *J. Biochem. Biophys. Meth.* **9**, 245-249
163. Takagaki, K., Nakamura, T., Majima, M., and Endo, M. (1985) *FEBS Lett.* **181**, 271-274
164. Takagaki, K., Nakamura,T., and Endo, M. (1988) *Biochim. Biophys. Acta* **966**, 94-98
165. Fukunaga, Y., Sobue, M., Suzuki, N., Kushida, H., and Suzuki, S. (1975) *Biochim. Biophys. Acta* **381**, 443-447
166. Takagaki, K., Nakamura, T., Kon, A., Tamura, S., and Endo, M. (1991) *J. Biochem.* **109**, 514-519
167. Takagaki, K,, Kon, A., Kawasaki, H., Nakamura, T., and Endo, M. (1989) *J. Biochem. Biophys. Meth.* **19**, 207-214
168. Takagaki, K., Kon, A., Kawasaki, H., Nakamura, T., Tamura, S., and Endo, M. (1990) *J. Biol. Chem.* **265**, 854-860
169. Kon, A., Takagaki, K., Kawasaki, H., Nakamura, T., and Endo, M. (1991) *J. Biochem.* **110**, 132-135
170. Takagaki, K., Kon, A., Kawasaki, H., Nakamura, T., Tamura, S., and Endo, M. (1990) *Biochem. Biophys. Res. Commun.* **169**, 15-21
171. Takagaki, K., Nakamura, T., Takeda, Y., Daidouji, K., and Endo, M. (1992) *J. Biol. Chem.* **267**, 18558-18563

172. Fukuda, M.N. and Matsumura, G. (1976) *J. Biol. Chem.* **251**, 6218-6225
173. Kitamikado, M., Ito, M., and Li, Y.-T. (1981) *J. Biol. Chem.* **256**, 3906-3909
174. DeGasperi, R., Li, Y.-T., and Li, S.-C. (1986) *J. Biol. Chem.* **261**, 5696-5698
175. Fushuku, N., Muramatsu, H., Uezono, M.M., and Muramatsu, T. (1987) *J. Biol. Chem.* **262**, 10086-10092
176. Yuki, H. and Fishman, W.H. (1963) *J. Biol. Chem.* **238**, 1877-1879
177. Thunberg, L., Bäckström, G., Wasteson, Å., Robinson, H.C., Ögren, S., and Lindahl, U. (1982) *J. Biol. Chem.* **257**, 10278-10282
178. Klein, U. and Von Figura, K. (1976) *Biochem. Biophys. Res. Commun.* **73**, 569-576
179. Öldberg, A., Heldin, C.H., Wasteson, A., Busch, C., and Höök, M. (1980) *Biochemistry* **19**, 5755-5762
180. Nakajima, M., Irimura, T., Di Ferrante, N., and Nicolson, G. L. (1984) *J. Biol. Chem.* **259**, 2283-2290
181. Yanagishita, M. (1985) *J. Biol. Chem.* **260**, 11075-11082
182. Takagaki, K., Nakamura, T., Majima, M., and Endo, M. (1988) *J. Biol. Chem.* **263**, 7000-7006
183. Milner, Y. and Avigad, G. (1967) *Carbohydr. Res.* **4**, 359-361
184. Majima, M., Takagaki, K., Igarashi, S., Nakamura, T., and Endo, M. (1984) *J. Biochem. Biophys. Meth.* **10**, 143-151
185. Hakomori, S., Handa, K., Iwabuchi, K., Yamamura, S., and Prinetti, A. (1998) *Glycobiology* **8**, xi-xix
186. Karlsson, K.A. (1989) *Annu. Rev. Biochem.* **58**, 309-350
187. Ito, M. and Yamagata, T. (1986) *J. Biol. Chem.* **261**, 14278-14282
188. Li, Y.-T., DeGasper, R., Muldrey, J.E., and Li, S.-C. (1986) *Biochem. Biophys. Res. Commun.* **141**, 346-352
189. Ito, M. and Yamagata, T. (1989) *J. Biol. Chem.* **264**, 9510-9519
190. Zhou, B., Li, S.-C., Laine, R.A., Huang, R.T.C., and Li, Y.-T. (1989) *J. Biol. Chem.* **264**, 12272-12277
191. Miura, Y., Arai, T., Ohtake, A., Ito, M., Yamamoto, K., and Yamagata, T. (1999) *Glycobiology* **9**, 957-960
192. Horibata, Y., Sakaguchi, K., Okino, N., Iida, H., Inagaki, M., Fujisawa, T., Hama, Y., and Ito, M. (2004) *J. Biol. Chem.* **279**, 29351-29358
193. Nakagawa, H., Yamada, T., Chien, J.J., Gardes, A., Kitamikado, M., Li, S.-C., and Li, Y.-T. (1980) *J. Biol. Chem.* **255**, 5955-5959
194. Nakamura, K. and Handa, S. (1984) *Anal. Biochem.* **142**, 406-410
195. Sakaguchi, K., Okino, N., Sueyoshi, N., Izu, H., and Ito, M. (2000) *J. Biochem.* **128**, 145-152
196. Ashida, H., Yamamoto, K., Kumagai, H., and Tochikura, T. (1992) *Eur. J. Biochem.* **205**, 729-735
197. Basu, S.S., Dastgheib-Hosseini, S., Hoover, G., Li, Z., and Basu, S. (1994) *Anal. Biochem.* **222**, 270-274
198. Li, Y.-T., Ishikawa, Y., and Li, S.-C. (1987) *Biochem. Biophys. Res. Commun.* **149**, 167-172
199. Horibata, Y., Okino, N., Ichinose, S., Omori, A., and Ito, M. (2000) *J. Biol. Chem.* **275**, 31297-31304
200. Izu, H., Izumi, Y., Kurome, Y., Sano, M., Kondo, A., Kato, I., and Ito, M. (1997) *J. Biol. Chem.* **272**, 19846-19850
201. Sakaguchi, K., Okino, N., Izu, H., and Ito, M. (1999) *Biochem. Biophys. Res. Commun.* **260**, 89-93
202. Wang, Q., Tull, D., Menke, A., Gilkes, N.R., Warren, R.A.J., Aebersold, R., and Wither, S.G. (1993) *J. Biol. Chem.* **268**, 14096-14102
203. Ito, M., Ikegami, Y., and Yamagata, T. (1991) *J. Biol. Chem.* **266**, 7919-7926
204. Kraulis, P.J. (1991) *J. Appl. Crystallog.* **24**, 946-950
205. Ito, M., Ikegami, Y., Tai, T., and Yamagata, T. (1993) *Eur. J. Biochem.* **218**, 637-643
206. Ito, M., Ikegami, Y., and Yamagata, T. (1993) *Eur. J. Biochem.* **218**, 645-649
207. Ito, M. and Komori, H. (1996) *J. Biol. Chem.* **271**, 12655-12660
208. Bruggemann, H., Henne, A., Hoster, F., Liesegang, H., Wiezer, A., Strittmatter, A., Hujer, S., Durre, P., and Gottschalk, G. (2004) *Science* **305**, 671-673
209. Higashi, H., Ito, M., Fukaya, N., Yamagata, S., and Yamagata, T. (1990) *Anal. Biochem.* **186**, 355-362
210. Shimamura, M., Hayase, T., Ito, M., Rasilo, M.-L., and Yamagata, T. (1988) *J. Biol. Chem.* **263**, 12124-12128
211. Hansson, G.C., Li, Y.-T., and Karlsson, H. (1989) *Biochemistry* **28**, 6672-6678

212. Li, Y.-T., Zhou, B., Rao, B.N.N., Schweinggruber, H., and Li, S.-C. (1991) *J. Biol. Chem.* **266**, 10723-10726
213. Ashida, H., Tsuji, Y., Yamamoto, K., Kumagai, H., and Tochikura, T. (1993) *Arch. Biochem. Biophys.* **305**, 559-562
214. Horibata, Y., Higashi, H., and Ito, M. (2001) *J. Biochem.* **130**, 263-268
215. Ji, L., Ito, M., Zhang, G., and Yamagata, T. (1995) *Glycobiology* **5**, 343-350
216. Muramoto, K., Kawahara, M., Kobayashi, K., Ito, M., Yamagata, T., and Kuroda, Y. (1994) *Biochem. Biophy. Res. Commun.* **202**, 398-402
217. Komori, H., Ichikawa, S., Hirabayashi, Y., and Ito, M. (1999) *J. Biol. Chem.* **274**, 8981-8987
218. Kasahara, K., Watanabe, K., Takeuchi, K., Kaneko, H., Oohira, A., Yamamoto, T., and Sanai, Y. (2000) *J. Biol. Chem.* **275**, 34701-34709
219. Kitamikado, M., Ueno, R., and Nakamura, T. (1970) *Bull. Jap. Soc. Sci. Fish.* **36**, 592-596
220. Kitamikado, M. and Ueno, R. (1972) *Bull. Jap. Soc. Sci. Fish.* **38**, 497-502
221. Hirano, S. and Meyer, K. (1973) *Connect. Tissue Res.* **2**, 1-10
222. Nakazawa, K. and Suzuki, S. (1975) *J. Biol. Chem.* **250**, 912-917
223. Ito, M., Hirabayashi, Y., and Yamagata, T. (1986) *J. Biochem.* **100**, 773-780
224. Scudder, P., Uemura, K., Dolby, J., Fukuda, M.N., and Feizi, T. (1983) *Biochem. J.* **213**, 485-494
225. Gooi, H.C., Feizi, T., Kapadia, A., Knowles, B.B., Solter, D., and Evans, M.J. (1981) *Nature* **292**, 156-158
226. Muramatsu, T., Gachelin, G., Dammorveli, M., Delaarbre, C., and Jacob, F. (1979) *Cell* **18**, 183-191
227. Watanabe, K. and Hakomori, S. (1976) *J. Exp. Med.* **144**, 645-653
228. Fukuda, M.N., Fukuda, M., and Hakomori, S. (1979) *J. Biol. Chem.* **254**, 3700-3703
229. Fukuda, M., Koeffler, H.P., and Minowada, J. (1981) *Proc. Natl. Acad. Sci. USA*, **78**, 6299-6303
230. Shur, B.D. and Hall, N.G. (1982) *J. Cell Biol.* **95**, 574-579
231. Eda, S., Yamanaka, M., and Beppu, M. (2004)) *J. Biol. Chem.* **279**, 5967-5974
232. Yamagishi, K., Suzuki, K., Imai, K., Mochizuki, H., Morikawa, K., Kyogashima, M., Kimata, K., and Watanabe, H. (2003) *J. Biol. Chem.* **278**, 25766-25772

3

Enzymatic Synthesis of Neo-*N*-glycans

3.1 Chemo-enzymatic Synthesis of Neoglycoconjugates Using Endo-*β*-*N*-acetylglucosaminidase

Progress in the field of glycobiology requires the synthesis and construction of glycoconjugates to elucidate the significance and functions of the oligosaccharides. The creation of new oligosaccharides with additional functions by modification of naturally occurring oligosaccharides and the addition of an oligosaccharide to a substance to give it a useful function are important subjects in "glycotechnology." Such attempts have been made using various methods, for example, the conjugation of a monosaccharide or oligosaccharide to a protein by a chemical method.[1] Although these methods are useful for attaching oligosaccharides to proteins, they cannot be used for attaching oligosaccharides to the original glycosylation sites of the native glycoproteins. However, there is an enzymatic method for doing this: an oligosaccharide is added to a substance using glycosyltransferases. This involves some difficulty since a large number of glycosyltransferases are required to construct an oligosaccharide with a complex structure. Moreover, such specific enzymes are hard to obtain because of the small quantities in cells. Therefore, it seems impossible to add an oligosaccharide to some substance using glycosyltransferases.

Recently, a practical method of synthesizing a glycopeptide has been developed combining the chemical synthesis of glycosyl peptides with the enzymatic method of transglycosylation of endoglycosidase, microbial endo-*β*-*N*-acetylglucosaminidase (Endo-*β*-GlcNAc-ase) (Fig. 3.1).

3.1.1 General Method for Chemo-enzymatic Synthesis of Glycopeptide

Although methods for the chemical synthesis of oligosaccharides in glycoconjugates have been developed, they are labor-intensive and involve complicated steps of protection and deprotection. However, several strategies for glycopeptide synthesis have been reported. These include the chemical addition of oligosaccharides to peptides by reductive amination[1] and solid-phase synthesis using gly-

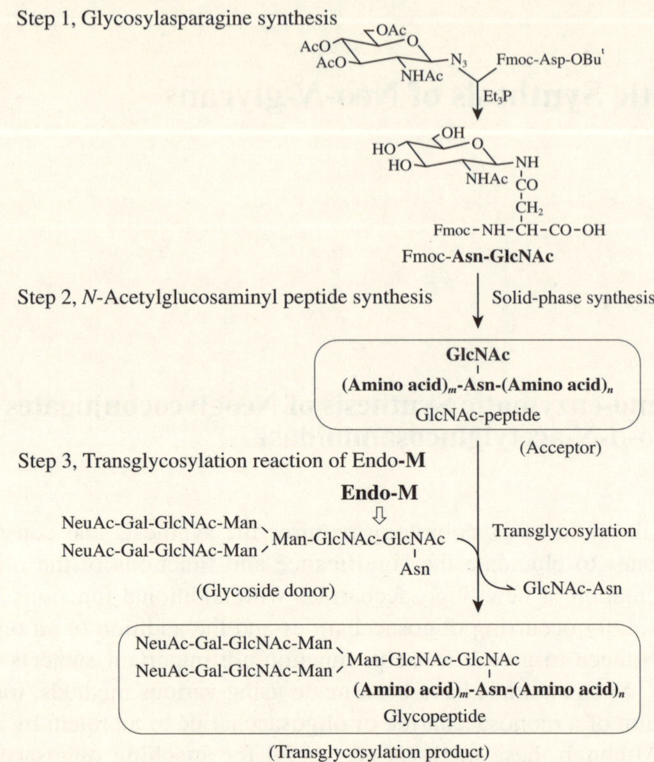

Fig. 3.1 Strategy of chemo-enzymatic synthesis of glycopeptide using transglycosylation activity of Endo-β-GlcNAc-ase. The procedure is a combination of three steps of chemical and enzymatic methods. Fmoc-Asp-OBu^t, Fmoc-aspartic acid α-*t*-butyl ester; Et_3P, triethylphosphine; Endo-M, Endo-β-GlcNAc-ase from *Mucor hiemalis*.

coamino acids as building blocks.[2] Although a peptide can be easily synthesized by a chemical method, it is almost impossible to synthesize a glycopeptide by a chemical method not only because the chemical synthesis of oligosaccharides involves complicated steps as noted above, but also because of the difficulty in binding with the peptide.

On the other hand, enzymatic methods of oligosaccharide synthesis have advantages because of their high stereo- and regio-selectivities. Thus the biosynthetic preparation of oligosaccharide is carried out using glycosyltransferases. However, many kinds of glycosyltransferases are required to synthesize oligosaccharides with complex structures. The enzymatic synthesis of glycopeptide using peptidases and glycosyltransferases has been reported.[3]

A. Transglycosylation Activity of Endo-β-GlcNAc-ase

It is well known that many glycosidases exhibit transglycosylation activities that involve the transfer of a carbohydrate moiety to the hydroxyl groups of various compounds, in addition to hydrolytic activities that involve the hydrolysis of glycosidic linkages. Transglycosylation activity of exoglycosidases is used for the enzymatic synthesis of various oligosaccharides.[4] In contrast to the transglycosylation activities of exoglycosidases, those of endoglycosidases are less known and not well studied. Transglycosylation is considered to be a special type of hydrolysis. In the hydrolysis of glycosidases, the carbohydrate moiety of the substrate is released and transferred to water. On the other hand, in the transglycosylation, the carbohydrate moiety released from the substrate is transferred to the hydroxyl groups of various compounds instead of water.[5] Therefore, the transglycosylation activity of endoglycosidases may be useful for transferring and adding an oligosaccharide (sugar chain) to various compounds having a hydroxyl group.

Hydrolysis reaction of endoglycosidase:
oligosaccharide-protein / lipid + HOH → oligosaccharide + protein / lipid
Transglycosylation reaction of endoglycosidase:
oligosaccharide-protein / lipid + R-OH → oligosaccharide-R + protein / lipid

This has attracted attention in the field of glycotechnology as being a very useful means of glycosylation.

Endo-β-GlcNAc-ase (EC 3.2.1.96) is a unique endoglycosidase that hydrolyzes *N,N'*-diacetylchitobiosyl linkages in oligosaccharides bound to Asn residues of various glycoproteins and glycopeptides and leaves one GlcNAc residue on the protein and peptide moieties. In 1986, Trimble *et al.* found that Endo-β-GlcNAc-ase of *Flavobacterium meningosepticum*, that is Endo-F,[6] had transglycosylation activity.[7] During studies on the substrate specificity of Endo-F, they observed that released oligosaccharides failed to incorporate tritium on reduction with NaB^3H_4. It seemed that the reducing end of the released oligosaccharide might be blocked by some compound. They found that glycerol added to the Endo-F preparation as a stabilizing reagent was put in the reaction mixture and the hydrolysis product of oligosaccharide was transferred to glycerol as the receptor by the transglycosylation activity of Endo-F during hydrolysis. In the structure of the transglycosylation product glycerol was glycosidically linked via its 1(or 3) carbon to the C1 of the reducing end of GlcNAc in the oligosaccharide. This finding suggested that Endo-β-GlcNAc-ases could transfer oligosaccharide to not only GlcNAc but also to other receptors having a hydroxyl group.

In 1988, a novel Endo-β-GlcNAc-ase was found in the culture fluid of *Mucor hiemalis* isolated from soil that could cleave not only the high-mannose type and hybrid type of Asn-linked oligosaccharides but also the complex type of oligosaccharide,[8] unlike other microbial Endo-β-GlcNAc-ases, which can act on only high-mannose and hybrid type oligosaccharides. This enzyme (named Endo-M after its source) showed transglycosylation activity and could transfer the oligosaccharides from glycopeptides to suitable acceptors with a GlcNAc residue during hydrolysis of the glycopeptide.[9] The transglycosylation activity of Endo-M catalyzes the following reaction (Fig. 3.2):

R-GlcNAc-GlcNAc-Asn (glycoside donor) + GlcNAc-Asn-R′ (acceptor) →
R-GlcNAc-GlcNAc-Asn-R′ (transglycosylation product) + GlcNAc-Asn

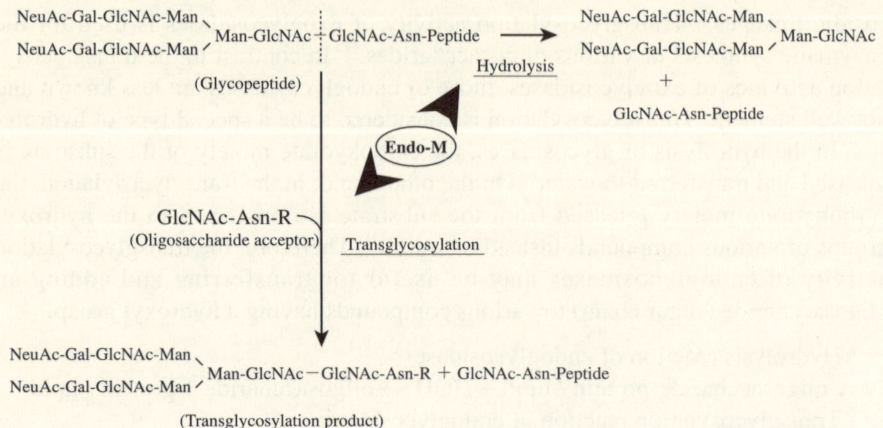

Fig. 3.2 Enzymatic action of Endo-M. NeuAc, *N*-acetylneuraminic acid (sialic acid).

(R: any type of oligosaccharide, R′: peptide or peptide derivatives).

These observations suggested that the transglycosylation of Endo-M might be useful for the synthesis of glycopeptides. This enzyme may also be useful for adding the oligosaccharide to various compounds and attracted attention in the fields of protein engineering and cell biology because there are almost no methods of glycosylating to proteins. The enzymatic method using the transglycosylation activity of endoglycosidase seems promising for glycosylation.

However, every Endo-*β*-GlcNAc-ases does not possess transglycosylation activity. So far as we know, Endo-M and Endo-A from *Arthrobacter protophormiae*[10] are the only microbial Endo-*β*-GlcNAc-ases having transglycosylation activity and useful for transferring oligosaccharides. Endo-H which is popular and commercially available does not have transglycosylation activity and cannot be used for the above purpose.

Concerning the mechanism of the transglycosylation activity of Endo-M and Endo-A, there is a report suggesting that their transglycosylation reactions proceeded via an oxazolinium ion intermediate.[11] Two possible mechanisms are proposed for the hydrolysis of the *N,N′*-diacetylchitobiose moiety in chitinase-type glycosidases: one is carried out via an oxocarbenium ion intermediate and the other via an oxazolinium ion intermediate.[12] According to the report, a novel disaccharide substrate of Man*β*1-4GlcNAc-oxazoline, the assumed intermediate of the transition state, was synthesized. This substrate was successful in transferring the oxazoline substrate to GlcNAc*β*1-*O*-*p*NP as an acceptor by Endo-M and Endo-A, resulting in the formation of the core trisaccharide derivative Man*β*1-4GlcNAc*β*1-4GlcNAc*β*1-*O*-*p*NP. On the other hand, Endo-H could not transfer the oxazoline substrate to GlcNAc*β*1-*O*-*p*NP. Based on these results, a new mechanism involving an oxazolinium ion intermediate has been proposed for the endoglycosidase-catalyzed transglycosylation (Fig. 3.3). It seems that only endoglycosidase which can catalyze the enzymatic reaction *via* the oxazolinium ion as an intermediate of

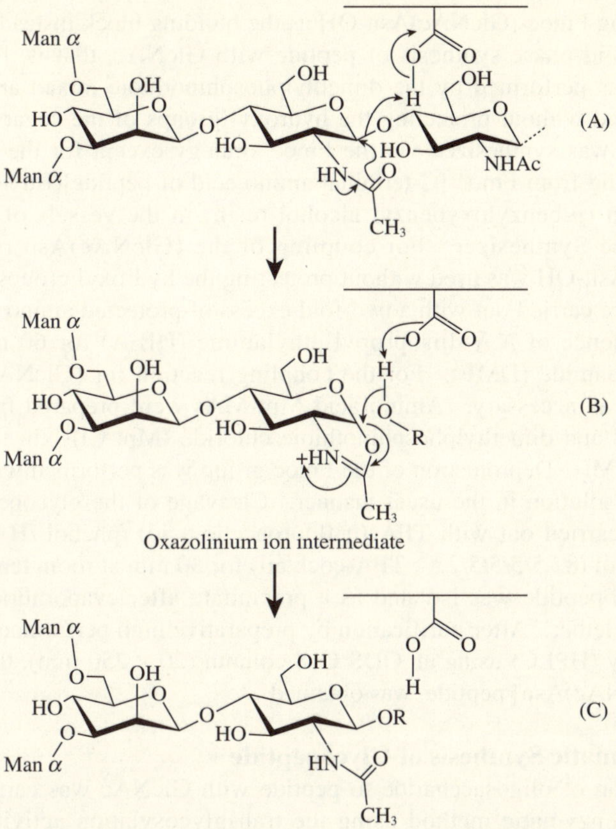

Fig. 3.3 Proposed reaction mechanism of Endo-M and Endo-A via oxazolinium ion intermediate.[11] (A) The oxygen of the glycosidic bond between two GlcNAc units is protonated by the carboxylic acid of an acidic amino acid followed by the formation of the oxazolinium ion intermediate by the nucleophilic attack of the amide carbonyl group to the anomeric center. (B) The oxazolinium ion intermediate is attacked by water or a glycosyl acceptor (R-OH). (C) The hydrolyzate or the transglycosylation product is formed.

the transition state has the transglycosylation activity.

B. Chemical Synthesis of Peptide Containing GlcNAc

As described above, Endo-M was able to transfer the oligosaccharides from glycopeptides to suitable acceptors with a GlcNAc residue. Inazu and Kobayashi developed a convenient synthetic method to prepare the peptide with a GlcNAc by solid-phase synthesis based on the Fmoc (9-fluorenylmethyloxycarbonyl) strategy in which Fmoc-(GlcNAc)Asn-OH is used instead of Fmoc-Asn-OH.[13] Chemical synthesis of the peptide with a GlcNAc was carried out as follows.[14] First, 2-acetamido-3,4,6-tri-*O*-acetyl-2-deoxy-D-glucopyranosyl azide and N^{α}-Fmoc-Asp-α-*t*-butyl ester were dissolved in dichloromethane at –78°C under an argon atmosphere. Next, triethylphosphine was added to the solution and Fmoc-N^{β}-(*N*-acetylglucosaminyl)-asparagine [Fmoc-(GlcNAc)Asn-OH] was obtained in about 77%

yield. Using Fmoc-(GlcNAc)Asn-OH as the building block instead of Fmoc-Asn-OH, the solid-phase synthesis of peptide with GlcNAc, that is, [(GlcNAc)Asn] peptide, was performed by the dimethylphosphinothioic mixed anhydride (Mpt-MA) method without protecting the hydroxyl groups of the sugar moiety as follows.[15] It was synthesized via the Fmoc strategy except for the coupling reactions, starting from Fmoc-[C-terminal amino acid of peptide](Bu^t)-OH coupled to Wang resin (*p*-benzyloxybenzyl alcohol resin) in the vessels of Multi-Peptide Solid-Phase Synthesizer. For coupling of the (GlcNAc)Asn residue, Fmoc-(GlcNAc)Asn-OH was used without protecting the hydroxyl groups. Coupling reactions were carried out with a two-fold excess of protected amino acid Mpt-MAs in the presence of *N*,*N*-diisopropyl ethylamine (DIEA) for 60 min in *N*,*N*-dimethylformamide (DMF). For the coupling reaction for (GlcNAc)Asn, double coupling was necessary. Amino acid Mpt-MAs were prepared from the Fmoc-amino acid and dimethylphosphinothioic chloride (Mpt-Cl)[16] in the presence of DIEA in DMF. Deprotection of the Fmoc group was performed with 20% piperidine/DMF solution in the usual manner. Cleavage of the glycopeptide from the resin was carried out with TFA (trifluoroacetic acid) /phenol /H_2O /thioanisol / ethanedithiol (82.5/5/5/5/2.5 ; TFA-cocktail) for 30 min at room temperature. The crude glycopeptide was isolated as a precipitate after evaporation and washing with diethylether. After purification by preparative high performance liquid chromatography (HPLC) using an ODS C18 column (20 × 250 mm), the desired peptide, [(GlcNAc)Asn] peptide, was obtained.

C. Enzymatic Synthesis of Glycopeptide

The addition of oligosaccharide to peptide with GlcNAc was carried out by the following enzymatic method using the transglycosylation activity of Endo-M. Endo-M was purified from the culture fluid of the fungus *Mucor hiemalis* strain No. 314 by various column chromatographies.[17] The enzyme preparation was almost free from other glycosidase and protease activities. The recombinant enzyme was also used.[18] Sialotransferrin glycopeptide as the donor of oligosaccharide was prepared by repeated exhaustive pronase digestion of human transferrin followed by Sephadex G-25 gel filtration.[19] Asialotransferrin glycopeptide was prepared from sialotransferrin glycopeptide through reaction with sialidase before the last pronase digestion. Both glycopeptides obtained were further purified by HPLC using a reversed-phase column before use. The donor of oligosaccharide was also obtained from a hen egg yolk[20]: a glycosyl asparagine having a disialobiantennary complex-type oligosaccharide, H-Asn[(NeuAc-Gal-GlcNAc-Man)$_2$-Man-(GlcNAc)$_2$]-OH was prepared by pronase digestion of a hen egg yolk glycopeptide having a disialobiantennary complex type oligosaccharide, H-Lys-Val-Ala-Asn[(NeuAc-Gal-GlcNAc-Man)$_2$-Man-(GlcNAc)$_2$]-Lys-Thr-OH.

Transglycosylation reaction of Endo-M was carried out in 10 μL of reaction mixture including 25 mM glycoside donor, 12.5 mM acceptor (GlcNAc-Asn- peptide), 60 mM potassium phosphate buffer (pH 6.0) and 40 milliunits of Endo-M solution (final 4 milliunits/mL). EDTA-3Na was added to the reaction mixture to a final concentration of 50 mM to protect the enzyme from the action of trace amounts of peptidase contamination in the enzyme preparation. The reaction mixture was incubated at 37°C for a suitable period. To terminate the reaction, the mixture was diluted 25-fold with cold distilled water and immediately frozen. Aliquots of the reaction mixture were analyzed by HPLC after filtration. Yield of the transglycosylation product was calculated by the following equation.[21)]

Yield (%) = (Area of the peak of the product / Initial area of the peak of the acceptor) × 100

HPLC analyses of peptides and glycopeptides were performed with a reversed-phase column (J'sphere ODS-M80, 4.6 × 250 mm). Elution was carried out with a linear acetonitrile gradient (0 - 10%) in 30 min at room temperature at a flow rate of 0.8 mL/min. Peptides and glycopeptides were detected by UV absorbance at 214 nm.

D. Typical Chemo-enzymatic Synthesis of Glycopeptide

A typical example of the chemo-enzymatic synthesis of a glycopeptide is described below.[22)] Peptide T is a bioactive peptide composed of eight amino acids of the sequence Ala-Ser-Thr-Thr-Thr-Asn-Tyr-Thr, and can block infection of human T cells by human immunodeficiency virus (HIV).[23)]

The first step of the chemo-enzymatic synthesis of glycosylated Peptide T containing *N*-linked oligosaccharides was the synthesis of *N*-acetylglucosaminyl Peptide T, which had GlcNAc attached to the Asn residue of the peptide, by a solid-phase method. The scheme of the solid-phase synthesis of *N*-acetylglucosaminyl Peptide T, [(GlcNAc)^{6}Asn] Peptide T, is shown in Fig. 3.4. The yield was about

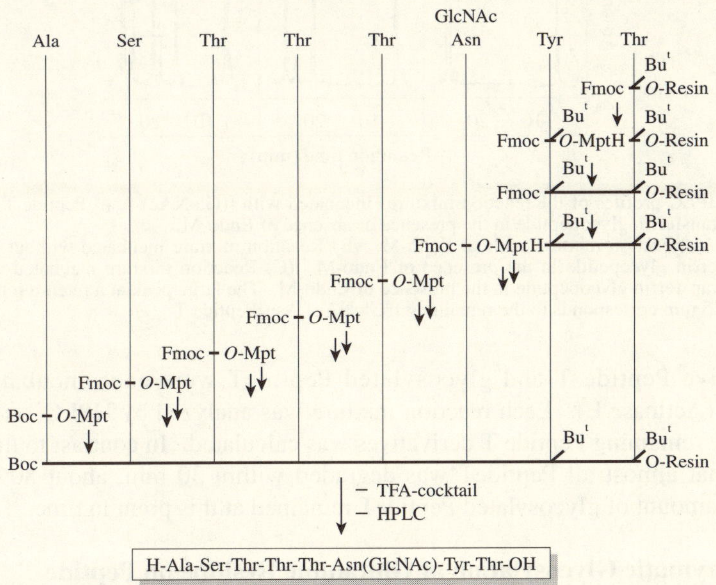

Fig. 3.4 Solid-phase synthesis of Peptide T containing GlcNAc attached to the Asn residue. Resin, *p*-alkoxybenzyl type; Mpt, dimethylphosphinothioic-; Fmoc, 9-fluorenylmethyloxycarbonyl- ; Boc, *t*-butyloxycarbonyl-.

39%. The structure of the [(GlcNAc)^{6}Asn] Peptide T obtained was identified by ^{1}H-NMR spectroscopy, MALDI-TOF MS spectrometry and amino acid analysis. The second step was the addition of oligosaccharide from glycopeptide to *N*-acetylglucosaminyl Peptide T, [(GlcNAc)^{6}Asn] Peptide T, using the transglycosylation activity of Endo-M. The reaction was carried out in the same manner using 25 mM sialotransferrin glycopeptide as the glycoside donor and 10 mM [(GlcNAc)^{6}Asn] Peptide T as the acceptor. After incubation for 4.5 h at 37°C, the reaction mixture was subjected to HPLC (Fig. 3.5). The newly occurring peaks were collected and analyzed by ESI mass spectrometry. In the spectrum of peak 1, a mass ion with a charge of –3, $[M\text{-}3H]^{3-}$ with *m/z* of 1019.4 was detected. The molecular mass of 3061.2 Da calculated from this value agreed with the theoretical value of a transglycosylation product in which sialobiantennary complex type oligosaccharide was transferred to [(GlcNAc)^{6}Asn] Peptide T (Mr 3063.9). The yield of the product was about 9%. The spectrum of peak 2 revealed a transglycosylation product in which a monosialobiantennary complex type oligosaccharide was transferred to [(GlcNAc)^{6}Asn] Peptide T (Mr 2772.6).

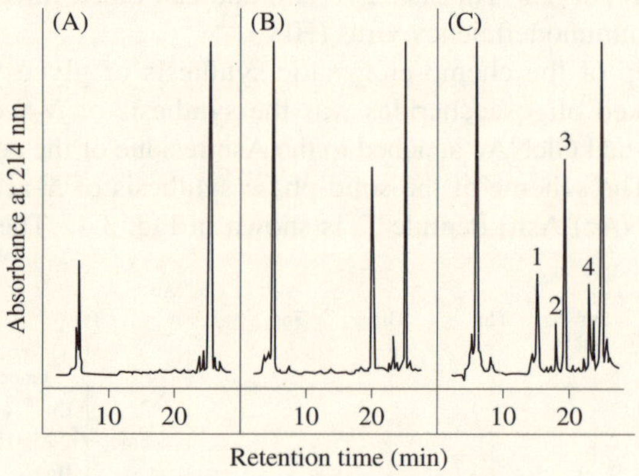

Fig. 3.5 HPLC profiles of the reaction mixtures incubated with [(GlcNAc)^{6}Asn] Peptide T and sialotransferrin glycopeptide in the presence or absence of Endo-M.
(A) Reaction mixture without Endo-M. (B) Reaction mixture incubated without sialotransferrin glycopetide in the presence of Endo-M. (C) Reaction mixture incubated with sialotransferrin glycopeptide in the presence of Endo-M. The large peak at a retention time of 24-25 min corresponds to the remaining [(GlcNAc)^{6}Asn]Peptide T.

Native Peptide T and glycosylated PeptideT were each incubated with Pronase (Actinase E). Each reaction mixture was analyzed by HPLC, and the ratio of the remaining Peptide T derivatives was calculated. In contrast to the observation that almost all PeptideT was degraded within 30 min, about 80% of the original amount of glycosylated PeptideT remained at this point in time.[22)]

E. Enzymatic Glycosylation of Glutamine Residue on Peptide

Oligosaccharides in glycoproteins are linked to the Asn residue in consensus sequence triplet Asn-Xaa-Ser/Thr, where Xaa is any amino acid except proline, of a protein as *N*-linked oligosaccharides. This triplet sequence is necessary for glycosylation. The biological addition of oligosaccharides to the Gln residues of peptides is impossible. However, using this chemo-enzymatic method, a glycopeptide having a Gln-linked oligosaccharide can be synthesized.

Substance P is a bioactive undecapeptide and a putative neurotransmitter that acts on the central and peripheral nervous systems.[24] This peptide has no Asn residue but it has two Gln residues as the fifth and sixth amino acid residues from the N-terminus. An attempt to introduce an *N*-linked oligosaccharide to the Gln residue of the peptide was made using a chemo-enzymatic method.[25] Two Substance P derivatives with GlcNAc attached to the fifth Gln residue and the sixth Gln residue were synthesized by the solid-phase method using the Fmoc strategy. A building block Fmoc-*N*-acetylglucosaminyl glutamine, Fmoc-(GlcNAc)Gln-OH, was used instead of Fmoc-Gln-OH at the fifth or the sixth amino acid by the Mpt-MA method. Thereafter, Substance P derivatives with GlcNAc were obtained. The transglycosylation of *N*-linked oligosaccharides to the (GlcNAc)Gln moieties of Substance P derivatives was performed by catalysis with Endo-M in the same manner as described above. The glycosylated Substance P was biologically active, although the activity was rather low, and stable against peptidase digestion.[25] The oligosaccharide moiety attached to the Gln residue of the peptide was not liberated by peptide-N^4-(*N*-acetyl-β-D-glucosaminyl) asparagine amidase F (PNGase F).

F. Chemo-enzymatic Synthesis of Glycopolymers Containing Multivalent Sialyl Oligosaccharides

This chemo-enzymatic method was applied to the synthesis of a glycopolymer containing multivalent oligosaccharides. It is well known that influenza viruses infect host cells through the binding of viral hemagglutinins (HAs) to sialyl glycoproteins or sialyl glycolipids as receptors on the host cell surface. The viruses recognize not only sialic acids in the receptors but also particular structures of the sugar chain. The above led to the attempt to chemo-enzymatically synthesize glycopolymers with polymeric compound backbones carrying multivalent sialooligosaccharides units as polymeric inhibitors of infection by human influenza viruses. Sialyl glycopeptide from hen egg yolk is a suitable oligosaccharide donor for the transglycosylation activity of Endo-M. It has sialyl (α2-6) *N*-acetyllactosamine residue in the nonreducing end, and this structure is known to be the hemagglutinin receptor of the A-type influenza virus.[26] There is a possibility that it might be used as an inhibitor against infection by influenza virus, but the binding ability of the sugar chain is not so strong. On the other hand, it is known that multivalent interaction occurs between viral hemagglutinins and cell surface sialic acid, and they should be amplified by the so-called glycoside cluster effects.[27] Various synthetic glycopolymers carrying multivalent sialic acid residues that target viral hemagglutinins have been prepared as influenza virus inhibitors using various polymers as backbones.[28] These conventional glycopolymers inhibit the binding of influenza viruses to host cell receptors with high affinity.

We synthesized glycopolymers that mimic these sialyl glycoproteins. In these glycopolymers, the sialooligosaccharides were clustered on a chitosan backbone

(Fig. 3.6). First, we prepared *p*-formylphenyl *β*-GlcNAc using a phase-transfer catalytic system,[29] then we transformed sialobianntenary complex type oligosaccharide of glycopeptide from hen egg yolk to this compound using Endo-M in a high yield (over 50%). This compound includes an aldehyde group which is able to condense the amino group by reductive amination in aqueous solution. Reductive amination of *p*-formylphenyl sialooligosaccharide compound with chitosan backbone was carried out in a good yield, and the glycopolymer was obtained. We carried out the hemagglutination inhibition test for influenza virus (New Caledonia/20/99 strain) using this glycopolymer. It was found that the ability for competitive inhibition with HRP-Fetuin is over 300-fold higher against fetuin. Thus the glycopolymer is a very good inhibitor for influenza virus hemagglutination.

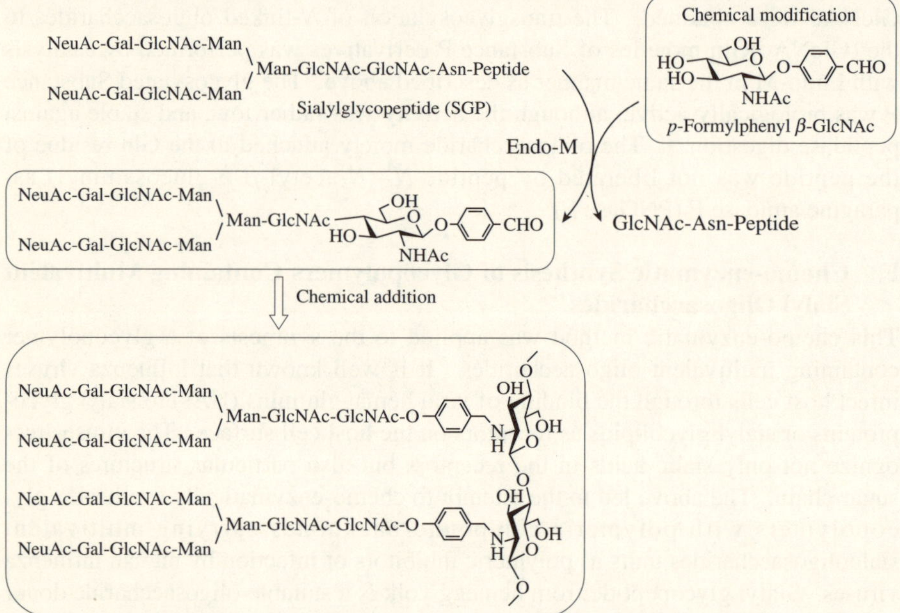

Fig. 3.6 Chemo-enzymatic synthesis of multiple sialooligosaccharide-binding polymers using transglycosylation activity of Endo-M.

A luminescent probe to detect influenza viruses was also synthesized using the transglycosylation activity of Endo-M. Tris-bipyridine ruthenium-complexes carrying disialocomplex type oligosaccharides were synthesized by the transglycosylation reaction of Endo-M. Their luminescence intensity is strongly depressed by virus binding.[30]

G. Improvement of Transglycosylation Reaction of Endo-M in Organic Solvents

In general, alcohols and acetonitrile are used for the suppression of hydrolysis and improvement of the transglycosylation reaction. Therefore, the transglycosylation activity of Endo-M in various organic solvents was examined.[31] The enzymatic reaction was carried out using *p*NP-β-GlcNAc as the acceptor for the oligosaccharide chain and sialyl glycopeptide from hen egg yolk as the donor of the oligosaccharide chain. The reaction mixture was incubated with Endo-M at 25°C for 3 h in various 30% organic solvents. In 30% acetone solution, the yield of the transglycosylation product was found to reach about 34% of the total amount of *p*NP-β-GlcNAc as the acceptor. Moreover, the yields of the transglycosylation products reached about 21% and 24% in dimethylsulfoxide (DMSO) and methanol, respectively (Table 3.1). On the other hand, the yield was only about 15% for the reaction in aqueous solution. Butanone, acetonitrile, dioxane, *N*,*N*-dimethylformamide (DMF), ethanol and tetrahydrofuran (THF) were not found to be effective. Concerning the effects of concentration of various organic solvents on the transglycosylation activity of Endo-M (the reaction was carried out for 2 h in the presence of 10-50% of various organic solvents), the yield of the transglycosylation product reached to about 40% in 30% acetone. Even in 40% DMSO and methanol, it reached 35-40%. It seemed that degradation of product barely occurred in the organic solvent even though it did so rapidly in aqueous solution. Thus, the transglycosylation activity of Endo-M was shown to be highly promoted in organic solvents, and Endo-M is likely to be extremely stable in these organic solvents.

Table 3.1 The yields of the transglycosylation product in reaction mixtures including various 30% organic solvents (v/v)

Organic solvent	Transglycosylation yield (%)
None	14.7
Acetone	34.1
Butanone	0
Acetonitrile	0
Dioxane	0
DMF	0
DMSO	21.3
Methanol	24.7
THF	0

Acceptor, *p*NP-β-GlcNAc; Donor, sialyl glycopeptide from hen egg yolk.

H. Substrate Specificity of Endo-M for Transglycosylation Reaction

As for the acceptors of the transglycosylation reaction by Endo-M, it was found that Endo-M transferred a sialocomplex type oligosaccharide from the glycopeptide to Glc and Man as well as GlcNAc (see Table 2.3). Moreover, not only the β-isomers but the α-isomers of these sugars also served as good glycosyl acceptors for the transglycosylation reaction. However, the transglycosylation activity of Endo-M on Xyl was very low and this enzyme did not transfer the oligosaccharide to sugars such as Gal, L-Fuc, GlcUA, cyclohexanol and cyclohexane-1,2-diol. These results suggest that Endo-M strictly recognizes the hydroxyl groups and their configurations at the C-4 and C-6 positions of the sugar acceptors: C-4 possesses the equatorial OH and the C-6 position should be $-CH_2OH$. The configurations at the C-1 and C-2 positions were hardly recognized by Endo-M.

3.1.2 Chemo-enzymatic Synthesis of Bioactive Glycopeptide Using Endoglycosidase

The present author and his research group have developed a chemo-enzymatic synthetic method for glycopeptide[32)] that combines the chemical synthesis of a peptide containing *N*-acetylglucosamine (GlcNAc) as a glycosylation tag (*N*-acetylglucosaminyl peptide)[13,15,33)] and the transglycosylation of the natural oligosaccharide block to the *N*-acetylglucosaminyl peptide catalyzed by Endo-*β*-GlcNAc-ase.[9)] Using this method, it becomes possible to introduce *N*-linked oligosaccharide into a peptide originally having no sugar chain.

We were interested in the remodeling of bioactive peptide or protein by the artificial addition of oligosaccharide and investigating the effect of *N*-glycosylation on the structure and activity of a bioactive peptide. Eel calcitonin (**1**) (see Fig.3.8) was used as the model peptide in this study. Calcitonin, a calcium metabolism-regulating hormone composed of 32 amino acids, exhibits hypocalcemic activity. It and its synthetic derivatives are of therapeutic use for hypocalcemia, Paget's disease and osteoporosis.[34)] It assumes a helical three-dimensional structure in a hydrophobic environment and exhibits extremely high biological activity. Although the third asparagine of calcitonin from the N-terminus (Asn-3) is a potential *N*-glycosylation site, it is not *N*-glycosylated in natural form. It has other chemo-enzymatically possible *N*-glycosylation sites such as aspartic acid (Asp-26) and two glutamine (Gln-14 and Gln-20) residues. We prepared several calcitonin derivatives *N*-glycosylated with various types of oligosaccharides at single or multiple sites. We found that the three-dimensional structure of calcitonin peptide backbone was scarcely affected by *N*-linked oligosaccharides. The effect of *N*-glycosylation on the biological activity of calcitonin is discussed.

A. Chemo-enzymatic Synthesis of Calcitonin Derivatives Containing *N*-linked Oligosaccharides

1) Chemical synthesis of *N*-acetylglucosaminyl calcitonin derivatives

Mizuno *et al.* synthesized calcitonin derivatives containing *N*-linked oligosaccharides at the Asn-3 residue following the synthetic scheme illustrated in Fig. 3.7.[35)]

The first step is the chemical synthesis of an *N*-acetylglucosaminyl calcitonin derivative whose Asn-3 is *N*-glycosylated with GlcNAc, [Asn(GlcNAc)[3]]-calcitonin (**2**). The synthesis was based on *i*) preparation of a building block *tert*-butyloxycarbonyl *N*-acetylglucosaminyl L-asparagine, Boc-Asn(GlcNAc)-OH, *ii*) synthesis of the N-terminal glycopeptide thioester segment, *iii*) synthesis of the C-terminal peptide segment, *iv*) condensation of these segment peptides by a thioester segment condensation method[36)] and *v*) deprotection and disufide bond formation between the Cys-1 and Cys-7 residues. Boc-Asn(GlcNAc)-OH was prepared using 2-acetamido-3,4,6-tri-*O*-benzyl-2-deoxy-*β*-D-glucopyranosyl azide

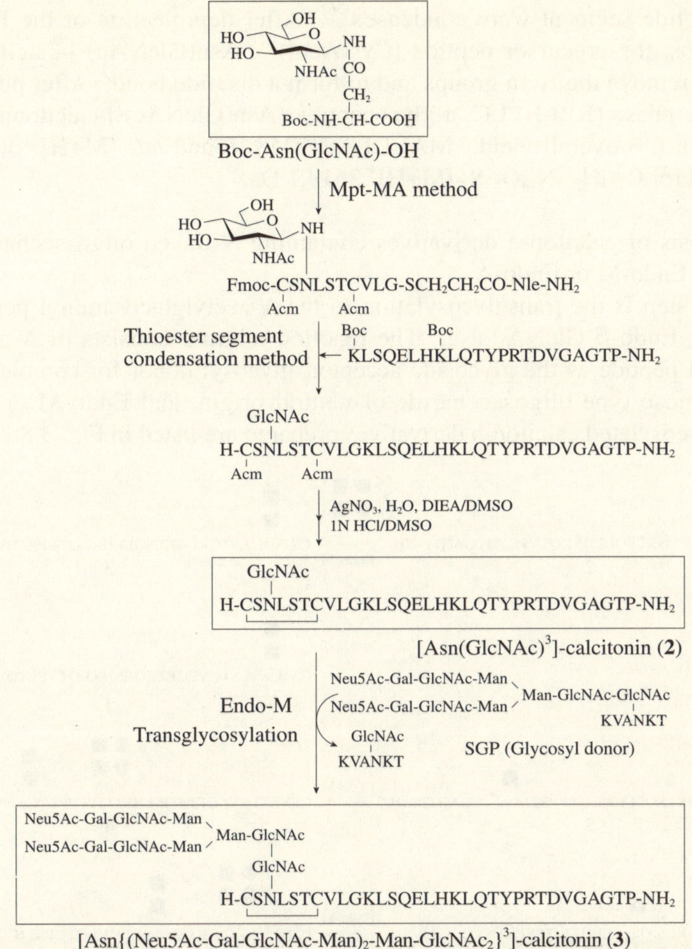

Fig. 3.7 Synthetic scheme of *N*-glycosylated calcitonin derivatives.

and Boc-L-aspartic acid α-benzyl ester (Boc-Asp-OBzl) in the presence of tri-*n*-butylphosphine (Bu_3P).[13,14] The N-terminal *N*-acetylglucosaminyl peptide thioester segment, Fmoc-[Cys(Acm)^{1,7},Asn(GlcNAc)^{3}]-calcitonin(1-10)-S-CH_2CH_2CO-Nle-NH_2 was prepared as follows. Peptide elongation of Boc-Gly-S-CH_2CH_2CO-Nle-MBHA-resin by the Boc solid-phase method gave a protected heptapeptide resin of calcitonin(4-10). Further, Asn-3(GlcNAc), Ser-2(Bzl) and Fmoc-Cys-1(Acm) residues were coupled manually by the corresponding dimethylphosphinothioic mixed anhydride (Mpt-MA)[33] without protection of hydroxyl functions of GlcNAc.[15] Another C-terminal peptide segment, calcitonin(11-32), was prepared by a Boc solid-phase method and the ε-amino group of lysine was protected by the Boc group to give [Lys(Boc)^{11,18}]-calcitonin(11-32). The N-terminal glycopeptide thioester segment and the C-terminal partially protected peptide segment were condensed.[36] After deprotection of the Fmoc and Boc groups, the precursor peptide [Cys(Acm)^{1,7}, Asn(GlcNAc)^{3}]-calcitonin was treated to remove the Acm groups and to form a disufide bond. After purification by reverse-phase (RP)-HPLC, a glycopeptide [Asn(GlcNAc)^{3}]-calcitonin (**2**) was obtained in 6% overall yield. MALDI-TOF MS: found *m/z* $[M+H]^+$ 3619.0 Da, calculated for $C_{154}H_{254}N_{44}O_{52}S_2$ $[M+H]^+$ 3619.1 Da.[35]

2) Synthesis of calcitonin derivatives containing *N*-linked oligosaccharides catalyzed by Endo-M or Endo-A
The final step is the transglycosylation to the *N*-acetylglucosaminyl peptide catalyzed by Endo-β-GlcNAc-ase. The reaction mixture consists of *N*-acetylglucosaminyl peptide as the glycoside acceptor, glycosyl donor for complex-type or high-mannose type oligosaccharide of natural origin, and Endo-M or Endo-A. The *N*-glycosylated calcitonin derivatives prepared are listed in Fig. 3.8.

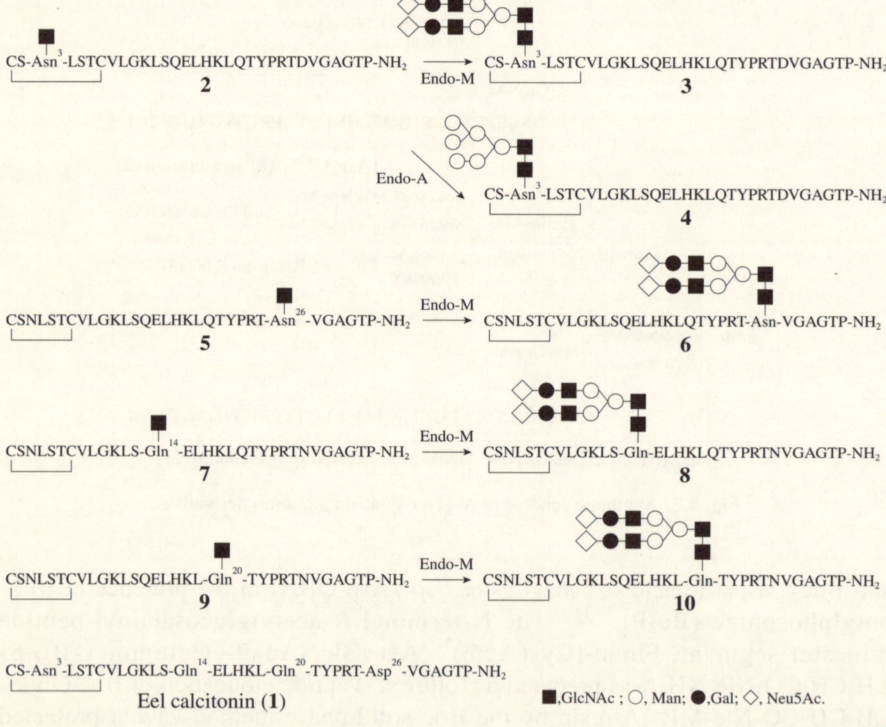

Fig. 3.8 Synthesis of calcitonin derivatives *N*-glycosylated at Asn-3, Asp-26, Gln-14 and Gln-20 sites.

General procedure for transglycosylation of complex-type oligosaccharide catalyzed by Endo-M: The purified recombinant Endo-M cloned in *Candida boidinii*[18] is used.[21] Sialyl glycopeptide (SGP)[20] prepared from hen egg yolk, H-Lys-Val-Ala-Asn[(Neu5Ac-Gal-GlcNAc-Man)$_2$-Man-(GlcNAc)$_2$]-Lys-Thr-OH, is used as the glycosyl donor of disialobiantennary complex-type oligosaccharide.

The excess use of SGP promotes the transglycosylation in a dose-dependent manner, and the transglycosylation yield to the glycoside acceptor added may exceed 50%.[37] The reaction mixture consists of 10 mM of *N*-acetylglucosaminyl peptide (glycoside acceptor), 100 mM of SGP (glycosyl donor) and 120 milliunits/mL of Endo-M in 60 mM potassium phosphate buffer of pH 6.25.[21] After incubation at 37°C for 20 to 30 min, the reaction is terminated by adding an equal volume of a cold aqueous solution of 0.5% trifluoroacetic acid (TFA), and the reaction mixture is analyzed by a RP-HPLC with a linear increase in acetonitrile concentration in a 0.1% TFA aqueous solution. The absorbance of the eluate is monitored at 214 nm. The transglycosylation yield (mol%) is calculated from the ratio of the peak area corresponding to the transglycosylation product to the initial peak of the acceptor without considering the remaining acceptor. The transglycosylation product is isolated by preparative RP-HPLC then freeze-dried.

Calcitonin derivative *N*-glycosylated with complex-type oligosaccharide at Asn-3: The transglycosylation reaction was performed using 22.1 mg of [Asn(GlcNAc)^3]-calcitonin (**2**) as the acceptor. The transglycosylation yield was 57% and 16.7 mg of the transglycosylation product was obtained (49% isolation yield without taking into consideration with the recovered acceptor). The product was identified as a calcitonin derivative containing a disialobiantennary complex-type oligosaccharide at the Asn-3 site, [Asn{(Neu5Ac-Gal-GlcNAc-Man)$_2$-Man-GlcNAc$_2$}^{3}]-calcitonin (**3**). MALDI-TOF MS: found *m/z* $[M+H]^+$ 5623.8 Da, calculated for $C_{230}H_{377}N_{49}O_{108}S_2$ $[M+H]^+$ 5621.9 Da.[21]

Calcitonin derivatives *N*-glycosylated with complex-type oligosaccharide at Asp-26: [Asn(GlcNAc)^{26}]-calcitonin (**5**), whose Asp-26 was replaced by Asn(GlcNAc)-OH, was prepared and used as the glycoside acceptor. The transglycosylation yield was 67%, and the transglycosylation product was identified as a calcitonin derivative containing a disialobiantennary complex-type oligosaccharide at the Asp-26 site, [Asn{(Neu5Ac-Gal-GlcNAc-Man)$_2$-Man-GlcNAc$_2$}^{26}]-calcitonin (**6**).[37]

Calcitonin derivatives *N*-glycosylated with complex-type oligosaccharide at Gln-14 or Gln-20: [Gln(GlcNAc)^{14}]-calcitonin (**7**) and [Gln(GlcNAc)^{20}]-calcitonin (**9**) whose Gln-14 and Gln-20 were replaced by *N*-acetylglucosaminyl L-glutamine, Gln(GlcNAc)-OH,[14] in which the reducing end of GlcNAc residue was bound to the N atom of the side chain amide of the glutamine residue in a *β*-anomeric configuration similar to Asn(GlcNAc)-OH, were prepared. Endo-M could recognize the GlcNAc moiety bound with Gln residue well,[25] and transglycosylated to them in yields of 60% and 62%, respectively. The transglycosylation products were identified as a calcitonin derivative containing a disialobiantennary complex-type oligosaccharide at the Gln-14 site, [Gln{(Neu5Ac-Gal-GlcNAc-Man)$_2$-Man-(GlcNAc)$_2$}^{14}]-calcitonin (**8**), and that at the Gln-20 site, [Gln{(Neu5Ac-Gal-GlcNAc-Man)$_2$-Man-(GlcNAc)$_2$}^{20}]-calcitonin (**10**), respectively.[37]

Calcitonin derivative *N*-glycosylated with a high-mannose type oligosaccharide at Asn-3: Endo-A,[38] which is specific for high-mannose type oligosaccharide, was used. The reaction mixture consisted of 10 mM of [Asn(GlcNAc)^3]-calcitonin (**2**), 10 mM of a high-mannosyl asparagine, Asn(Man_6-$GlcNAc_2$)-OH (M6-GP), prepared from ovalbumin[39] as the glycosyl donor, 50 milliunits/mL of a recombinant Endo-A[40] and 15% acetone in 25 mM ammonium acetate buffer, pH 6.0, and was incubated at 37°C for 90 min. A calcitonin derivative containing a high-mannose type oligosaccharide [Asn(Man_6-$GlcNAc_2$)^3]-calcitonin (**4**) was obtained in 33% yield.[41]

3) Synthesis of plurally *N*-glycosylated calcitonin derivatives

Glycoproteins are generally modified by plural oligosaccharides. As a model of such glycoproteins, we attempted to add plural oligosaccharides to any two sites among Asn-3, Asp-26, Gln-14 and Gln-20 of calcitonin. Our strategy is *i*) chemical synthesis of *N*-acetylglucosaminyl calcitonin derivatives containing two Asn(GlcNAc)-OH and/or Gln(GlcNAc)-OH moieties, *ii*) addition of plural complex-type oligosaccharides onto two *N*-acetylglucosaminyl sites using Endo-M and *iii*) addition of both complex-type and high-mannose type oligosaccharides using Endo-M and Endo-A. The scheme for plural addition of *N*-linked oligosaccharides onto Asn-3 and Asp-26 is shown in Fig. 3.9.

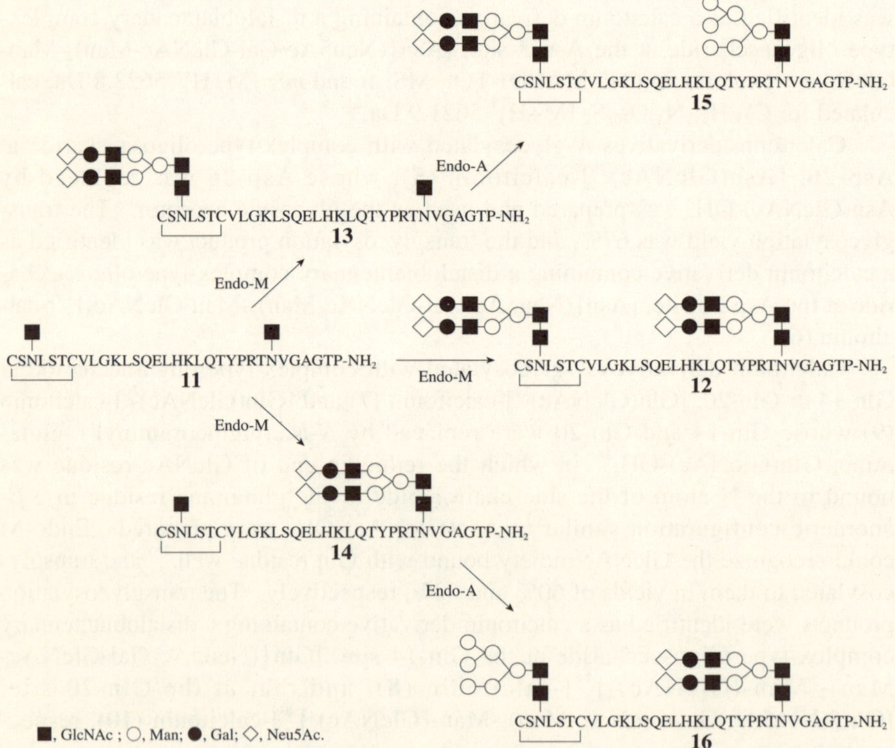

Fig. 3.9 Synthetic scheme of calcitonin derivatives *N*-glycosylated at two sites.

[Asn(GlcNAc)^{3,26}]-calcitonin (**11**), whose Asn-3 and Asp-26 residues were replaced by Asn(GlcNAc)-OH, was chemically synthesized and used as an acceptor for the transglycosylation by Endo-M. The reaction mixture consisted of 10 mM of [Asn(GlcNAc)^{3,26}]-calcitonin (**11**), 150 mM of SGP and 120 mU/mL of Endo-M in 60 mM potassium phosphate buffer of pH 6.25 and was incubated at 37°C for 30 min. The transglycosylation to each GlcNAc moiety of the acceptor proceeded well. As shown in the HPLC profile of the reaction mixture (Fig. 3.10), three peaks of products, P1, P2 and P3, were detected in each yield of 44%, 24% and 21% in front of the remaining acceptor (11%). The main product P1 was identified as [Asn{(Neu5Ac-Gal-GlcNAc-Man)$_2$-Man-(GlcNAc)$_2$}^{3,26}]-calcitonin (**12**), whose two GlcNAc sites were transglycosylated with disialobiantennary complex-type oligosaccharides. The other products, P2 and P3, were identified as [Asn{(Neu5Ac-Gal-GlcNAc-Man)$_2$-Man-(GlcNAc)$_2$}^{3}, Asn(GlcNAc)^{26}]-calcitonin (**13**) and [Asn(GlcNAc)^3, Asn{(Neu5Ac-Gal-GlcNAc-Man)$_2$-Man-(GlcNAc)$_2$}^{26}]-calcitonin (**14**).[42] Each had one GlcNAc site which was transglycosylated with a disialobiantennary complex-type oligosaccharide.

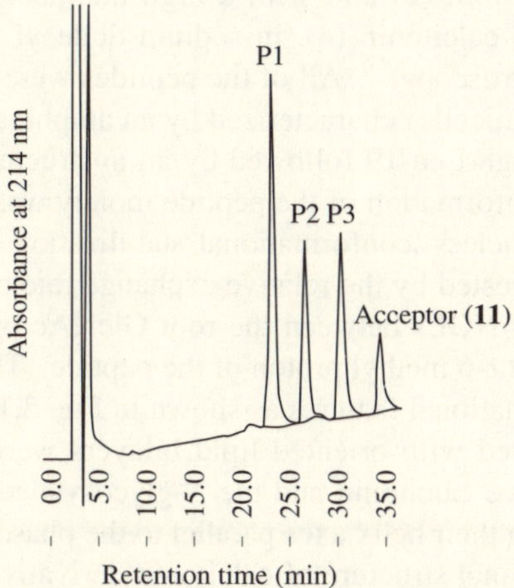

Fig. 3.10 HPLC profile of the reaction mixture for transglycosylation to [Asn(GlcNAc)^{3,26}]-calcitonin (**11**).
The reaction mixture consisted of 10 mM of [Asn(GlcNAc)^{3,26}]-calcitonin (**11**), 150 mM SGP and 120 mU/mL of Endo-M in 60 mM potassium phosphate buffer, pH 6.25. After incubation at 37°C for 30 min, the reaction mixture was analyzed by RP-HPLC (Linear increase in 27.5 to 37.5% acetonitrile concentration in 0.1% TFA over 40 min at a flow rate of 0.8 mL/min using $\phi 4.6 \times 250$ mm ODS column).[42]

Next, using the above N-glycosylated calcitonin derivative **13** or **14** as the glycoside acceptor and M6-GP as the glycosyl donor, the transglycosylation of a high-mannose type oligosaccharide to the free GlcNAc site was performed by Endo-A. Endo-A transglycosylated a high-mannose type oligosaccharide to the free GlcNAc moiety (48% and 41% yield, respectively) without hydrolysis of the already attached complex-type oligosaccharide, and the calcitonin derivative containing complex-type and high-mannose type oligosaccharides, [Asn{(Neu5Ac-

Gal-GlcNAc-Man)$_2$-Man-(GlcNAc)$_2$}^{3}, Asn(Man$_6$-GlcNAc$_2$)^{26}]-calcitonin (**15**) and [Asn(Man$_6$-GlcNAc$_2$)^3, Asn{(Neu5Ac-Gal-GlcNAc-Man)$_2$-Man-GlcNAc$_2$}^{26}]-calcitonin (16) were prepared.[42)]

B. Effect of *N*-linked Oligosaccharide on the Three-dimensional Structure of Calcitonin

Calcitonin derivatives are known to assume a helical three-dimensional structure in a hydrophobic environment as elucidated by NMR or circular dichroism (CD). The Asn-3, which is a potential *N*-glycosylation site, of calcitonin is in the loop-like structure induced by the disulfide bond between Cys-1 and Cys-7 at the N-terminus. Both Gln-14 and Gln-20 are assumed to be in the α-helix structure, and Asp-26 in the random coil region at the C-terminus. We studied the effect of *N*-glycosylation on the three-dimensional structure of calcitonin peptide backbone using the glycosyl calcitonin derivatives prepared.

Hashimoto *et al.* determined the three-dimensional structures of calcitonin (**1**) and the two derivatives *N*-glycosylated at the Asn-3 site with GlcNAc, [Asn(GlcNAc)^3]-calcitonin (**2**) and with a high-mannose type oligosaccharide, [Asn(Man$_6$-GlcNAc$_2$)^3]-calcitonin (**4**), in sodium dodecyl sulfate (SDS) micelles by solution NMR spectroscopy.[43)] All of the peptides were found to have an identical conformations in micelles characterized by an amphipathic α-helix consisting of residues Ser-5 through Leu-19 followed by an unstructural region at the C-terminus. The overall conformation of the peptide moiety was not affected by the *N*-glycosylation. Nevertheless, conformational stabilization of the α-helix with *N*-glycosylation was suggested by the relative exchange rate of Leu-12 amide proton and by the presence of NOEs between the root GlcNAc proton (H1,2,3,5) of the carbohydrate and the Thr-6 methyl proton of the peptide. The calculated structures show the same conformational features as shown in Fig. 3.11.[43)] The topologies of these peptides associated with oriented lipid bilayers were determined by solid-state NMR. Both native calcitonin and the *N*-glycosylated calcitonin were found to have orientation with their helix axes parallel to the phase of the lipid bilayers.[43)]

The three-dimensional structure of calcitonin derivatives was also studied using CD spectra in trifluoroethanol (TFE)-H_2O.[44)] Calcitonin assumed a random coil structure in aqueous solution and gradually accumulated helical content as the TFE concentration increased. At 40% TFE concentration, the helical content almost reached saturation with about 60%. The calcitonin derivatives *N*-glycosylated with GlcNAc and with a high-mannose type oligosaccharide (compounds **2** and **4**) showed CD spectra quite similar to those of calcitonin at 40% TFE solution.[44)] Then the calcitonin derivatives *N*-glycosylated at Asn-3, Asp-26, Gln-14 or Gln-20 with GlcNAc (compounds **2**, **5**, **7** and **9**) and with disialobiantennary complex-type oligosaccharides (compounds **3**, **6**, **8** and **10**) were examined. All of these calcitonin derivatives showed CD spectra quite similar to those of calcitonin, with the only exception of compound **10** in which a small reduction in helical content was detected.[37)] Thus, the *N*-glycosylation scarcely affected the three-dimensional structure of the calcitonin peptide backbone irrespective of the attachment sites and the structures of oligosaccharides.

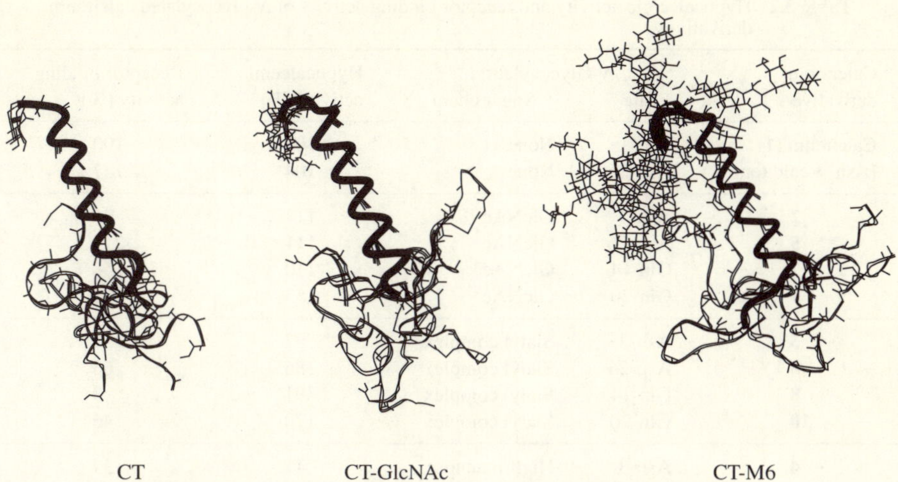

Fig. 3.11 Three-dimensional structure of calcitonin derivatives in sodium dodecyl sulfate micelles. CT, calcitonin (**1**); CT-GlcNAc, [Asn(GlcNAc)^3]-calcitonin (**2**); CT-M6, [Asn(Man$_6$-GlcNAc$_2$)^3]-calcitonin (**4**)
Figures are shown as a superposition of the backbone atoms of the 10 lower energy structures. The structures are aligned at residues 5 - 19.[43)]
[Reprinted with permission from Hashimoto, Y. *et al.* (1999) *Biochemistry* **38**, 8381, Copyright (1999) American Chemical Society]

C. Effect of *N*-linked Oligosaccharide on the Biological Activity of Calcitonin

Effect of *N*-linked oligosaccharide on hypocalcemic activity: Calcitonin is known to inhibit osteoclast bone resorption through receptors that are abundantly expressed on the plasma membrane of osteoclasts, and expresses hypocalcemic activity. This activity is analyzed by a method using male rats *in vivo* based on the measurement of free calcium concentration in blood.[45)] The hypocalcemic activity of the *N*-glycosylated calcitonin derivatives is shown as the relative value to that of calcitonin in Table 3.2 (second column from right).[37)] The replacement of Asp-26 to asparagine ([Asn^{26}]-calcitonin) did not change the activity. The addition of GlcNAc to any site scarcely affected the activity except that to Gln-20 (compound **9**) where higher activity was observed. The addition of a sialocomplex type oligosaccharide at Asn-3 or Gln-14 (compounds **3** and **8**) scarcely changed the activity, but that at Asn-26 and Gln-20 sites (compounds **6** and **10**) significantly enhanced the activity.[37)] The derivative containing a high-mannose type oligosaccharide at Asn-3 (compound **4**) showed lower activity.[44)] Thus, the *N*-glycosylation of calcitonin exhibited different effects on biological activity depending on the type of oligosaccharide and attachment site, but not on the difference in amino acid residues (asparagine or glutamine).

Effect of *N*-linked oligosaccharide on receptor binding activity: The receptor binding activity of the calcitonin derivatives was measured with a direct competition assay using mouse osteoclast-like cells and ^{125}I-labeled elcatonin, a synthetic analogue of eel calcitonin that shows receptor binding and hypocalcemic activity

Table 3.2 Hypocalcemic activity and receptor binding activity of *N*-glycosylated calcitonin derivatives

Calcitonin derivative	*N*-Glycosylation Site	Sugar chain	Hypocalcemic activity (%)*	Receptor binding activity (%)*
Calcitonin (**1**)		None	100	100
[Asn[26]]-calcitonin	Asp-26	None	104	187
2	Asn-3	GlcNAc	114	86
5	Asp-26	GlcNAc	111	214
7	Gln-14	GlcNAc	110	58
9	Gln-20	GlcNAc	157	86
3	Asn-3	Sialyl complex[#1]	97	6
6	Asp-26	Sialyl complex	186	66
8	Gln-14	Sialyl complex	91	3
10	Gln-20	Sialyl complex	170	46
4	Asn-3	High-mannose[#2]	41	39

* Hypocalcemic activity and receptor binding activity are expressed as a value relative to that of calcitonin.[37,44)]

#1 (Neu5Ac-Gal-GlcNAc-Man)$_2$-Man-(GlcNAc)$_2$

#2 (Man)$_6$-(GlcNAc)$_2$

comparable to those of calcitonin.[34,44)] From the IC$_{50}$ values of the competition of calcitonin and its derivatives with ^{125}I-elcatonin binding to calcitonin receptor on the osteoclast-like cells, the receptor binding activity of *N*-glycosylated calcitonin derivatives relative to that of calcitonin was calculated (Table 3.2, right column).[37)] A general trend for the longer oligosaccharide to have smaller binding activity was found. Among these, the derivatives *N*-glycosylated with a sialocomplex type oligosaccharide at Asn-3 or Gln-14 (compounds **3** and **8**) had a dramatically decreased binding activity of less than one tenth that of calcitonin. [Asn[26]]-calcitonin and its *N*-acetylglucosaminyl derivative (compound **5**) showed the highest activity of about twice the activity of calcitonin.

Effect of *N*-linked oligosaccharide on biodistribution: The calcitonin derivative *N*-glycosylated with a sialyl complex-type oligosaccharide at Asn-3 (compound **3**) showed nearly the same hypocalcemic activity as that of calcitonin, even though it had only one tenth the receptor binding activity. For a better understanding of the discrepancy between the hypocalcemic activity and the receptor binding activity of the *N*-glycosylated calcitonin derivatives, the biodistribution or metabolism of the derivatives in living organisms needs to be considered.

Using radioisotope-labeled calcitonin derivatives whose Tyr-22 was labeled with ^{125}I, their biodistribution was studied in mice measuring the radioactivity in blood, bone, liver and kidney.[44)] The derivatives *N*-glycosylated with a sialocomplex type oligosaccharide at Asn-3 or Asp-26 (compounds **3** and **6**) exhibited higher blood concentration than calcitonin. They accumulated to a great extent in the kidney and less in the liver.[37)] They may escape liver trap and hence accumulate in relative advance in the kidney, which controls electrolyte balance. The decrease in receptor binding activity of these calcitonin derivatives *N*-glycosylated with sialocomplex type oligosaccharide may be compensated by this biodistribution profile. This may explain the fact that their hypocalcemic activity was not dramatically decreased and exhibited a moderate activity (compound **3**) or even enhanced activity in the case of compound **6**.

3.1.3 Enzymatic Synthesis of Neoglycoconjugates Using Oligosaccharide-transfer Activity of Endo-A

A. Enzymatic Synthesis of Neooligosaccharides by Transglycosylation Activity of Endo-A

The present author and his co-workers examined the enzymatic synthesis of novel oligosaccharides that are attached to mono- or disaccharides at the reducing-end *N*-acetylglucosamine by Endo-A. We succeeded in synthesizing novel oligosaccharides, $(Man)_6$GlcNAcGlc, $(Man)_6$GlcNAcMan, $(Man)_6$GlcNAc-$(GlcNAc)_2$ and $(Man)_6$GlcNAc$(Glc)_2$ by the enzyme.[10,46] The susceptibility of these novel oligosaccharides to Endo-β-GlcNAc-ase was examined. Endo-β-GlcNAc-ases such as Endo-H, Endo-F and Endo-Flavo degraded these pyridylaminated (PA-) oligosaccharides, $(Man)_6$GlcNAcGlc-PA and $(Man)_6$GlcNAc$(Glc)_2$-PA, to PA-glucose and PA-$(Glc)_2$.[10] Substrate specificities of Endo-β-GlcNAc-ases have been studied with heterogeneous oligosaccharides.[47,48] However, these studies were mainly limited to the aglycon specificity of the enzyme because modification of the aglycon moiety does not generally occur in glycoproteins. There are some reports on the aglycon specificity of Endo-β-GlcNAc-ases.[49-51] Endo-β-GlcNAc-ase from *Diplococcus* (Endo-D) cleaves sugar chains that are linked to *N*-acetylglucosamine, GlcNAc-Asn, or (Fuc)GlcNAc-Asn, but Endo-CII from *Clostridium* cannot act on sugar chains linked with (Fuc)GlcNAc-Asn. Using the novel oligosaccharides synthesized by transglycosylation activity of Endo-A, we showed that Endo-β-GlcNAc-ase degrades not only GlcNAc-GlcNAc linkages but also GlcNAc-Man and GlcNAc-$(Glc)_2$ linkages. These results suggest that the aglycon specificity of Endo-β-GlcNAc-ase is very broad and that the enzyme recognizes only one GlcNAc residue of the disaccharide moiety in the *N*-linked oligosaccharides.[10,46]

B. Enzymatic Synthesis of Substrates for Colorimetric Detection of Endo-β-GlcNAc-ase Activity

Endo-β-GlcNAc-ase has been measured with radioactive asparagine oligosaccharide $(Man)_5(GlcNAc)_2$Asn-[^{3}H]dansyl (DNS) and its *N*-[^{14}C]acetyl derivatives.[52,53] Iwase *et al.* measured Endo-β-GlcNAc-ase activity by HPLC with DNS-asparaginyl oligosaccharides as substrates.[54] Cumbersome methods are needed for the separation of products GlcNAc-Asn-[^{3}H]DNS, GlcNAc-Asn-*N*-[^{14}C]acetyl and GlcNAc-Asn-DNS by high-voltage paper electrophoresis or HPLC.

We examined the enzymatic synthesis of a novel oligosaccharide, $(Man)_6$GlcNAc-Glc-*p*NP.[55] In aqueous medium, Endo-A produced predominantly the hydrolytic product $(Man)_6$GlcNAc and the transglycosylation product

$(Man)_6$GlcNAc-Glc-*p*NP was readily degraded to $(Man)_6$GlcNAc and Glc-*p*NP to give a low yield of the transglycosylation product. Transglycosylation activity of Endo-A can be enhanced markedly in media containing organic solvents such as acetone, dioxane and dimethyl sulfoxide (DMSO)[56] (see Chapter 2, section 2.1.1). In the DMSO-containing medium, suppression of the hydrolytic activity increased as DMSO concentration was increased to 20-30%, where transglycosylation was maximum. In order to prepare a larger quantity of $(Man)_6$GlcNAc-Glc-*p*NP, we used the 20% DMSO-containing medium and a 45-min reaction at 37°C for the transglycosylation. Under these conditions, $(Man)_6$GlcNAc-Glc-*p*NP was synthesized with a reaction yield of approximately 65% relative to the acceptor added.

We examined the hydrolysis of $(Man)_6$GlcNAc-Glc-*p*NP by various Endo-β-GlcNAc-ases and found that the enzymes degraded $(Man)_6$GlcNAc-Glc-*p*NP to $(Man)_6$GlcNAc and Glc-*p*NP, as expected. We have established a new procedure for the colorimetric detection of Endo-β-GlcNAc-ase activity using $(Man)_6$GlcNAc-Glc-*p*NP, which is as simple as that for other exoglycosidase assays with *p*NP-glycosides as substrates (Fig. 3.12A). $(Man)_6$GlcNAc-Glc-*p*NP (20 nmol) was incubated with 0.5 milliunits of Endo-Flavo (*Flavobacterium* Endo-β-GlcNAc-ase) in 30 μL of 200 mM acetate buffer (pH 6.0) and the mixture was incubated for an appropriate time at 37°C. The enzymatic reaction was stopped by boiling and 20 milliunits of yeast α-glucosidase was added. After incubation for 30 min at 37°C, 450 μL of 0.2 M borate buffer (pH 9.8) was added, and the liberated *p*-nitrophenol was determined spectrophotometrically at 405 nm.

Fig. 3.12A shows the analyses of the time course of the hydrolysis of $(Man)_6$GlcNAc-Glc-*p*NP by Endo-Flavo enzyme. The relationship between the reaction time and the color intensity of *p*NP liberated was shown to be linear. We

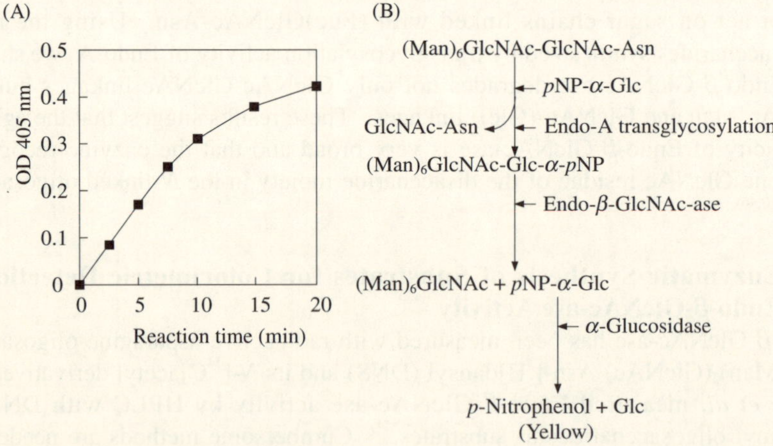

Fig. 3.12 (A) Time course of products formed from $(Man)_6$GlcNAc-Glc-*p*NP digested with Endo-β-GlcNAc-ase and α-glucosidase.
(B) Detection system for Endo-β-GlcNAc-ase with a synthetic oligosaccharide $(Man)_6$GlcNAc-Glc-*p*NP.
[Reprinted from *Anal. Biochem.* **257**, Takegawa, K. *et al.*, 222, Copyright (1998), with permission from Elsevier]

found that the results obtained by the colorimetric assay system using $(Man)_6$GlcNAc-Glc-*p*NP were very close to those using the conventional methods using $(Man)_6(GlcNAc)_2$-Asn-DNS as substrate.[54)] Our method has several advantages over these techniques. The assay of Endo-β-GlcNAc-ase activity is a simple procedure, like other exoglycosidase assays with *p*NP-derivatives as substrates and does not require specialized equipment. The method presented here is practical and useful for the routine assay of Endo-β-GlcNAc-ase (Fig. 3.12B).

As an indicator of galactosidase activity, the chromogenic substrate 5-bromo-4-chloro-3-indolyl β-D-galactopyranoside (X-Gal) is most commonly used for blue-white color selection in genetic engineering. Therefore transglycosylation activity of Endo-A was used for the enzymatic synthesis of the novel oligosaccharide $(Man)_6$GlcNAc-5-bromo-4-chloro-3-indolyl-β-glucoside (Glc-β-X). We observed that in media containing organic solvent, the hydrolytic activity of Endo-A was suppressed and the transglycosylation activity enhanced.[56)] We have taken advantage of this finding and synthesized $(Man)_6$GlcNAc-Glc-β-X by Endo-A transglycosylation.[57)]

We attempted a coexpression of β-glucosidase with Endo-β-GlcNAc-ase. The transformant harboring Endo-A and β-glucosidase genes formed blue colonies in the presence of $(Man)_6$GlcNAc-Glc-β-X. Other transformants, except those harboring the β-glucosidase or Endo-β-GlcNAc-ase gene, formed white colonies as expected. These results revealed that Endo-β-GlcNAc-ase activity can be directly detected by plate assay using $(Man)_6$GlcNAc-Glc-β-X (Fig. 3.13). The plate assay can simultaneously detect thousands of Endo-β-GlcNAc-ase-containing bacterial colonies. Therefore, the method presented here will be useful for the rapid detection of Endo-β-GlcNAc-ase activity.[57)]

C. Chemo-enzymatic Synthesis of a High-mannose Type Glycopeptide Analog Containing a Glc-Asn Linkage

Glycopeptidase [peptide-N^4-(*N*-acetyl-β-D-glucosamininyl)asparagine amidase, also called *N*-glycanase, peptide *N*-glycanase and glycoamidase] is found in a number of organisms across different kingdoms.[58-61)] The enzyme catalyzes the release of intact oligosaccharides from *N*-linked glycopeptides and glycoproteins via hydrolysis of the β-amide of the linking asparagine. A number of natural and unnatural mono- and disaccharide glycopeptides were synthesized and tested for substrate specificity. Cellobiose and lactose glycopeptides, in which GlcNAc-GlcNAc linked to Asn is replaced by a disaccharide of non-aminosugars, are not substrates for glycopeptidases.[62)] To further prove the mechanism and substrate requirements for glycopeptidases, an non-natural product, $(Man)_9$GlcNAcGlc-pentapeptide, was synthesized by transglycosylation of Endo-A using a synthetic pentapeptide containing a Glc-Asn linkage[63)] (Fig. 3.14). The Glc-Asn linkage has been found in Archaebacteria[64)] and laminin,[65)] although the structure of the oligosaccharide beyond the linking glucose has not been elucidated. The synthetic glycopeptide may serve further as a model to study this class of glycoconjugates.

Glucose-containing peptide [Tyr-Ile-(Glc)Asn-Ala-Ser] was chemically synthesized and Endo-A transfer of the $(Man)_9$GlcNAc structure performed. The syn-

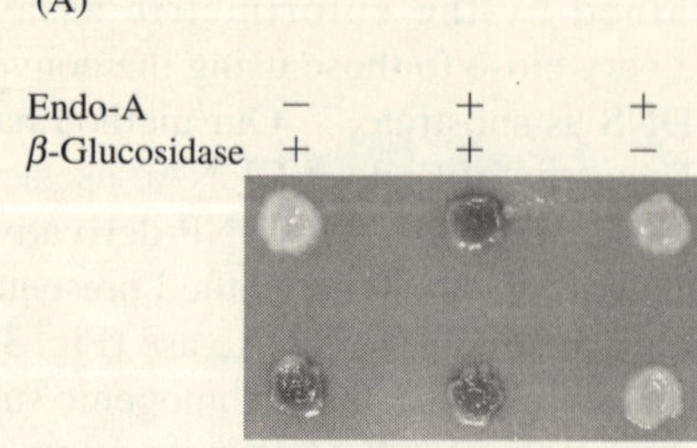

(B)

Man Man Man Man Man Man GlcNAc-O (Man)$_6$GlcNAc-Glc-β-X

Endo-β-GlcNAc-ase

Glc-β-X

β-Glucosidase

X

Blue

Fig. 3.13 (A) A plate assay for the detection of Endo-β-GlcNAc-ase using (Man)$_6$GlcNAc-Glc-β-X. The recombinant *E. coli* XL1-Blue colonies were cultured on LB agar plates containing 100 μg/mL ampicillin, 50 μg/mL chloramphenicol, and 1 mM IPTG for 12 h at 37°C in the presence or absence of Endo-A and β-glucosidase genes, as indicated. Substrates are A, Glc-β-X (35 nmol/cm^2) and B, (Man)$_6$GlcNAc-Glc-β-X (35 nmol/cm^2).
(B) Detection system for Endo-β-GlcNAc-ase using the synthetic oligosaccharide (Man)$_6$GlcNAc-Glc-β-X.

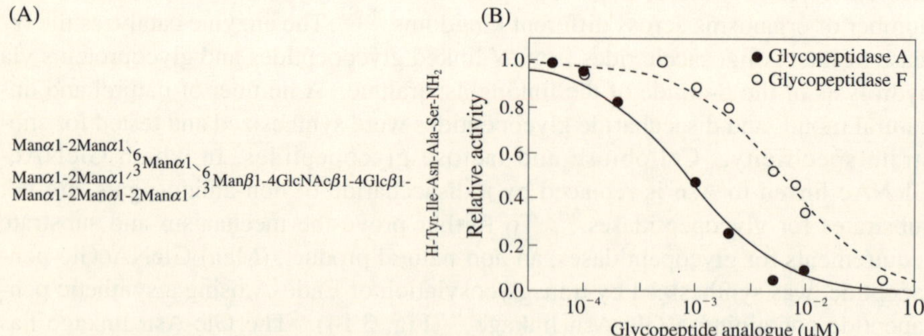

Fig. 3.14 (A) Structure of a high-mannose type glycopeptide analogue containing a Glc-Asn linkage.
(B) Inhibition of glycopeptidases by (Man)$_9$GlcNAc-Glc-peptide.
[Reprinted from *Bioorg. Med. Chem. Lett.* **8**, Deras, I.L. *et al.*, 1765, Copyright (1998), with permission from Elsevier]

thetic product $(Man)_9$GlcNAcGlc-peptide was obtained as a white solid in 39% yield.[63] When the synthetic neoglycopeptide was incubated with glycopeptidases, a number of detectable hydrolytic products were observed.[63] The synthetic product was then tested as inhibitor for glycopeptidase A from almond[66,67] or glycopeptidase F from *Flavobacterium meningosepticum.*[68,69] Using glycopeptide isolated from thermolysin digest of bovine asialofetuin as substrate, glycopeptidase was incubated in the presence of 0.5-1000 μM of $(Man)_9$GlcNAcGlc-pentapeptide, and the reaction mixture was then analyzed by HPLC. The glycopeptide shows clear inhibitory activity toward both glycopeptidases A and F, and an IC_{50} of 8 μM for glycopeptidase A and 62 μM for glycopeptidase F (Fig. 3.14). These results indicate that the non-natural product $(Man)_9$GlcNAcGlc-pentapeptide was not hydrolyzed by glycopeptidases from plant and bacterial sources, but it inhibited both enzymes in the micromolar range.[62]

D. Enzymatic Synthesis of Neoglycopolymers

The transglycosylation activity of Endo-A can be enhanced dramatically by the inclusion of organic solvent in the reaction mixture.[53] This finding was extended to the synthesis of important intermediates for the preparation of neoglycoconjugates. The transglycosylation by Endo-A using $(Man)_9(GlcNAc)_2$Asn as donor to some functionalized *N*-acetylglucosamine glycosides was tested. When acceptor concentration was 0.2 M, Endo-A transferred $(Man)_9$GlcNAc to GlcNAc-*O*-$(CH_2)_6NH_2$ (93% of the converted substrate), GlcNAc-*O*-$CH_2CH{=}CH_2$ (99%), GlcNAc-*O*-$(CH_2)_3CH{=}CH_2$ (90%) and GlcNAc-*O*-$(CH_2)_3NHCOCH{=}CH_2$ (78%) with yields of 81, 81, 84 and 70% of the starting substrate, respectively.[70] The thioglycosides of GlcNAc are good acceptors for Endo-A transglycosylation. When GlcNAc-*S*-CH_2CN, GlcNAc-*S*-$(CH_2)_3CH_3$ and GlcNAc-*S*-$CH_2CONHCH_2CH(OMe)_2$ were used as acceptors at 0.2 M, the transglycosylation indices were 88, 86 and 95%, with yields of 83, 78 and 81%, respectively.[70]

We have synthesized $(Man)_9(GlcNAc)_2$-NAP [3-(*N*-acryloylamino)-proryl] by Endo-A transglycosylation, and a glycopolymer was obtained from $(Man)_9(GlcNAc)_2$-NAP and acrylamide using TEMED and ammonium persulfate as catalysts (Fig. 3.15). The fractions containing the polymer eluted at the void volume of the Sephadex G-50 column were pooled and lyophilized. The sugar content of the polymer was estimated to be 37% by the phenol-H_2SO_4 method. From these results, the ratio of sugar side chains to acrylamide residues was estimated to be 1.44.[70] The molecular weight of the glycopolymer was estimated to be between 1,500,000 and 2,000,000.

Clustering of monosaccharides by attaching them to simple branched peptides enhances inhibitory potencies for some C-type lectins.[71,72] Formation of glycopolymers is a convenient way to provide glycoside clustering.[73] Therefore, we examined the inhibitory effects of the glycopolymer on mannose-binding proteins (MBPs). A solid-phase assay was carried out on serum- and liver MBP-carbohydrate recognition domain (CRD), using the $(Man)_9$GlcNAc-glycopolymer and soybean agglutinin (SBA), which contains the same $(Man)_9(GlcNAc)_2$ as sugar chains. A dramatic increase in the inhibition of MBP-CRDs compared with that

(A)

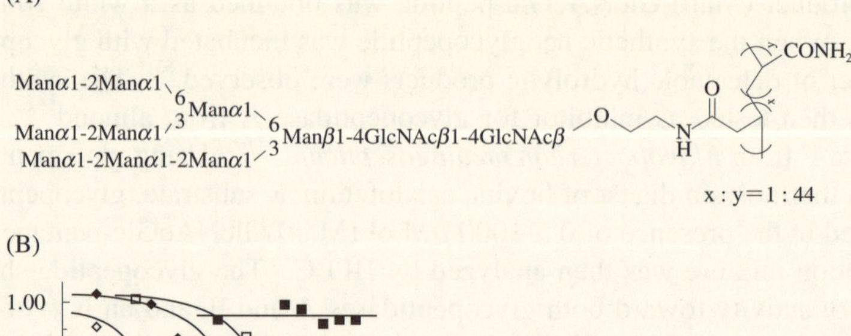

(B)

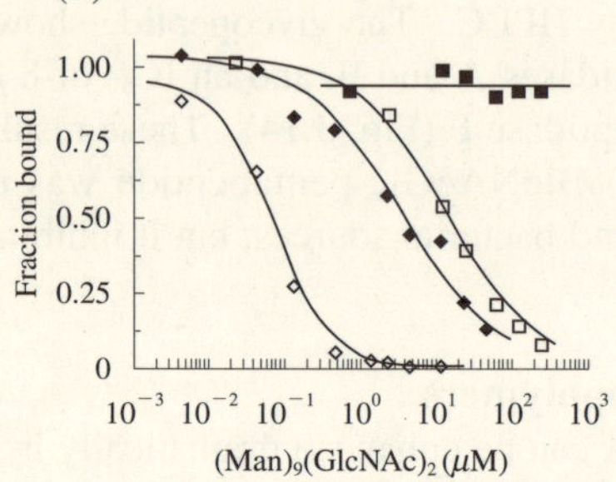

Fig. 3.15 (A) Structure of a neoglycopolymer having the $(Man)_9(GlcNAc)_2$ sugar chain. (B) Inhibition of binding by serum- and liver-MBP-CRDs by the glycopolymer. Concentrations of SBA and glycopolymer are expressed on the bases of $(Man)_9(GlcNAc)_2$. ■, SBA + serum MBP-CRD; □, SBA + liver MBP-CRD; ◆, glycopolymer + serum MBP-CRD; ◇, glycopolymer + liver MBP-CRD.
[Reprinted with permission from Fan, J.-Q. *et al.* (1995) *J. Biol. Chem.* **270**, 17735]

by the native glycoprotein (SBA) which contains the same $(Man)_9(GlcNAc)_2$ oligosaccharide was demonstrated. In the case of the liver MBP-CRD, an approximately 180-fold enhancement of inhibitory potency over the native SBA was attained by the glycopolymers (Fig. 3.15). Although no significant inhibition of the serum MBP-CRD was observed for SBA, the glycopolymer demonstrated surprisingly strong inhibitory potence (I_{50} = 3.5 mM).[70)]

3.2 Remodeling of Sugar Chains in Glycoproteins Using Endo-β-N-acetylglucosaminidase

3.2.1 Remodeling of Sugar Chains by Endo-A

Asparagine (*N*)-linked glycosylation begins with the transfer of the precursor oligosaccharide $(Glc)_3(Man)_9(GlcNAc)_2$ to a nascent polypeptide, and the precursor oligosaccharide of the glycoprotein is modified by Golgi-localized enzymes, which generate a heterogeneous assortment of oligosaccharide structures.[74-76)] Therefore, naturally occurring glycoproteins have a high degree of heterogeneity in their oligosaccharide moiety. For example, human serum immunoglobulin G contains over 30 varieties of biantennary *N*-linked sugar chains.[77)] It is difficult to study the biological roles of individual oligosaccharides because they are highly heterogeneous in glycoproteins.

There are several approaches to the synthesis of neoglycoproteins. Besides conventional analysis with specific glycosidases,[47)] site-directed mutagenesis can be used for the synthesis of neoglycoproteins (introduction of Asn-X-Ser/Thr site for *N*-glycosylation) and identification of the role of individual *N*-linked sugar chains in glycoproteins.[78,79)] However, these DNA-mediated studies call for the isolation and preparation of the genes that code for glycoproteins, and the structures of the *N*-linked oligosaccharides vary with the host cells. Moreover, *N*-linked sugar chains are often crucial during glycoprotein folding and oligomerization. When glycosylation is inhibited using tunicamycin (an inhibitor of dolichol oligosaccharide precursor synthesis), or by mutational elimination of consensus sequences, the folding of numerous glycoproteins fails. Typically, the proteins aggregate soon after synthesis and associate noncovalently with Bip, an abundant ER chaperone known to bind a variety of misfolded and incompletely folded proteins.[80)]

We have developed a novel method for the enzymatic synthesis of neoglycoproteins using oligosaccharide-transferring activity with Endo-A. A method for the conversion of heterogeneous to homogeneous *N*-linked sugar chains in glycoproteins, utilizing the transglycosylation activity of Endo-A, is illustrated in Fig. 3.16. The method is suitable and practical for the improved synthesis of neoglycopeptides and neoglycoproteins and useful for attaching the same *N*-linked sugar chains to all the original *N*-glycosylation sites of glycoprotein molecules.

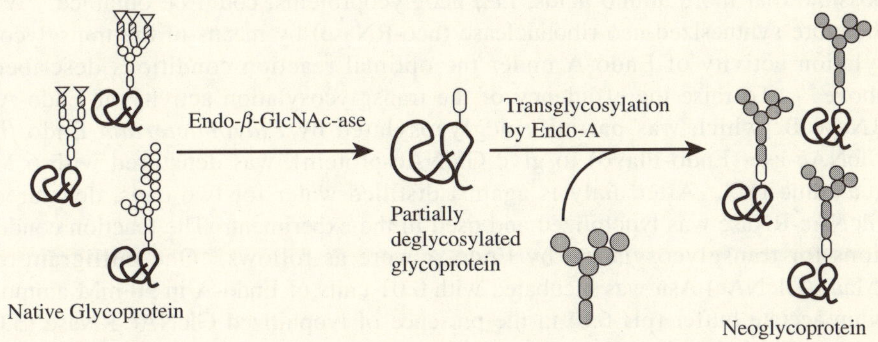

Fig. 3.16 Enzymatic synthesis of neoglycoproteins by Endo-A transglycosylation.

The present author and his group found that Endo-A transferred $(Man)_6GlcNAc$ released from $(Man)_6(GlcNAc)_2$ to 4-L-aspartylglycosylamine (GlcNAc-Asn).[81)] Therefore we expected Endo-A to be useful for synthesizing not only novel oligosaccharides but also neoglycoproteins. We examined the transglycosylation activity catalyzed by Endo-A toward GlcNAc-peptides (10 amino acids) prepared from ovalbumin. Partially deglycosylated glycopeptide (GlcNAc-peptide) from ovalbumin was isolated by pepsin digestion and Endo-β-GlcNAc-ase treatments, and the GlcNAc-containing glycopeptide was used as the acceptor of the transglycosylation reaction.[81)] Reaction mixtures containing 4 milliunits of Endo-A, 100 nmol of GlcNAc-peptide and 25 mM ammonium acetate buffer (pH 6.0) were incubated for 10 min at 37°C. The reaction was started by adding 50 μg of $(Man)_6(GlcNAc)_2$Asn and stopped by boiling for 3 min at 100°C. The reaction mixture was then directly analyzed by HPLC. Only one major peak was obtained without $(Man)_6(GlcNAc)_2$Asn in the reaction mixture. However, another peptide peak was generated in the presence of $(Man)_6(GlcNAc)_2$Asn. This peak was collected, hydrolyzed with 6 M HCl or 2 M trifluoroacetic acid and the amino acid or amino sugar contents were investigated using an amino acid analyzer. The results suggested that the transfer of $(Man)_6$GlcNAc to the GlcNAc-peptide occurred through the transglycosylation of Endo-A. To further confirm the structure of $(Man)_6$GlcNAc-GlcNAc-peptide, we analyzed the peak by means of ion mass spectrometry, and this yielded molecular ion peaks of 2,573, which corresponds to the calculated value of $(Man)_6(GlcNAc)_2$-peptide. These results indicated that a neoglycopeptide had been synthesized by the transglycosylation of Endo-A.[81)]

Chemo-enzymatic addition of a high-mannose type oligosaccharide to eel calcitonin (32 amino acids), a calcium-regulating hormone, was also examined. A glycosylated calcitonin derivative was efficiently and practically synthesized with a reaction yield of 33% relative to the concentration of the acceptor added.[41)]

A. Remodeling of Sugar Chains in Ribonuclease

The above results showed that Endo-A can synthesize neoglycopeptide, and it was possible that more amino acids, *i.e.*, neoglycoproteins, could be obtained. We therefore synthesized neo-ribonuclease (neo-RNase) by means of the transglycosylation activity of Endo-A under the optimal reaction conditions described above.[81)] To raise the efficiency of the transglycosylation activity of Endo-A, RNase B [which was partially deglycosylated by *Flavobacterium* Endo-β-GlcNAc-ase (Endo-Flavo) to give GlcNAc-protein], was denatured with 6 M guanidine-HCl. After dialysis against distilled water for two days, denatured GlcNAc-RNase was lyophilized and used in the experiment. The reaction conditions for transglycosylation by Endo-A were as follows. One milligram of $(Man)_6(GlcNAc)_2$Asn was incubated with 0.01 units of Endo-A in 50 mM ammonium acetate buffer (pH 6.0) in the presence of lyophilized GlcNAc-RNase (3.0 mg, total volume, 50 μL). After incubating the reaction mixture for 10 min at 37°C, the reaction mixture was dialyzed against 10 mM ammonium acetate buffer (pH 6.0) and placed on a column (1.5 × 6 cm) containing ConA-Sepharose 4B at 4°C. Most of the GlcNAc-RNase passed through the ConA column. However, some of it bound to the column, but it was eluted with buffer containing 1 M methyl α-mannoside. To characterize the Con A-bound RNase further, both forms were pooled and analyzed by SDS-PAGE. RNase that did not bind to the Con A column had the same molecular weight as GlcNAc-RNase (Fig. 3.17A). The molecular weight of the bound RNase was increased by about 2,000 and it migrated to the position of native RNase B (Fig. 3.17A). When bound RNase was incubated with 0.01 units of Endo-Flavo, it was deglycosylated again and migrated to the position of GlcNAc-RNase (Fig. 3.17A). These results showed that Endo-A transferred the oligosaccharide $(Man)_6$GlcNAc, from

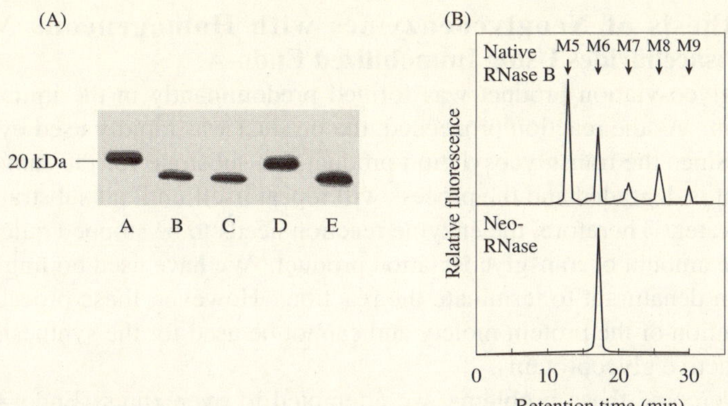

Fig. 3.17 (A) SDS-PAGE of the reaction products produced by Endo-A. $(Man)_6GlcNAc$ was incubated with Endo-A in the presence of GlcNAc-RNase. The reaction was eluted through a Con A-Sepharose column, and bound RNase was then eluted with 1 M methyl α-mannoside. Each protein peak was subjected to SDS-PAGE. Lane A, native RNase B; lane B, GlcNAc-RNase by Endo-β-GlcNAc-ase digestion; lane C, unbound RNase; lane D, bound RNase; lane E, bound RNase digested by Endo-β-GlcNAc-ase.
(B) Comparison of the *N*-linked oligosaccharides in native and neo-RNases. The *N*-linked sugar chains were released from native and neo-RNases by hydrazinolysis and derivatized with PA. A portion of each PA-oligosaccharide was analyzed by HPLC with the Takara PALPAK N column. The arrows indicating the elution positions of standard PA-oligosaccharides M5-M9 correspond to $(Man)_{5\text{-}9}(GlcNAc)_2$-PA, respectively.
[Reprinted with permission from Takegawa, K. *et al.* (1995) *J. Biol. Chem.* **270**, 3098]

$(Man)_6(GlcNAc)_2$Asn to GlcNAc-RNase. Therefore, we called the bound RNase "neo-RNase" and set out to find whether $(Man)_6GlcNAc$ was actually linked to it.

The *N*-linked sugar chains were compared between the native and the neo-RNase. One of the two forms of RNase (500 μg) was incubated with 0.1 units of Endo-Flavo at 37°C for 12 h. After the *N*-linked sugar chains were partially released from the protein moiety, the reaction mixtures were pyridylaminated (PA) and analyzed by HPLC. PA-oligosaccharides from the native RNase B separated into five peaks [peaks $(Man)_5GlcNAc$-PA to $(Man)_9GlcNAc$-PA] (Fig. 3.17B). Liang *et al.*[82] have reported that the *N*-linked sugar chains of RNase B are $(Man)_{5\text{-}8}(GlcNAc)_2$. However, we found a small amount of $(Man)_9(GlcNAc)_2$. The *N*-linked sugar chain of neo-RNase consisted entirely of $(Man)_6(GlcNAc)_2$ (Fig. 3.17B). These results showed that the heterogeneous *N*-linked sugar chains, $(Man)_{5\text{-}9}(GlcNAc)_2$, of the RNase B were converted into homogeneous $(Man)_6(GlcNAc)_2$ by the oligosaccharide-transferring mechanism of Endo-A.

Our data showed that the GlcNAc-peptides derived from ovalbumin (10 amino acids) and eel calcitonin (32 amino acids) are better acceptors than bovine RNase B (GlcNAc-protein) for the transglycosylation reaction of Endo-A. For example, Endo-A can readily deglycosylate RNase B but shows less oligosaccharide-transferring activity toward the native form of partially deglycosylated RNase B. This may be considered a result of the accessibility of glycosylation sites to Endo-A, based on the stereostructure of the protein.

B. Synthesis of Neoglycoenzymes with Homogeneous *N*-linked Oligosaccharides Using Immobilized Endo-A

The transglycosylation product was formed predominantly in the initial stage of the reaction. As the reaction proceeded, the product was rapidly used by the same enzyme. Since the transglycosylation product is a substrate for the same enzyme, the product is degraded and the process will repeat itself until all substrate is transferred to water. Therefore, the enzyme reaction needs to be stopped quickly to obtain a large amount of transglycosylation product. We have used boiling treatment and protein denaturant to terminate the reaction. However, these procedures lead to denaturation of the protein moiety and cannot be used for the synthesis of enzymatically active glycoproteins.

To overcome these problems, we attempted to overexpress Endo-A as a fusion protein linked to glutathione *S*-transferase (GST). Interestingly, we found that not only GST-Endo-A fusion protein but also the immobilized GST-Endo-A retained the transglycosylation activity. Taking advantage of this finding, we reported a new method for the enzymatic synthesis of active neoglycoenzymes attached to homogeneous oligosaccharides by immobilized GST-Endo-A fusion protein.[83)]

Next we examined the transglycosylation to the partially deglycosylated RNase B without denaturation of the protein moiety. After incubation for the indicated time at 37°C, the reaction mixture was analyzed by SDS-PAGE (Fig. 3.18). One protein band was newly generated with an increase of 2,000 as compared with the deglycosylated RNase B, and it migrated to the position of the native RNase B.[81)] The transglycosylation product peaked in content at 10 min, and was then gradually hydrolyzed by the same enzyme. The reaction mixture was stained by lectin blotting using HRP-Con A (Fig. 3.18). The transglycosylation product with the same size as native RNase B was stained (Fig. 3.18, lane 3), but GlcNAc-RNase was not (Fig. 3.18, lane 2). The *N*-linked oligosaccharides of native RNase B consist of $(Man)_{5\text{-}9}(GlcNAc)$.[81)] Neo-RNase was purified on a Con A-agarose

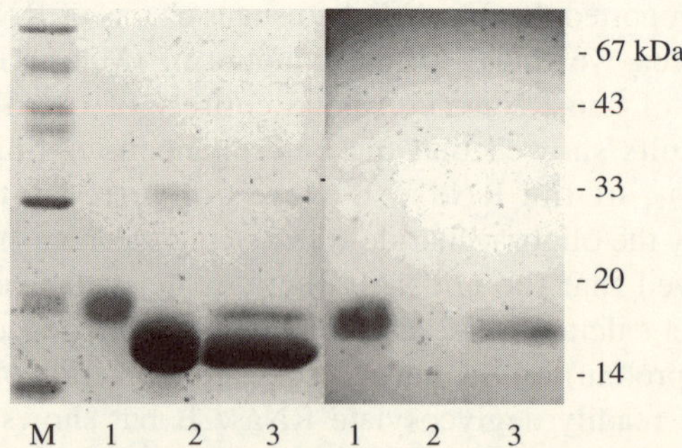

Fig. 3.18 Lectin blotting of neo-RNase. The samples were loaded onto a 15% gel. (Left) Stained with Coomassie Brilliant Blue R-250. (Right) Analyzed by lectin blotting using HRP-Con A. Lane M, molecular weight markers; lane 1, native RNase B; lane 2, GlcNAc-RNase B digested with Endo-*β*-GlcNAc-ase; lane 3, reaction mixture of neo-RNase.
[Reprinted from *Biochem. Biophys. Res. Commun.* **267**, Fujita, K. *et al.*, 137, Copyright (2000), with permission from Elsevier]

column in the same way as RNase B, and the *N*-linked oligosaccharides of neo-RNase were confirmed to be modified to $(Man)_6(GlcNAc)_2$ analyzed by HPLC.[81] We examined the enzymatic activity of neo-RNase in comparison with that of RNase B toward cytidine 2′,3′-cyclic monophosphate. They showed similar levels of activity and pH-activity profiles, indicating that $(Man)_6$GlcNAc was transferred to the deglycosylated RNase B without a change in enzymatic properties.

The method presented here is suitable for the synthesis of neoglycoenzymes without denaturation of the protein moiety and useful for attaching the same *N*-linked sugar chains to all the original glycosylation sites of glycoprotein molecules.

Transglycosylation of Endo-A is an attractive method for remodeling glycoproteins, because *N*-linked sugar chains can be transferred to a glycoside acceptor *en bloc*. Moreover, Endo-A has broad ranges of optimum pH and organic solvent stabilities, and does not require nucleotide sugars for the reactions. Because of the substrate specificity of the enzyme, however, the conversion of the *N*-linked sugar chains by Endo-A is limited to high-mannose type oligosaccharides.[84] In contrast, Endo-β-GlcNAc-ase from *Mucor hiemalis* (Endo-M) could transfer complex-type oligosaccharides to various glycoproteins (See Section 3.1.1). Therefore, a combination of Endo-A and Endo-M will enable us to design more complex neoglycoproteins with different *N*-linked oligosaccharides moieties.

3.2.2 Remodeling of Sugar Chains by Endo-M

The sugar chain in glycoproteins plays important roles in biological processes such as cell-cell interactions and substrate-receptor recognition. Although prokaryotes do not express *N*-glycans, all animals, plants and eukaryotes have the *N*-glycan synthesis pathway including processing and trimming. However, yeasts and vegetative molds do not appear to complete the trimming of sugar residues and are unable to generate typical complex type sugar chains that are generally found in animal glycoproteins. Yeast often further specialize their high-mannose glycans by extending them into large mannans.[85] Therefore, when human glycoprotein is produced by recombinant yeast that is constructed using gene-engineering techniques, it has yeast-specific and large mannans, not human-compatible oligosaccharide. Such large mannans are strong antigens in humans. Using the transglycosylation activity of Endo-M, we attempted to change the high-mannose type of oligosaccharide in glycoproteins to a complex type of oligosaccharide that is a human compatible-oligosaccharide. Fig. 3.19 shows a method for remodeling the sugar chain in glycoproteins.

Bovine pancreas ribonuclease B has one sugar chain of the high-mannose type.[82] The transfer of the complex type oligosaccharides from sialoglycopeptide to the ribonuclease treated by Endo-β-GlcNAc-ase was carried out as follows. First, 50 μg of ribonuclease B and 0.1 milliunits of Endo-H in a total volume of 500 μL with 50 mM sodium phosphate buffer (pH 5.5) was incubated overnight at 37°C to obtain GlcNAc-ribonuclease B, which contains one GlcNAc residue and is partially deglycosylated. Without heat treatment, the reaction mixture was im-

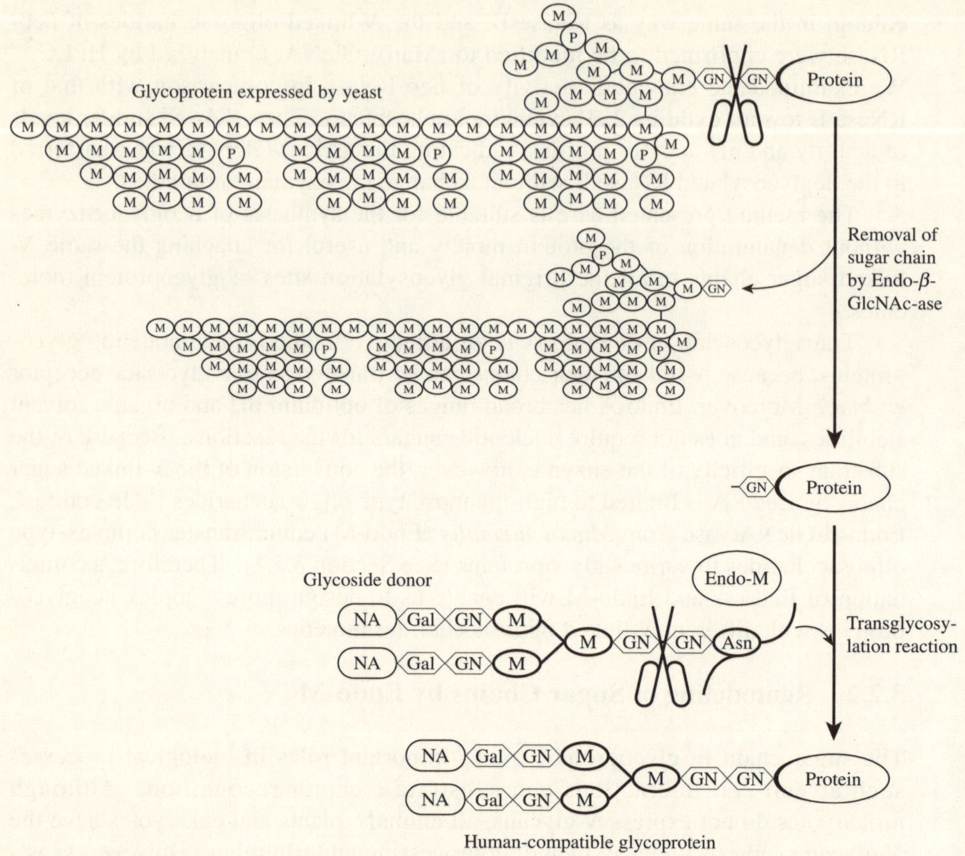

Fig. 3.19 Remodeling of sugar chain on glycoprotein using transglycosylation activity of Endo-M. M, mannose; GN, *N*-acetylglucosamine; Gal, galactose; NA, *N*-acetylneuraminic acid; P, phosphate.

mediately loaded onto a gel filtration column (Superdex 75 HR) equilibrated with 50 mM sodium phosphate buffer (pH 7.0) with 0.15 M NaCl and eluted using an AKTA FPLC system (Amersham Pharmacia Biotech.). The fractions including GlcNAc-ribonuclease B were concentrated and desalted on by Ultrafree-15 centrifugal filters 10K (Millipore). Next, the transglycosylation reaction by Endo-M was carried out as follows: 0.5 milliunits of recombinant Endo-M was incubated with 3.5 mM sialyl glycopeptide from hen egg yolk, which contains sialobiantennary complex type oligosaccharide, in 40 μL of 25 mM sodium acetate buffer (pH 6.0) in the presence of 40 μg of GlcNAc-ribonuclease B. After incubation of the mixture for 60 min at 30°C, the reaction was stopped by boiling for 3 min. The reaction mixtures were analyzed by SDS-PAGE. As shown in Fig. 3.20 A, one protein band was newly generated with an increase of 2,000 Da by 60 min-incubation (lane 3) compared with GlcNAc-ribonuclease B (lane 2). The reaction mixture was also subjected to lectin blotting using horseradish peroxidase (HRP) con-

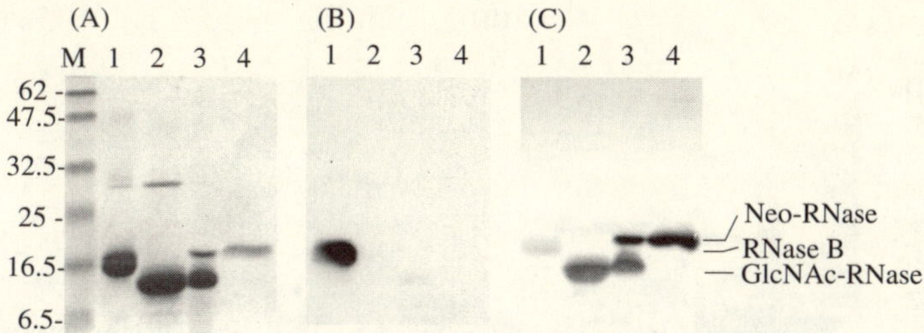

Fig. 3.20 Lectin blotting analysis of new RNase B the oligosaccharide of which was changed to a complex type oligosaccharide.
The samples were analyzed on SDS-PAGE. (A) Stained by Coomassie Brilliant Blue R-250. (B) Lectin blotting with HRP-Con A. (C) Lectin blotting with biotinyl-SSA. Lane M, prestained molecular markers; lane 1, native RNase B; lane 2, Endo-H-treated RNase B (GlcNAc-RNase); lane 3, reaction mixture; lane 4, purified new RNase having complex type oligosaccharide (Neo-RNase).

jugated-Con-A, which is specific for mannose (Fig. 3.20 B) and biotinylated *Sambucus sieboldiana* agglutinin (SSA) which is specific for sialic acid (Fig. 3.20 C). The Neo-ribonuclease was detected by SSA blotting (Fig. 3.20 C, lane 3) but was not detected by Con-A blotting (Fig. 3.20 B, lane 3). These results indicated that the high-mannose type oligosaccharides of ribonuclease B had been converted to the complex type oligosaccharides.

We examined the enzymatic activity of Neo-ribonuclease in comparison with that of original ribonuclease B toward cytidine 2′,3′-cyclic monophosphate. Both ribonucleases showed almost the same activity.

Partially deglycosylated invertase (GlcNAc-invertase) and yeast carboxypeptidase (GlcNAc-CPY) treated by Endo-H were also used as the acceptor for the oligosaccharide-transferring reaction by Endo-M. The enzymatic reactions were carried out with the same composition and conditions of the reaction for GlcNAc-ribonuclease B as described above, except for the acceptor. After incubation, the reaction mixtures were analyzed by SDS-PAGE (Fig. 3.21 A) and lectin blotting using biotinylated SSA (Fig. 3.21 B). The reaction mixtures of GlcNAc-invertase and GlcNAc-CPY generated a new protein band (Fig. 3.21 A, lane 2 and lane 4, respectively) and were also detected with SSA blotting (Fig. 3.21 B, lane 2 and lane 4, respectively). However, GlcNAc-invertase and GlcNAc-CPY were not detected (Fig. 3.21 B, lane 1 and lane 3, respectively). These results showed that the complex type oligosaccharides were bound to these proteins.

The above results suggest the possibility of remodeling or exchanging the high-mannose type of oligosaccharides in glycoproteins produced by yeasts or molds to the complex types that are not synthesized by these microorganisms. The simplest approach for the synthesis of glycoproteins is preparation by cultured recombinant cells. The yeast *Saccharomyces cerevisiae* is a useful organism for the production of recombinant glycoproteins containing *N*-linked oligosaccharides. However, the mannan glycans on the recombinant proteins are strong antigens for mammals, as noted above. Chiba *et al.*[86] constructed a *S. cerevisiae* mutant to convert the mannan-type oligosaccharides of glycoproteins to

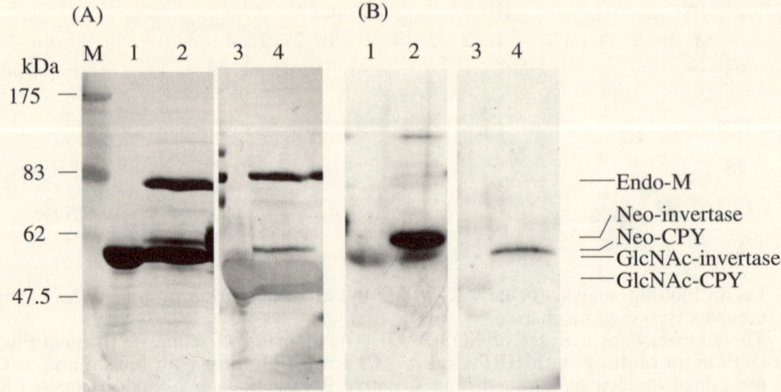

Fig. 3.21 Lectin blotting analyses of new invertase and yeast carboxypeptidase (CPY) the oligosaccharides of which were changed to complex type oligosaccharides.
The samples were analyzed on SDS-PAGE. (A) Stained by Coomassie Brilliant Blue R-250. (B) Lectin blotting with biotinyl-SSA. Lane 1, Endo-H-treated invertase (GlcNAc-invertase); lane 2, transglycosylation reaction mixture of Endo-M with invertase; lane 3, Endo-H-treated CPY (GlcNAc-CPY); lane 4, transglycosylation reaction mixture of Endo-M with CPY; lane M, molecular weight markers.

$(Man)_5(GlcNAc)_2$. However, the synthesis of glycoproteins containing complex type oligosaccharides has not been successful.

Sears and Wong[87] reviewed the biochemical approaches to the synthesis of homogeneous glycoproteins by means of native peptide ligation, glycosyltransferase and endoglycosidase-catalyzed transglycosylation. Until now, our approach using Endo-M is the simplest for the synthesis of neoglycoprotein and has the potential for remodeling mannan glycans into complex type oligosaccharides. This conversion system of oligosaccharides will also help to provide homogeneous glycoproteins to study the effects of oligosaccharides on glycoproteins. As Endo-M has wide substrate specificity for high-mannose type, hybrid type and biantennary complex type oligosaccharides, the transglycosylation reaction catalyzed by Endo-M will become a useful technique in protein engineering.

3.3 Synthesis of *β*-Mannosides Using Endo-*β*-mannosidase

3.3.1 Synthesis of the Man*β*1-4GlcNAc Structure

β-Mannosyl linkage is one of the glycosyl linkages that are difficult to chemically synthesize because a hydroxyl group at an anomer carbon and C-2 of Man*β* is located in the gauche conformation.[88] The Man*β*1-4GlcNAc structure appears in the core structure of the *N*-linked sugar chain. Therefore it is crucial to establish a routine synthetic method to form the Man*β*1-4GlcNAc structure for the synthesis of *N*-linked sugar chains.

Various efforts for improving enantio-selectivity of the *β*-mannosyl linkage in chemical synthesis have been made and some practical methods for the synthesis of Manβ1-4GlcNAc have been reported.[88-93] Several methods have been applied to synthesize the core trisaccharide of the *N*-linked sugar chain Manβ1-4GlcNAcβ1-4GlcNAc[94,95] and the core pentasaccharide Manα1-6(Manα1-3)Manβ1-4GlcNAcβ1-4GlcNAc.[96,97] However these methods require separation from Manα-derivatives produced during chemical reactions and much labor to prepare glycosyl donors and acceptors for coupling and their products are protected forms.

An enzymatic method for the preparation of the Manβ1-4GlcNAc linkage using mannosidases specific for the β-linkage is attractive. Transglycosylation activities of β-mannanase from *Aspergillus niger* and β-mannosidases from *Aspergillus oryzae* and *Helix pomatia* have been applied to the synthesis of the core trisaccharide Manβ1-4GlcNAcβ1-4GlcNAc.[98-100] In addition, the chemo-enzymatic synthesis of Manβ1-4GlcNAcβ1-4GlcNAc using β-mannosyltransferase has been reported,[101] but the preparation of the acceptor substrate (glycolipid) is a laborious process.

Recently, endo-β-mannosidase was discovered from plant species and purified from lily flowers as described in Chapter 2 section 2.1.2.[102,103] The purified lily endo-β-mannosidase hydrolyzes the β-mannosyl linkage of oligomannose-type *N*-linked sugar chains, $(\mathrm{Man})_n$Manα1-6Manβ1-4GlcNAcβ1-4GlcNAc (n = 0~2) and produces $(\mathrm{Man})_n$Manα1-6Man and GlcNAcβ1-4GlcNAc.[102,103] This enzyme hydrolyzes free *N*-linked sugar chains as well as their pyridylaminated derivatives and their corresponding glycopeptides.[104] Because of its substrate specificity, endo-β-mannosidase is expected to have unique characteristics in the glycosidase-catalyzed synthesis of the Manβ1-4GlcNAc structure. Oligomannose is expected to be a building block for the formation of oligomannose-type *N*-linked sugar chains.

3.3.2 Discovery of Transglycosylation Activity of Endo-β-mannosidase

As the first step for this application of this endoglycosidase, activities for the reverse reaction of hydrolysis (condensation reaction) and/or the transglycosylation reaction of this enzyme were investigated. In the presence of 90 mM Manα1-6Man and 140 mM GN2-PA (abbreviations for PA-sugars are shown in Table 3.3), which are the hydrolysis products of M2B-PA by digestion with endo-β-mannosidase, the enzyme did not generate any product, meaning that the reverse reaction of the hydrolysis with endo-β-mannosidase does not occur under the conditions used. However, in the presence of 90 mM M2B-peptide as the donor substrate and 140 mM GN2-PA as the acceptor substrate, incubation with the purified lily enzyme at 37°C for 10 h produced a substantial amount (67% based on a donor substrate) of the newly generated product[105] (Fig. 3.22). The structure of this product was identified as M2B-PA by two-dimensional sugar mapping[106] coupled with the α-mannosidase and the β-mannosidase digestion of the product and by matrix-assisted laser desorption ionization/time-of-flight mass spectrometry.

Table 3.3 Abbreviations of PA-sugars

Abbreviation	Structure
GN2-PA	GlcNAcβ1-4GlcNAc-PA
M1-PA	Manβ1-4GlcNAcβ1-4GlcNAc-PA
M2B-PA	Manα1 6Manβ1-4GlcNAcβ1-4GlcNAc-PA
M3B-PA	Manα1 6 3Manβ1-4GlcNAcβ1-4GlcNAc-PA Manα1
M3C-PA	3Manα1 Manα1 6Manβ1-4GlcNAcβ1-4GlcNAc-PA
M4B-PA	Manα1 6 3Manα1 Manα1 6Manβ1-4GlcNAcβ1-4GlcNAc-PA
M5A-PA	Manα1 6 3Manα1 Manα1 6 3Manβ1-4GlcNAcβ1-4GlcNAc-PA Manα1

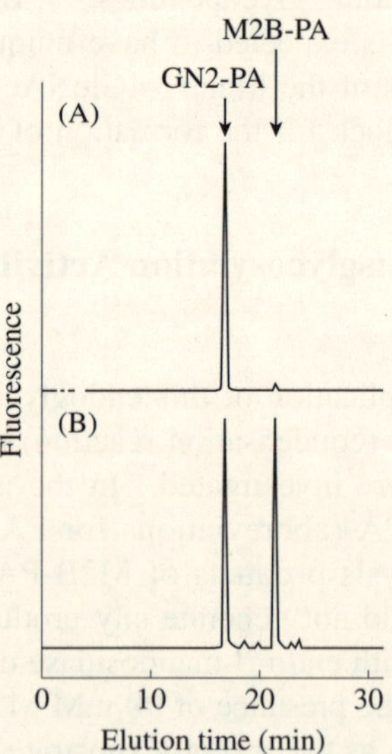

Fig. 3.22 Size-fractionation HPLC of transglycosylation products with lily endo-β-mannosidase. Lily endo-β-mannosidase (40 mU) was incubated with 90 mM M2B-peptide and 140 mM GN2-PA at 37˚C for 0 h (A) and 10 h (B).
[Reprinted with permission from Sasaki, A. *et al.* (2005) *FEBS J.* **272**, 1662]

This product was produced by transfer of Manα1-6Man from M2B-peptide to GN2-PA with β-linkage. These observations show that the lily endo-β-mannosidase has transglycosylation activity.

3.3.3 Transfer of Mannose to GlcNAc by β1-4 Linkage Using Endo-β-mannosidase

The regio- and stereoselectivity of transglycosylation of lily endo-β-mannosidase was investigated. In the presence of M1-peptide as the donor substrate and *p*NP-β-GlcNAc as the acceptor substrate, by incubation (at 37°C for 10 h) with endo-β-mannosidase, two products were generated (Fig. 3.23). The major product (Arrow 2) was analyzed by methylation analysis and the structure of the minor product (Arrow 1) is described in section 3.3.5. Permethylated alditol acetates derived from the major product were analyzed by gas chromatography-mass spectrometry and 1,5-di-*O*-acetyl-2,3,4,6-tetra-*O*-methyl-mannitol and 2-deoxy-2-(*N*-methyl)acetamido-1,4,5-tri-*O*-acetyl-3,6-di-*O*-methyl-glucitol were detected in an equimolar ratio. 4,6-Di-*O*-methyl- and 3,4-di-*O*-methyl *N*-acetylhexosamine derivatives were not detected, showing that the mannose residue transferred is linked to the C-4 position of *p*NP-β-GlcNAc. In addition, digestion of this product by *Achatina fulica* β-mannosidase gave *p*NP-β-GlcNAc, indicating that this enzyme synthesizes the β-linked mannose structure. Thus, lily endo-β-mannosidase synthesizes Manβ1-4GlcNAc structure by its transglycosylation activity.

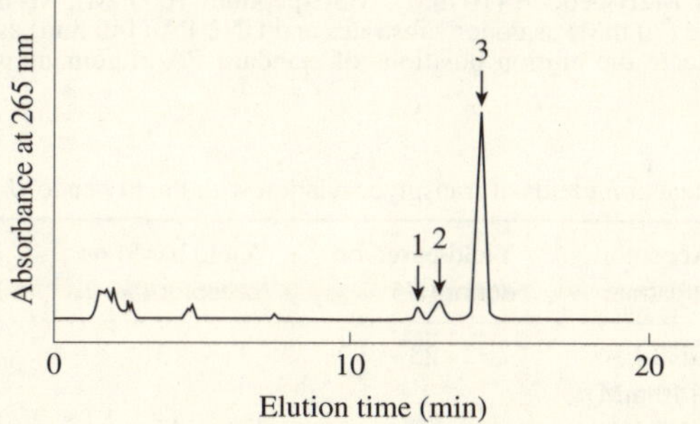

Fig. 3.23 Reverse-phase HPLC of transglycosylation products with lily endo-β-mannosidase using 80 mM M1-peptide and 140 mM *p*NP-β-GlcNAc at 37°C for 10 h. Arrows 1 and 2 indicate the reaction products Manβ1-Manβ1-4GlcNAc-*p*NP and Manβ1-4GlcNAc-*p*NP, respectively. Arrow 3 indicates the elution position of *p*NP-β-GlcNAc.
[Reprinted with permission from Sasaki, A. *et al.* (2005) *FEBS J.* **272**, 1662]

3.3.4 Transfer of Oligomannose to GN2 Using Endo-β-mannosidase

M2B-peptide is shown to be a good donor substrate for transglycosylation of endo-β-mannosidase to transfer Manα1-6Man to GN2-PA with β-linkage as described in section 3.3.2. A mixture of M2B-peptide, M3C-peptide, M4B-peptide and M5A-peptide as donor substrates and GN2-PA as the acceptor substrate were incubated with the enzyme.[105] The first three glycopeptides above in a mixture worked as donor substrates for transglycosylation to generate their corresponding pyridylaminated derivatives (Fig. 3.24; Table 3.4). However, the last glycopeptide did not work as a donor substrate. In addition, M3B-peptide did not work as a donor substrate (Table 3.4). The glycopeptides containing the Manα1-3Manβ structure could not be used as donor substrate for either transglycosylation or hydrolysis[102,103] of the lily endo-β-mannosidase.

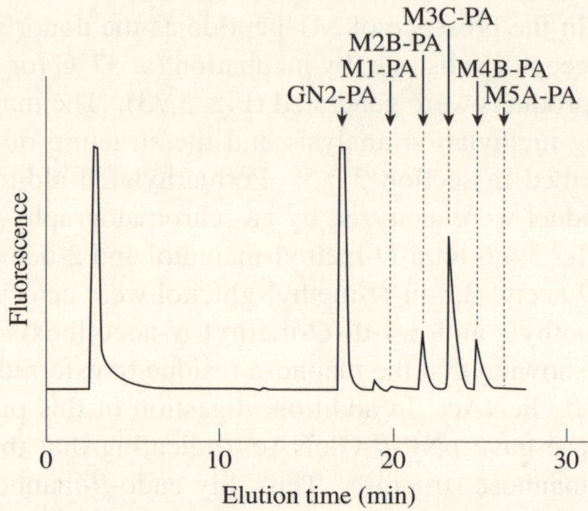

Fig. 3.24 Size-fractionation HPLC of transglycosylation products with lily endo-β-mannosidase using a mixture of M2B-peptide (10 mM), M3C-peptide (100 mM), M4B-peptide (60 mM) and M5A-peptide (70 mM) as donor substrates and GN2-PA (140 mm) as the acceptor substrate. Arrows indicate the elution positions of standard PA-oligomannose-type *N*-linked sugar chains.

Table 3.4 Reaction yields of transglycosylation with the lily endo-β-mannosidase

Donor substrate	Acceptor substrate	Yield based on donor (%)	Yield based on acceptor (%)	Relative hydrolysis rate
M1-peptide (90 mM)	GN2-PA (140 mM)	23	15	4
M2B-peptide (90 mM)	GN2-PA (140 mM)	67	43	100
M3B-peptide (90 mM)	GN2-PA (140 mM)	0	0	< 0.01
M3C-peptide (100 mM)	GN2-PA (140 mM)	19	13	48
M4B-peptide (60 mM)	GN2-PA (140 mM)	11	5	42
M5A-peptide (70 mM)	GN2-PA (140 mM)	0	0	< 0.01

Even though the concentration of the donor substrate (M2B-peptide) was changed in the presence of the same concentration of the acceptor substrate, reaction yields based on donor substrate did not change much (Table 3.5).

Table 3.5 Dependence of the concentrations of a donor substrate (M2B-peptide) on transglycosylation to 140 mM GN2-PA as an acceptor substrate with the lily endo-β-mannosidase

Amount of M2B-peptide used as donor substrate	Yield based on donor (%)	Yield based on acceptor (%)
10 mM	67	5
50 mM	64	23
90 mM	67	43

3.3.5 Transfer of Mannose to Various Monosaccharides Using Endo-β-mannosidase

Transglycosylation of endo-β-mannosidase using various monosaccharide derivatives as acceptor substrates was investigated. To simplify the analysis of the transglycosylation products, M1-peptide was used as the donor substrate in a series of reactions and *p*NP-β-GlcNAc, *p*NP-β-GalNAc, *p*NP-β-Glc and *p*NP-β-Man were used as acceptor substrates. In the presence of 80 mM M1-peptide and 140 mM of each acceptor substrate, transmannosylation to an acceptor saccharide occurred as shown in Table 3.6. *p*NP-β-Glc, *p*NP-β-GlcNAc and *p*NP-β-Man worked as acceptor substrates in transglycosylation of Man by the β-linkage, while *p*NP-β-GalNAc did not. Among four acceptor substrates, only GalNAc has an axial hydroxyl group at C-4, indicating that the β-mannosyl transfer by the endo-β-mannosidase requires an equatorial hydroxyl group at C-4 of the acceptor substrate. The structural difference among Glc, GlcNAc and Man, which work as acceptor substrates for transglycosylation, is found at C-2; the kind of substituent and its orientation at C-2 are different from each other, showing that transglycosylation by endo-β-mannosidase does not require structural strictness on C-2 of acceptor monosaccharides, but structure of a substituent on C-2 are partly influenced on transglycosylation because the reaction yields against each *p*NP derivative differed from each other (Table 3.6). Successive transfer of two mannoses was observed when *p*NP-β-Glc and *p*NP-β-GlcNAc were used.

Table 3.6 Transglycosylation using 80 mM M1-peptide as the donor substrate and several *p*NP-monosaccharides as acceptor substrates with the lily endo-β-mannosidase

Acceptor substrate (140 mM)	Product	Yield based on donor (%)	Yield based on acceptor (%)
*p*NP-β-GlcNAc	Man-GlcNAc-*p*NP	21	12
	Man-Man-GlcNAc-*p*NP	11	6
*p*NP-β-GalNAc		0	0
*p*NP-β-Glc	Man-Glc-*p*NP	40	23
	Man-Man-Glc-*p*NP	14	8
*p*NP-β-Man	Man-Man-*p*NP	17	10

3.3.6 Future Prospects

The lily endo-β-mannosidase has been shown to possess transglycosylation activity. This endoglycosidase transfers mannose and oligomannose to acceptor sugars by β-linkage to synthesize various sugar chains containing the Manβ structure. The structures of the *N*-linked sugar chains synthesized depend on substrate specificity for hydrolysis activity of endo-β-mannosidase, *i.e.*, oligomannosides containing the Manα1-3Manβ structure can not be used as donor substrates. Not only GlcNAcβ1-4GlcNAc-PA but *p*NP derivatives of GlcNAc, Glc and Man have also been shown to work as acceptor substrates. Monosaccharides, which have an equatorial hydroxyl group at their C-4 position, *e.g.*, xylose and rhamnose, may also work as acceptor substrates. Thus, the characteristics of this enzyme can be applied to the synthesis of various types of oligosaccharides containing the Manβ linkage such as *N*-linked sugar chains and arthro-series glycosphingolipids.[107] The donor substrate for mannosyltransfer used in our laboratory is M1-peptide. This glycopeptide is easily prepared on a synthetic scale by complete α-mannosidase digestion of glycopeptides obtained by pronase digestion of glycoproteins. In addition, the expression system of the *Arabidopsis* endo-β-mannosidase in the heterologous cells has been constructed.[103] Therefore, large quantities of endo-β-mannosidase are available for large-scale synthesis of β-mannosides.

References

1. Lee, Y.C. and Lee, R.T. (1992) in *Glycoconjugates* (Allen, H.J. and Kisailus, E.C., eds.) pp. 121-165, Marcel Dekker, Inc., N.Y.
2. Medal, M. and Bock, K. (1994) *Glycoconj. J.* **11**, 59-63
3. Schuster, M., Wang, P., Paulson J.C., and Wong, C.H. (1994) *J. Am. Chem. Soc.* **116**, 1135-1136
4. Cote, G.L. and Tao, B.Y. (1990) *Glycoconj. J.* **7**, 145-162
5. Edelman, J. (1956) *Adv. Enzymol.* **17**, 189-232
6. Elder, J.H. and Alexander, S. (1982) *Proc. Natl. Acad. Sci. USA* **79**, 4540-4544
7. Trimble, R.B., Atkinson, P.H., Tarentino, A.L., Plummer, T.H., Maley, F., and Tomer, K.B. (1986) *J. Biol. Chem.* **261**, 12000-12005
8. Kadowaki, S., Yamamoto. K., Fujisaki, M., Kumagai H., and Tochikura, T. (1988) *Agric. Biol. Chem.* **52**, 2387-2389
9. Yamamoto, K., Kadowaki, S., Watanabe, J., and Kumagai, H. (1994) *Biochem. Biophys. Res. Commun.* **203**, 244-252
10. Takegawa, K., Yamaguchi, S., Kondo, A., Kato, I., and Iwahara, S. (1991) *Biochem. Int.* **25**, 829-835
11. Fujita, M., Shoda, S., Haneda, K., Inazu, T., Takegawa, K., and Yamamoto, K. (2001) *Biochim. Biophys. Acta* **1528**, 9-14
12. Brameld, K.A., Shrader, W.D., Imperiali, B., and Goddard, W.A. (1998) *J. Mol. Biol.* **280**, 913-923
13. Inazu, T. and Kobayashi, K. (1993) *Synlett* 1993, 869-870
14. Mizuno, M., Muramoto, I., Kobayashi, K., Yaginuma, H., and Inazu, T. (1999) *Synthesis* 1999, 162-165
15. Inazu, T., Mizuno, M., Kohda, Y., Kobayashi, K., and Yaginuma, H. (1996) in *Peptide Chemistry 1995* (Nishi, N., ed.), pp.61-64, Protein Research Foundation, Osaka
16. Inazu, T., Mizuno M., Maegami, T., and Haneda, K. (1997) in *Peptide Chemistry 1996* (Kitada, C., ed.), pp. 41-44, Protein Research Foundation,Osaka
17. Kadowaki, S., Yamamoto, K., Fujisaki, M., Izumi, K., Tochikura, T., and Yokoyama, T. (1988) *Agric. Biol. Chem.* **54**, 97-106
18. Fujita, K., Kobayashi, K., Iwamatsu, A., Takeuchi, M., Kumagai, H., and Yamamoto, K. (2004)

Arch. Biochem. Biophys. **432**, 41-49

19. Huang, C.-C., Mayer, H.B., and Montogomery, R. (1970) *Carbohyd. Res.* **13**, 127-137
20. Seko, A., Koketsu, M., Nishizono, M., Enoki, Y., Ibrahim, H.R., Juneja, L.R., Kim, M., and Yamamoto, T. (1997) *Biochim. Biophys. Acta* **1335**, 23-32
21. Haneda, K., Inazu, T., Mizuno, M., and Yamamoto, K. (2003) *Methods Enzymol.* **362** (Lee, Y.C. and Lee, R.C., eds.) pp. 74-85, Academic Press, Oxford
22. Yamamoto, K., Fujimori, K., Haneda, K., Mizuno, M., Inazu, T., and Kumagai, H. (1998) *Carbohyd. Res.* **305**, 415-422
23. Pert, C.B., Hill, J.M., Ruff, M.R., Berman, R.M., Robey, W.G., Arthur, L.O., Ruscetti, F.W., and Farrar,W.L. (1986). *Proc. Natl. Acad. Sci. USA* **83**, 9254-9258
24. Marx, J.L. (1979) *Science* **205**, 886-889
25. Haneda, K., Inazu, T., Mizuno, M., Iguchi, R., Tanabe, H., Fujimori, K., Yamamoto, K., Kumagai, H., Tsumori, K., and Munekata, E. (2001) *Biochim. Biophys. Acta* **1526**, 242-248
26. Suzuki, Y. (1994) *Prog. Lipid Res.* **33**, 429-457
27. Sigal, G.B., Mammen, M., Dahmann, G., and Whitesides, G.M. (1996) *J. Am. Chem. Soc.* **118**, 3789-3800
28. Furuike, T., Aiba, S., Suzuki, T., Takahashi, T., Suzuki, Y., and Yamada, K. (2000) *J. Chem. Soc., Perkin Trans.* **1**, 3000-3005
29. Roy, R., Tropper, F.D., Romanowska, A., Letellier, M., Cousineau, L., Meunier, S.J., and Boratynski, J. (1991) *Glycoconj. J.* **8**, 75-81
30. Kojima, S., Hasegawa, T., Yonemura, T., Sasaki, K., Yamamoto, K., Makimura, Y., Takahashi, T., Suzuki, T., Suzuki, Y., and Kobayashi, K. (2003) *Chem. Commun.* **11**, 1250-1251
31. Akaike, E., Tsutsumida, M., Osumi, K., Fujita, M., Yamanoi, T., Yamamoto, K., and Fujita, K. (2004) *Carbohyd. Res.* **339**, 719-722
32. Haneda, K., Inazu, T., Yamamoto, K., Kumagai, H., Nakahara, Y., and Kobata, A. (1996) *Carbohydr. Res.* **292**, 61-70
33. Inazu, T., Kobayashi, K., and Yaginuma, H. (1994) *Peptide Chem. 1993*, pp. 101-104, Protein Research Foundation, Osaka
34. Azria, M. (1989) *The Calcitonins: Physiology and Pharmacology*, Karger, Basel
35. Mizuno, M., Haneda, K., Iguchi, R., Muramoto, I., Kawakami, T., Aimoto, S., Yamamoto, K., and Inazu, T. (1999) *J. Am. Chem. Soc.* **121**, 284-290
36. Hojo, H. and Aimoto, S. (1996) *Bull. Chem. Soc. Jpn.* **69**, 3331-3338
37. Haneda, K., Tagashira, M., Yoshino, E., Takeuchi, M., Inazu, T., Toma, K., Iijima, H., Isogai, Y., Hori, M., Takamatsu, S., Hujibayashi, Y., Kobayashi, K., Takeuchi, M., and Yamamoto, K. (2004) *Glycoconj. J.* **21**, 377-386
38. Takegawa, K., Yamaguchi, S., Kondo, A., Iwamoto, H., Nakoshi, M., Kato, I., and Iwahara, S. (1991) *Biochem. Int.* **24**, 849-855
39. Tai, T., Yamashita, K., Ogata-Arakawa, M., Koide, N., Muramatsu, T., Iwashita, S., Inoue, Y., and Kobata, A. (1975) *J. Biol. Chem.* **250**, 8569-8575
40. Takegawa, K., Yamabe, K., Fujita, K., Tabuchi, M., Mita, M., Izu, H., Watanabe, A., Asada, Y., Sano, M., Kondo, A., Kato, I., and Iwahara, S. (1997) *Arch. Biochem. Biophys.* **338**, 22-28
41. Yamamoto, K., Haneda, K., Iguchi, R., Inazu, T., Mizuno, M., Takegawa, K., Kondo, A., and Kato, I. (1999) *J. Biosci. Bioeng.* **87**, 175-179
42. Haneda, K., Takeuchi, M., Inazu, T., Toma, K., Tagashira, M., Kobayashi, K., Yamamoto, K., and Takegawa, K. (2002) *Peptide Science 2001*, pp. 89-92, Protein Research Foundation, Osaka
43. Hashimoto, H., Toma, K., Nishikido, J., Yamamoto, K., Haneda, K., Inazu, T., Valentine, K.G., and Opella, S.J. (1999) *Biochemistry* **38**, 8377-8384
44. Tagashira, M., Tanaka, A., Hisatani, K., Isogai, Y., Hori, M., Takamatsu, S., Fujibayashi, Y., Yamamoto, K., Haneda, K., Inazu, T., and Toma, K. (2001) *Glycoconj. J.* **18**, 449-455
45. Otani, M., Kitazawa, S., Yamauchi, H., Meguro, T., and Orimo, H. (1978) *Horm. Metab. Res.* **10**, 252-256
46. Fujita, K., Miyamura, T., Sano, M., Kato, I., and Takegawa, K. (2002) *J. Biosci. Bioeng.* **93**, 614-617
47. Kobata, A. (1979) *Anal. Biochem.* **100**, 1-14
48. Tachibana, Y., Yamashita, K., and Kobata, A. (1982) *Arch. Biochem. Biophys.* **214**, 199-210
49. Tarentino, A.L. and Maley, F. (1975) *Biochem. Biophys. Res. Commun.* **67**, 455-462
50. Arakawa, M. and Muramatsu, T. (1974) *J. Biochem.* **76**, 307-317
51. Muramatsu, T., Koide, N., and Maeyama, K. (1978) *J.Biochem.* **83**, 363-370
52. Tarentino, A.L. and Maley, F. (1974) *J. Biol. Chem.* **249**, 811-817
53. Koide, N. and Muramatsu, T. (1974) *J. Biol. Chem.* **249**, 4897-4904

54. Iwase, H., Morinaga, T., Li, Y.-T., and Li, S.-C. (1981) *Anal. Biochem.* **113**, 93-95
55. Takegawa, K., Fujita, K., Fan, J.-Q., Tabuchi, M., Tanaka, N., Kondo, A., Iwamoto, H., Kato, I., Lee, Y.C., and Iwahara, S. (1998) *Anal. Biochem.* **257**, 218-223
56. Fan, J.-Q., Takegawa, K., Iwahara, S., Kondo, A., Kato, I., Abeygunawardana, C., and Lee, Y.C. (1995) *J. Biol. Chem.* **270**, 17723-17729
57. Fujita, K., Asada, Y., Yamamoto, K., and Takegawa, K. (2000) *J. Biosci. Bioeng.* **90**, 462-464
58. Suzuki, T., Kitajima K., Emori, Y., Inoue, Y., and Inoue, S. (1997) *Proc. Natl. Acad. Sci. USA* **94**, 6244-6249
59. Berger, S., Menudier, A., Julien, R., and Karamanos, Y. (1995) *Biochimie* **77**, 751-760
60. Kitajima, K., Suzuki, T., Kouchi, Z., Inoue, S., and Inoue, Y. (1995) *Arch. Biochem. Biophys.* **319**, 393-401
61. Seko, A., Kitajima, K., Iwamatsu, T., Inoue, Y., and Inoue, S. (1999) *Glycobiology* **9**, 887-895
62. Fan, J.-Q. and Lee, Y.C. (1997) *J. Biol. Chem.* **272**, 27058-27064
63. Deras, I.L., Takegawa, K., Kondo, A., Kato, I., and Lee, Y.C. (1998) *Bioorg. Med. Chem. Lett.* **8**, 1763-1766
64. Messner, P. (1997) *Glycoconj. J.* **14**, 3-11
65. Schreiner, R., Schnabel, E., and Wieland, F. (1994) *J. Cell Biol.* **124**, 1071-1081
66. Takahashi, N. and Nishibe, H. (1978) *J. Biochem.* **84**, 1467-1473
67. Plummer, T.H., Jr., Phelan, A.W., and Tarentino, A.L. (1987) *Eur. J. Biochem.* **163**, 167-1173
68. Plummer, T.H., Jr. and Tarentino, A.L. (1981) *J. Biol. Chem.* **256**, 10243-10246
69. Plummer, T.H., Jr., Elder, J.H., Alexander, S., Phelan, A.W., and Tarentino, A.L. (1984) *J. Biol. Chem.* **259**, 10700-10704
70. Fan, J.-Q., Quesenberry, M.S., Takegawa, K., Iwahara, S., Kondo, A., Kato, I., and Lee, Y.C. (1995) *J. Biol. Chem.* **270**, 17730-17735
71. Lee, R.T. and Lee Y.C. (1987) *Methods Enzymol.* **138**, 424-429
72. Lee, R.T., Ichikawa, Y., Kawasaki, T., Drickamer, K., and Lee, Y.C. (1992) *Arch. Biochem. Biophys.* **299**, 129-136
73. Lee, Y.C. (1993) *Biochem. Soc. Trans.* **21**, 460-463
74. Dean, N. (1998) *Biochim. Biophys. Acta* **1426**, 309-322
75. Silberstein, S. and Gilmore, R. (1996) *FASEB J.* **10**, 849-858
76. Lerouge, P., Cabanes-Macheteau, M., Rayon, C., Fischette-Laine, A.C., Gomord, V., and Faye, L. (1998) *Plant Mol. Biol.* **38**, 31-48
77. Parekh, R.B., Dwek, R.A., Sutton, B.J., Fernandes, D.L., Leung, A.K., Takeuchi, F., Nagano, Y., Miyamoto, T., and Kobata, A. (1985) *Nature* **316**, 452-457
78. Matzuk, M.M. and Boime, I. (1988) *J. Cell Biol.* **106**, 1049-1059
79. Matzuk, M.M., Keene, J.F., and Biome, I. (1989) *J. Biol. Chem.* **264**, 2409-2414
80. Gething, M.J. (1999) *Semin. Cell Dev. Biol.* **10**, 465-472
81. Takegawa, K., Tabuchi, M., Yamaguchi, S., Kondo, A., Kato, I., and Iwahara, S. (1995) *J. Biol. Chem.* **270**, 3094-3099
82. Liang, C.-J., Yamashita, K., and Kobata, A. (1980) *J. Biochem.* **88**, 51-58
83. Fujita, K., Tanaka, N., Sano, M., Kato, I., Asada, Y., and Takegawa, K. (2000) *Biochem. Biophys. Res. Commun.* **267**, 134-138
84. Takegawa, K., Nakoshi, M., Iwahara, S., Yamamoto, K., and Tochikura, T. (1989) *Appl. Environ. Microbiol.* **55**, 3107-3112
85. Freeze, H.H. (1999) in *Essentials of Glycobiology* (Varki, A. *et al.*, eds.) pp. 286-289, Cold Spring Harbor Laboratory Press, NY
86. Chiba, Y., Suzuki, M., Yoshida, S., Yoshida, A., Ikenaga, H., Takeuchi, M., Jigami, Y., and Ichishima, E. (1998) *J. Biol. Chem.* **273**, 26298-26304
87. Sears, P. and Wong, C.H. (2001) *Science* **291**, 2344-2350
88. Ito, Y. and Ohnishi, Y. (2001) in *Glycoscience: Chemistry and Chemical Biology.* (Fraser-Reid, B.O., Tatsuta, K., and Thiem, J., eds.) pp. 1589-1619, Springer-Verlag, Berlin
89. Shaban, M.A.E. and Jeanloz, R.W. (1976) *Carbohyd. Res.* **52**, 115-127
90. Barresi, F. and Hindsgaul, O. (1991) *J. Am. Chem. Soc.* **113**, 9376-9377
91. Stork, G. and Kim, G. (1992) *J. Am. Chem. Soc.* **114**, 1087-1088
92. Ito, Y. and Ogawa, T. (1994) *Angew. Chem. Int. Ed. Engl.* **33**, 1765-1767
93. Crich, D. and Sun, S. (1997) *J. Org. Chem.* **62**, 1198-1199
94. Augé, C.D., Warren, R.W., Jeanloz, M., Kiso, M., and Anderson, L. (1980) *Carbohydr. Res.* **82**, 85-95
95. Günther, W. and Kunz, H. (1990) *Angew. Chem. Int. Ed. Engl.* **29**, 1050-1051
96. Paulsen, H. and Lebuhn, R. (1984) *Carbohydr. Res.* **130**, 85-101

97. Dan, A., Ito, Y., and Ogawa, T. (1995) *J. Org. Chem.* **60**, 4680-4681
98. Usui, T., Suzuki, M., Sato, T., Kawagishi, H., Adachi, K., and Sano, H. (1994) *Glycoconj. J.* **11**, 105-110
99. Singh, S., Scigelova, M., and Crout, D.H.G. (1996) *Chem. Commun.* **8**, 993-994
100. Scigelova, M., Singh, S., and Crout, D.H.G. (1999) *J. Chem. Soc., Perkin Trans. 1.* **1999**, 777-782
101. Watt, G.M., Revers, L., Webberley, M.C., Wilson, I.B.H., and Flitsch, S.L. (1998) *Carbohydr. Res.* **305**, 533-541
102. Sasaki, A., Yamagishi, M., Mega, T., Norioka, S., Natsuka, S., and Hase, S. (1999) *J.Biochem.* **125**, 363-367
103. Ishimizu, T., Sasaki, A., Okutani, S., Maeda, M., Yamagishi, M., and Hase, S. (2004) *J.Biol. Chem.* **279**, 38555-38562
104. Sasaki, A., Ishimizu, T., and Hase, S. (2005) *J. Biochem.* **137**, 87-93
105. Sasaki, A., Ishimizu, T., Geyer, R., and Hase, S. (2005) *FEBS J.* **272**, 1660-1668
106. Hase, S. (1994) *Methods Enzymol.* **230**, 225-237
107. Dennis, R.D. and Wiegandt, H. (1993) *Adv. Lipid Res.* **26**, 321-351

4

Enzymatic Synthesis of Neo-*O*-glycans

4.1 Enzymatic Synthesis of *O*-Linked Glycopeptides Using Endo-α-*N*-acetylgalactosaminidase

The Galβ1-3GalNAc disaccharide, *O*-linked to either serine or threonine in mucin type glycoproteins, is a basic structure in mucin type sugar chains. Most other *O*-linked sugar chains are in fact constructed using Galβ1-3GalNAc, which is hence designated as a core 1 molecule in these structures. This disaccharide is also known as the Thomsen-Friedenreich antigen (T-antigen) and has been shown to have a role in the development of carcinoma[1, 2] and is found on the cell surface of myelogenous leukocytes.[3] Sialylated T-antigen is also found in mucins expressed by human breast cancer cell lines.[4] Synthetic glycopeptides bearing this disaccharide, including sialylated species, have the potential, therefore, to be effective immunogens in the stimulation of an immune response against cancer cells that express it on their surface.

Because of the clinical importance of Galβ1-3GalNAc, many organic[5] and enzymatic[6] reactions that synthesize T-antigen-containing glycopeptides have now been described. Organic chemical synthesis, however, requires special knowledge and technical expertise and therefore only experienced organic chemists have the ability to synthesize target glycopeptides.[7-9] In contrast, enzymatic synthesis does not need such expertise, and most biologists could replicate enzymatic synthesis procedures from published protocols in the literature. The present author has therefore summarized the reported enzymatic methods for the synthesis in this chapter. The enzymatic synthesis of glycopeptides can be classified into three different approaches: (1) the use of transferases,[10] (2) the use of exoglycosidases[6] and (3) the use of endoglycosidases.[11-13] The procedures that use either transferases or exoglycosidases are stepwise protocols, whereas, in contrast, endoglycosidase methods can produce glycopeptides in single-step reactions. Each method has both advantages and disadvantages. The most significant disadvantage of the method using transferase or exoglycosidase is that one must first synthesize a GalNAc-linked peptide before commencing the enzymatic elongation reaction using transferase or glycosidase. GalNAc-linked peptides can be generated by chemical synthesis, which is generally beyond the expertise of most biochemists. In contrast, the method using endoglycosidase shown below does not require the use of GalNAc-linked peptide as an acceptor, but free peptide alone can be used. Therefore, we believe that the method using endo-α-*N*-acetylgalactosaminidase (Endo-α-GalNAc-ase) for the synthesis of *O*-linked glycopeptides is the most acceptable to biochemists.

4.1.1 One-step Synthesis of *O*-Linked Glycopeptides Using Endo-α-GalNAc-ase

Among the methods employed in the synthesis of glycopeptides, strategies that use endo-glycosidases are the most attractive since entire sugar chains can be transferred in one-step reactions as opposed to the stepwise protocols required for chemoenzymatic synthesis. From this point of view, enzymatic transglycosylations have been studied using Endo-β-GlcNAc-ase A (Endo-A)[14-20] and M (Endo-M)[21] in the synthesis of *N*-linked glycopeptides. The syntheses of *N*-glycans using these enzymes are reviewed in Chapter 3 section 3.2, and in addition, the endo-β-xylosidase isolated from the mollusk *Patiopecten* has now been successfully used for the attachment of glycosaminoglycan chains to peptides.[22] This method successfully attached glycosaminoglycan chains directly to the target peptides using the transglycosylation activity of the enzyme, whereas transglycosylation using Endo-M requires GlcNAc-linked peptide as the acceptor.

In this chapter, one-step synthesis of *O*-glycans using Endo-α-GalNAc-ase is described. Very few reports have so far been published on synthetic approaches to the generation of glycopeptides using Endo-α-GalNAc-ase. Although several Endo-α-GalNAc-ases have been isolated and characterized for their substrate specificity, the corresponding transglycosylation reactions for the synthesis of *O*-linked glycopeptides have not yet been critically examined.

Transglycosylation using Endo-α-GalNAc-ase from *Streptococcus pneumoniae* and *Bacillus* sp. has, however, been reported by two groups and discussed below.

4.1.2 Transglycosylation Using Endo-α-GalNAc-ase from *Streptococcus pneumoniae*

Endo-α-GalNAc-ase from *Streptococcus pneumoniae* was originally characterized as an enzyme with specificity for the Galβ1-3GalNAc disaccharide, but recently this enzyme has been shown to have a much broader specificity[23] than previously thought.[24,25] In 1989, transfer activity of the Galβ1-3GalNAc residue was demonstrated by Bardales and Bhavanandan.[26] Treatment of asialoglycoproteins, having Galβ1-3GalNAcα1-Ser/Thr linkages, with Endo-α-GalNAc-ase-containing glycerol resulted in the formation of mainly Galβ1-3GalNAcα1-1glycerol. In this transglycosylation reaction, the anomeric configuration was retained and the same type of anomeric linkage was also formed as the one cleaved in the substrate, α-linkage in this instance. Similar enzymatic transglycosylation has also been observed with compounds having primary hydroxyl groups such as *p*-nitrophenol (*p*NP), threonine, serine, glucose, galactose and fucose. There was, however, no description of whether the hydroxyl groups of either serine or threonine residues in peptides could act as acceptors in the transglycosylation reactions.

4.1.3 Transglycosylation Using Endo-α-GalNAc-ase from *Bacillus* sp.

Ashida *et al.* isolated Endo-α-GalNAc-ase from the soil bacteria, *Bacillus* sp. A198 strain.[27] This enzyme also showed high specificity for Galβ1-3GalNAc and demonstrated aglycon specificity for *p*NP, benzyl alcohol and the serine/threonine residues of several proteins. When transglycosylation was conducted using Galβ1-3GalNAcα1-*p*NP as the donor and glucose as the acceptor, the reaction yielded the compounds Galβ1-3GalNAcα1-2Glc and Galβ1-3GalNAcα1-1α/βGlc. The enzyme also showed transfer activity for Galβ1-3GalNAc residues to the hydroxyl groups of various compounds including 1-alkanols,[28] glucose, galactose, mannose, sorbitol, maltose and sucrose in yields of 14-36%.

The representative transglycosylation procedure is as follows: 2 mM Galβ1-3GalNAcα1-*p*NP and 100 mU/mL of enzyme are incubated with 15% (v/v) 1-hexanol as the acceptor in the presence of 0.2% (v/v) sodium cholate. After incubation at 37°C for 3 h, the reaction mixture is analyzed by TLC using chloroform/ methanol/ water (65/35/8) as the developing solvent. The anomeric configurations of GalNAc were confirmed to be α-linkages by sequential digestion using β-galactosidase isolated from jack-bean and with α-GalNAc-ase from *Acremonium* sp. The transglycosylation product was shown to be completely hydrolyzed, and the anomeric structure of the transglycosylation product was confirmed to be of α-configuration.

It was not, however, shown whether the enzyme could transfer the disaccharides to the hydroxyl groups of either serine or threonine residue in peptides.

4.1.4 Transglycosylation Using Endo-α-GalNAc-ase from *Streptomyces* sp.

Endo-α-GalNAc-ase from *Streptomyces* sp. has been isolated and purified from the culture medium of *Streptomyces* sp. OH-11242. This enzyme was found to be capable of liberating not only Galβ1-3GalNAc but other larger oligosaccharides through the hydrolysis of the *O*-glycosidic linkage between GalNAc and Ser/Thr residues in peptides.[29-31] The substrate specificity of this enzyme was shown to be different from that of other Endo-α-GalNAc-ases such as those from *S. pneumoniae* and *Bacillus* sp. The details of the substrate specificity of this enzyme are described in Chapter 2 section 2.2.1.

The *p*NP glycoside of Galβ1-3GalNAc is a good substrate of this enzyme, and Galβ1-3GalNAcα1-*p*NP can consequently be used as a donor in the transglycosylation reaction. Galβ1-3GalNAcα1-*p*NP can be enzymatically synthesized quite easily *via* transglycosylation using Galβ1-*p*NP as the donor and GalNAcα1-*p*NP as the acceptor with the aid of recombinant β1-3galactosidase from *B. circulans,* as shown in Fig. 4.1.

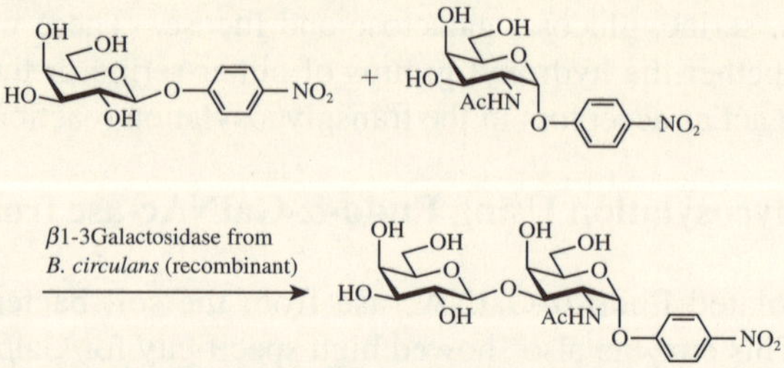

Fig. 4.1 Synthesis of Galβ1-3GalNAcα1-*p*NP by an enzymatic transglycosylation reaction using recombinant β1-3galactosidase from *B. circulans.*

Using Galβ1-3GalNAcα1-*p*NP as the donor, the transfer reaction of Galβ1-3GalNAc residues to the hydroxyl groups of serine or threonine residues in peptides was examined (Fig. 4.2).[12] A hexapeptide, Leu-Ser-Gln-Val-His-Arg, was used as the acceptor. This peptide is identical to the C-terminal sequence of FGF-5S (short form of Fibroblast Growth Factor) and has a stimulating effect, similar to that of FGF itself, on DNA synthesis in fibroblast cell lines.[32] The benzyloxycarbonyl (Cbz) group was attached to the *N*-terminal amino group in order to increase the hydrophobicity for the isolation of the reaction products with ODS-column attached HPLC.

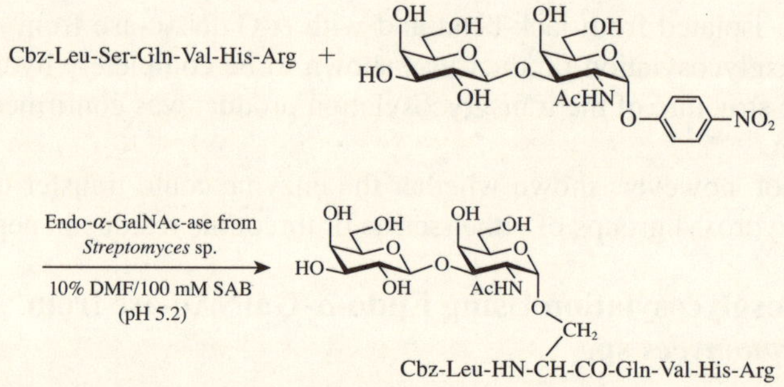

Fig. 4.2 Enzymatic synthesis of Galβ1-3GalNAc-linked hexapeptide *via* transglycosylation using Endo-α-GalNAc-ase from *Streptomyces* sp.

A. Method

Galβ1-3GalNAcα1-*p*NP (210 mg) and Cbz-linked FGF-fragment hexapeptide (Cbz-Leu-Ser-Gln-Val-His-Arg, 100 mg) were dissolved in 25 mL of 0.1 M sodium acetate buffer (pH 5.2) containing 10% DMF, and 10 units of partially purified Endo-α-GalNAc-ase from *Streptomyces* sp. (31.3 units/mL) was then added to the mixture. The solution was incubated at 37°C and reverse phase HPLC was performed at the appropriate time intervals. A representative HPLC chart is shown in Fig. 4.3. The reaction was stopped after 2 h by heating the solution in a boiling water bath for 3 min. The reaction mixtures were then applied to a preparative HPLC column (Mightysil RP-18, Kanto Chemicals Co., Tokyo, Japan, ϕ 3.5× 25 cm) and eluted with a gradient from 10% to 50% acetonitrile (120 min, 10

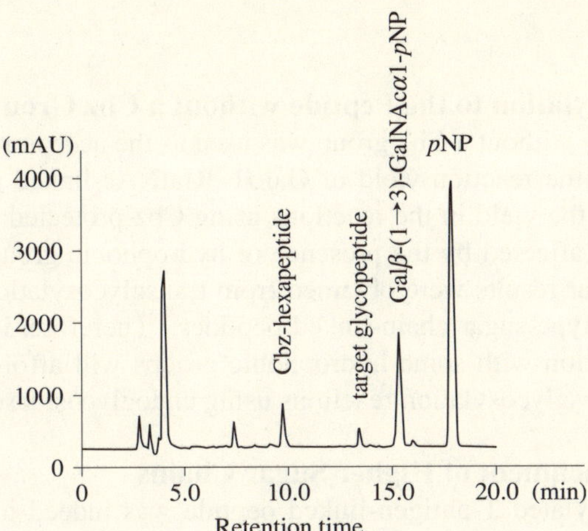

Fig. 4.3 HPLC chart of the transglycosylation reaction using Galβ1-3GalNAcα1-*p*NP and Cbz-hexapeptide in the presence of Endo-α-GalNAc-ase from *Streptomyces* sp.

mL/min) containing 0.1% trifluoroacetic acid to give 20 mg of Galβ1-3GalNAc-linked hexapeptide (yield; 13%).

B. Structure Determination

The structure of the transglycosylated product was confirmed by NMR spectroscopy. Each of the proton signals from both sugars and amino acids in the 500 MHz ^{1}H-NMR spectrum was assigned by a TOCSY spectrum initially and all of the cross peaks in the HSQC spectrum (Fig. 4.4) were then assigned. As $J_{1,2}$ of the GalNAc residue was 4.09 Hz and the C-1 chemical shift of GalNAc was 98 ppm (from N3 in Fig 4.4), the anomeric configuration was confirmed to be an α-

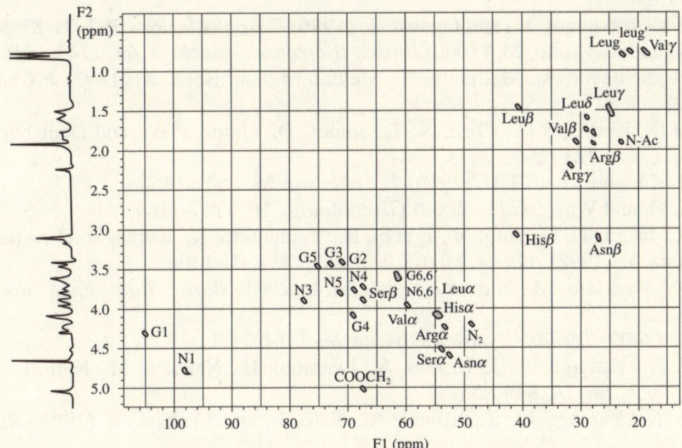

Fig. 4.4 HSQC spectrum of Cbz-Leu-(Galβ1-3GalNAcα1-)Ser-Gln-Val-His-Arg measured by a Varian Inova 500 MHz ^{1}H-NMR spectrometer. The axes of the NMR spectrum indicate the chemical shift of the ^{1}H (F_2) and ^{13}C (F_1) nuclei.

linkage.

C. Transglycosylation to the Peptide without a Cbz Group

When free peptide without a Cbz group was used as the acceptor in the transglycosylation reaction, the reaction yield of Galβ1-3GalNAc-linked peptide was 11%. As shown above, the yield in the reactions using Cbz-protected peptide was 13%. This may also be affected by the presence of hydrophobic groups in the acceptor molecules. Similar results were obtained from transglycosylations using Endo-M to form complex type sugar chain-linked peptides. Therefore, it can be generally stated that protection with some hydrophobic groups will afford better results in the enzymatic transglycosylation reactions using endoglycosidases.

D. For the Attachment of Higher Sugar Chains

Although the sialylated T-antigen-linked peptide was indeed obtained, the yield proved to be very low. The reason for this is not clear, but it may be that the reaction conditions are not yet fully optimized. Another reason for the low yield may be due to the difficulty in purifying the final product.

From the practical point of view, in order to successfully synthesize a sialyl T-antigen-linked peptide, it would be more fruitful to first synthesize a Galβ1-3GalNAc-linked peptide, then attach a sialyl group by transglycosylation using commercially available recombinant sialyltransferase from rat liver.

References

1. Yamashita, Y., Chung, Y., Horie, R., Kannagi, R., and Sowa, M.J. (1995) *J. Natl. Cancer Inst.* **87**, 441-446
2. Kurosaka, A., Nakajima, H., Funakoshi, J., Matsuyama, M., Nagayo, T., and Yamashina, I. (1983) *J. Biol. Chem.* **258**, 11594-11598
3. Fukuda, M., Carlsson, S.R., Klock, J.C., and Dell, A. (1986) *J. Biol. Chem.* **261**, 12796-12806
4. Hull, S.R., Bright, A., Carraway, K.L., Abe, M., Hayes, D.F., and Kufe, D.W. (1989) *Cancer Commun.* **1**, 261-267
5. Nakahara, Y., Nakahara, Y., and Ogawa, T. (1996) *Carbohydr. Res.* **292**, 71-81
6. Ajisaka, K. and Miyasato, M. (2000) *Biosci. Biotechnol. Biochem.* **64**, 1743-1746
7. Paulsen, H., Schleyer, A., Mathieux, N., Meldal, M., and Bock, K. (1997) *J. Chem. Soc., Perkin Trans.* **1**, 281-293
8. Schwarz, J.B., Kuduk, S.D., Chen, X.-T., Sames, D., Glunz, P.W., and Danishefsky, S.J. (1999) *J. Am. Chem. Soc.* **121**, 2662-2673
9. Osborn, H.M.I. and Khan, T.H. (1999) *Tetrahedron* **55**, 1807-1850
10. Koeller, K.M. and Wong, C.H. (2000) *Glycobiology* **10**, 1157-1169
11. Haneda, K., Inazu, T., Mizuno, M., Iguchi, R., Yamamoto, K., Kumagai, H., Aimoto, S., Suzuki, H., and Noda, T. (1998) *Bioorg. Med. Chem. Lett.* **8**, 1303-1306
12. Ajisaka, K., Miyasato, M., and Ishii-Karakasa, I. (2001) *Biosci. Biotechnol. Biochem.* **65**, 1240-1243
13. Ajisaka, K. (2002) *Trends Glycosci. Glycotechnol.* **14**, 1-11
14. Takegawa, K., Yamaguchi, S., Kondo, A., Iwamoto, H., Nakoshi, M., Kato, I., and Iwahara, S. (1991) *Biochem. Int.* **24**, 849-855
15. Takegawa, K., Yamaguchi, S., Kondo, A., Kato, I., and Iwahara, S. (1991) *Biochem. Int.* **25**, 829-835
16. Fan, J.-Q., Takegawa, K., Iwahara, S., Kondo, A., Kato, I., Abeygunawardana, C., and Lee, Y.C. (1995) *J. Biol. Chem.* **270**, 17723-17729
17. Fan, J.-Q., Quesenberry, M. S., Takegawa, K., Iwahara, S., Kondo, A., Kato, I., and Lee, Y.C. (1995) *J. Biol. Chem.* **270**, 17730-17735

18. Fan, J.-Q., Huynh, L.H., Reinhold, B.B., Reinhold, V.N., Takegawa, K., Iwahara, S., Kondo, A., Kato, I., and Lee, Y.C. (1996) *Glycoconj. J.* **13**, 643-652
19. Deras, I.L., Takegawa, K., Kondo, A., Kato, I., and Lee, Y.C. (1998) *Bioorg. Med. Chem. Lett.* **8**, 1763-1766
20. Takegawa, K., Fujita, K., Fan, J.-Q., Tabuchi, M., Tanaka, N., Kondo, A., Iwamoto, H., Kato, I., Lee, Y.C., and Iwahara, S. (1998) *Anal. Biochem.* **257**, 218-223
21. Haneda, K., Inazu, T., Yamamoto, K., Kumagai, H., Nakahara, Y., and Kobata, A. (1996) *Carbohydr. Res.* **292**, 61-70
22. Ishido, K., Takagaki, K., Iwafune, M., Yoshihara, S., Sasaki, M., and Endo, M. (2002) *J. Biol. Chem.* **277**, 11889-11895
23. Brooks, M.M. and Savage, A.V. (1997) *Glycoconj. J.* **14**, 183-190
24. Bhavanandan, V.P., Umemoto, J., and Davidson, E.A. (1976) *Biochem. Biophys. Res. Commun.* **70**, 738-745
25. Fan, J.-Q., Yamamoto, K., Hirabayashi, Y., Kumagai, H., and Tochikura, T. (1990) *Biochem. Biophys. Res. Commun.* **169**, 751-757
26. Bardales, R. and Bhavanandan, V.P. (1989) *J. Biol. Chem.* **264**, 19893-19897
27. Ashida, H., Yamamoto, K., Murata, T., Usui, T., and Kumagai, H. (2000) *Arch. Biochem. Biophys.* **373**, 394-400.
28. Ashida, H., Yamamoto, K., and Kumagai, H. (2001) *Carbohydr. Res.* **330**, 487-493
29. Ishii-Karakasa, I., Iwase, H., Hotta, K., Tanaka, Y., and Omura, S. (1992) *Biochem. J.* **288**, 475-482.
30. Ishii-Karakasa, I., Iwase, H., and Hotta, K. (1997) *Eur. J. Biochem.* **247**, 709-715
31. Tanaka, Y., Takahashi, Y., Shinose, M., Omura, S., Ishii-Karakasa, I., Iwase, H., and Hotta, K. (1998) *J. Ferment. Bioeng.* **85**, 381-387.
32. Zhan, X., Bates, B., Hu, X., and Goldfarb, M. (1988) *Mol. Cel. Biol.* **8**, 3487-3495

5

Enzymatic Synthesis of Neoproteoglycans

Proteoglycans are glycoconjugates whose core proteins have one to scores of high molecular glycosaminoglycan (GAG) chains attached with a covalent bond to the serine (Ser) residue. These macromolecules exist in cell surfaces and extracellular matrix (ECM), and have not only functions derived from the core protein but also various biological activities derived from this GAG structure.

GAGs, the sugar chains of proteoglycans, are known to have domain structures involved in various biological functions such as anticoagulation, cell proliferation and infection by *Plasmodium*.[1-4] Rapid progress has been made in the chemical synthesis of GAGs.[5, 6] The existing techniques, however, cannot synthesize a long GAG chain which shows biological activity. Recombinant protein produced by genetic engineering has no sugar chain or imperfect or different sugar chains from those of native protein.[7] Even if chemical synthesis could generate exactly the same sugar chains as the native sugar chains, there is no technique to introduce them into the recombinant protein. However, some techniques for introducing sugar chains into protein enzymatically have been developed.[8, 9]

It is important to prepare many kinds of GAG oligosaccharides as a library to elucidate correlations between biological functions and domain structures mentioned above. It is known that glycosidase has transglycosylation activity as a reverse reaction of the hydrolysis reaction. Recently, this activity has been used for the synthesis of oligosaccharides *in vitro*.[10, 11] Endoglycosidases release oligosaccharides by hydrolysis reaction, and transfer simultaneously these oligosaccharides to other molecule by transglycosylation reaction. In this case, a particular endoglycosidase acts like a restriction enzyme which recognizes specific linkages in sugar chains for cleavage of oligosaccharides and concurrently acts like a synthase or ligase which forms a specific bond.[12] The transglycosylation makes reconstruction of the sugar chain possible. Few endoglycosidases act on proteoglycans. At present, testicular hyaluronidase that acts on GAG chains and endo-β-xylosidase that acts on linkage structures between GAG chains and peptide in proteoglycans have been identified.[13-15] This chapter outlines the technique for synthesis of artificial chondroitin sulfate (ChS) oligosaccharides using the transglycosylation activity of testicular hyaluronidase, which is an endo-β-*N*-acetylhexosaminidase, and for attaching GAG chains into core proteins in proteoglycans using endo-β-xylosidase.

5.1 Enzymatic Synthesis of Neoglycans Using Hyaluronidase

5.1.1 Hydrolysis Reaction of Hyaluronidase

Testicular hyaluronidase (EC3.2.1.35, hyaluronidase) is an endo-*β*-*N*-acetylhexosaminidase that acts on the *N*-acetylglucosaminyl-glucuronic acid (GlcNAc*β*1-4GlcUA) structure in hyaluronan (HA) with tetrasaccharides and hexasaccharides as major terminal products[16, 17] (Fig. 5.1). This enzyme basically releases the disaccharides whose structure is GlcUA*β*1-3GlcNAc (or GalNAc, *N*-acetylgalactosamine) from the nonreducing end of long GAG chains by hydrolysis activity. When HA longer than hexasaccharides having a glucuronic acid (GlcUA) at the nonreducing end are used as the substrate for this enzyme, *N*-acetylglucosaminide (GlcNAc) bonds are hydrolyzed sequentially to disaccharides from the nonreducing end.[16] On the other hand, when HA longer than heptasaccharides having GlcNAc at the nonreducing end are used as the substrate, only the second *N*-acetylglucosaminide bond from the nonreducing end is hydrolyzed to selectively release trisaccharides.[16] Hyaluronidase also acts on the GalNAc*β*1-4GlcUA structure in chondroitin (Ch), chondroitin 4-sulfate (Ch4S) and chondroitin 6-sulfate (Ch6S)[18] (Table 5.1). In this case, this enzyme releases the disaccharides whose

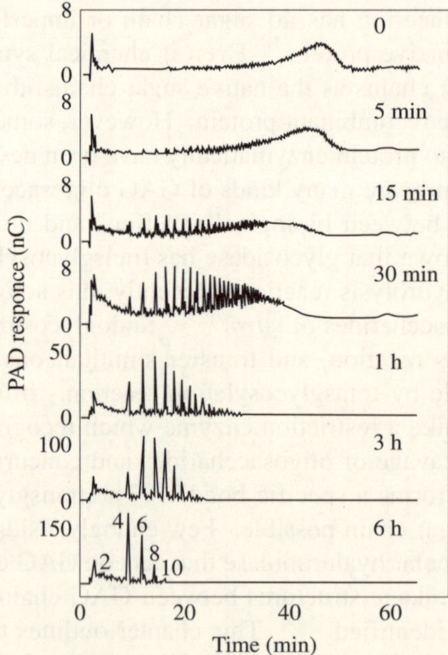

Fig. 5.1 High-performance anion-exchange chromatography with pulsed amperometric detection (HPAEC-PAD) of hydrolysis reaction products of hyaluronidase. HA was digested with bovine testicular hyaluronidase for the indicated times (0-6 h) at 37°C. The number for peaks 2, 4, 6, 8 and 10 indicates di-, tetra-, hexa-, octa-, and decasaccharides, respectively.

Table 5.1 The sites of hydrolysis and transglycosylation in GAG chain with hyaluronidase

I. Hyaluronan
  *GlcUAβ1-3GlcNAcβ1↓4GlcUAβ1-3GlcNAcβ1-.....
II. Chondroitin
  *GlcUAβ1-3GalNAcβ1↓4GlcUAβ1-3GalNAcβ1-.....
  Chondroitin 4-sulfate
  *GlcUAβ1-3GalNAc(4-OSO_3^-)β1↓4GlcUAβ1-3GalNAc(4-OSO_3^-)β1-.....
  Chondroitin 6-sulfate
  *GlcUAβ1-3GalNAc(6-OSO_3^-)β1↓4GlcUAβ1-3GalNAc(6-OSO_3^-)β1-.....
III. Dermatan sulfate
  IduUAα1-3GalNAc(4-OSO_3^-)β1-4IduUAα1-3GalNAc(4-OSO_3^-)β1-.....
  Desulfated dermatan sulfate
  *IduUAα1-3GalNAcβ1↓4IduUAα1-3GalNAcβ1-.....
IV. Chondroitin sulfate D
  *GlcUA(2-OSO_3^-)β1-3GalNAc(6-OSO_3^-)β1-4GlcUA(2-OSO_3^-)β1-3GalNAc(6-OSO_3^-)β1-.....
V. Chondroitin sulfate E
  *GlcUAβ1-3GalNAc(4-OSO_3^-, 6-OSO_3^-)β1↓4GlcUAβ1-3GalNAc(4-OSO_3^-, 6-OSO_3^- or none-sulfate)β1-.....

Vertical arrows show the sites of the hydrolysis of hyaluronidase. Asterisks show the regions that can be acceptors of the transglycosylation of hyaluronidase. Disaccharides released from the nonreducing end of GAG chains by hyaluronidase digestion are shown in gothic, and the sugar chains that can be transferred as donors are shown in hatched boxes.
[Reprinted with permission from Endo, M. *et al.* (2001) *Kikan Kagaku Sousetsu* **48**, 25]

structure is GlcUAβ1-3GalNAc from the nonreducing end of long GAG chains by hydrolysis. Hyaluronidase also hydrolyzes dermatan sulfate (DS), although this activity is low. Optimal conditions for hydrolysis reaction of hyaluronidase are pH 4.0-5.0 with NaCl.

5.1.2 Transglycosylation with Hyaluronidase as an Endoglycosidase

Simultaneously with this hydrolysis reaction, this enzyme can also transfer new disaccharides to GlcUA at the nonreducing end of the acceptor, GAG, by forming β1-4 bonds.[16, 19, 20] Based on this activity, hyaluronidase can sequentially transfer new disaccharides at the nonreducing end of GAG chains, and as a result, GAG chains with biological function can be restructured (Fig. 5.2).

A. Detection of the Reaction Products by Hydrolysis and Transglycosylation of Hyaluronidase

The product of the transglycosylation reaction, as well as that of hydrolysis, can be easily detected by gel filtration HPLC, and also be analyzed with a mass spectrometer.[16, 20, 21] Fig. 5.3 and Table 5.2 show one example of HPLC. Labeling with 2-aminopyridine (PA) at the reducing end of the GAG oligosaccharide[22-24] makes it easy to detect transglycosylation products, respectively identify donors, acceptors and products, and effectively prevent side reaction. Structures at the nonreducing end of donors and acceptors can be confirmed by digestion with exoglycosidases such as β-glucuronidase, α-L-iduronidase, or β-*N*-acetylhexosaminidases,[20] and can be analyzed with a mass spectrometer.[25] One example of

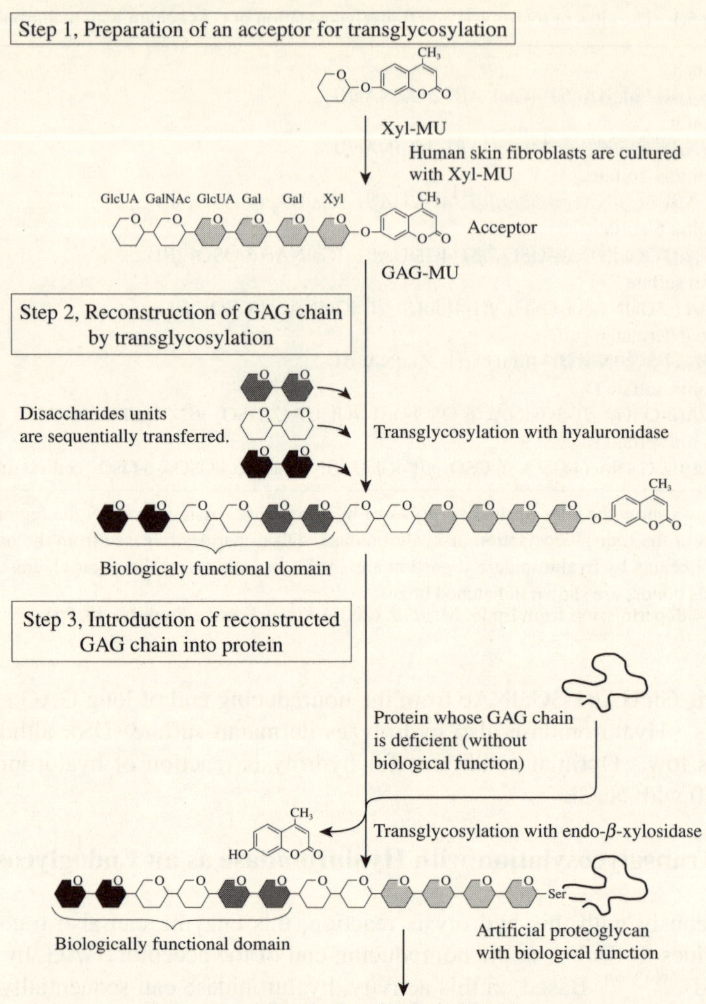

Fig. 5.2 Strategy for reconstruction of the glycosaminoglycan chain and introducing it into protein. Step 1 (Preparation of an acceptor for transglycosylation): GAG chain attached to MU is elongated when Xyl-MU as an initiator for GAG synthesis is added in the culture conditioned medium of cultured human skin fibroblasts. Step 2 (Reconstruction of GAG chain by transglycosylation): Purified GAG-MU as an acceptor and any kind of GAG as a donor is incubated with hyaluronidase, then the disaccharide unit is sequentially transferred to the acceptor. Step 3 (Introduction of reconstructed GAG chain into protein): Transglycosylated product is introduced into Ser residues in protein whose GAG chain is deficient. MU, 4-methylumbelliferone; Xyl, xylose; PA, 2-aminopyridine; GAG, glycosaminoglycan.
[Reprinted with permission from Endo, M. *et al.* (2001) *Kikan Kagaku Sousetsu* **48**, 26]

the quantitative change of the hydrolysis product analyzed by a mass spectrometer is shown (Fig. 5.3B and C). Structural information on the newly transglycosylated GAG chain and its linkage can be confirmed by HPLC analysis of the unsaturated disaccharide generated by eliminase digestion with chondroitin AC-II lyase or ABC lyase, *etc.*

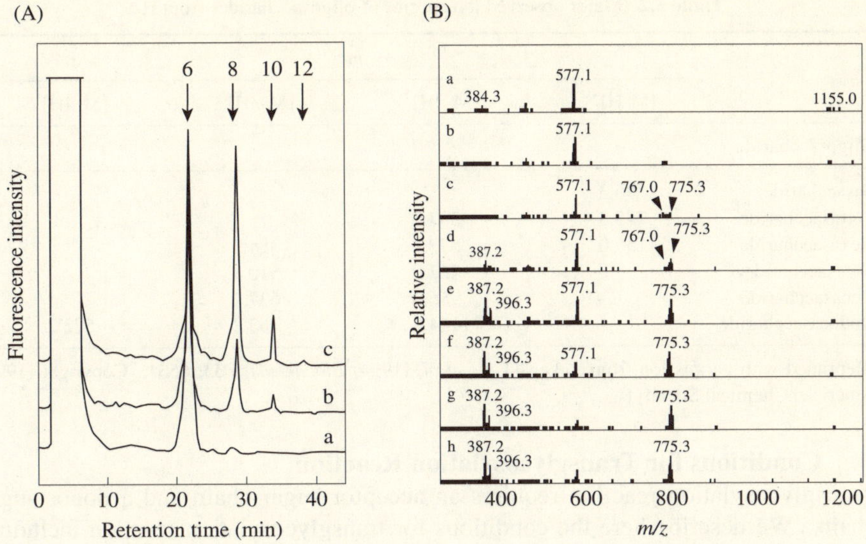

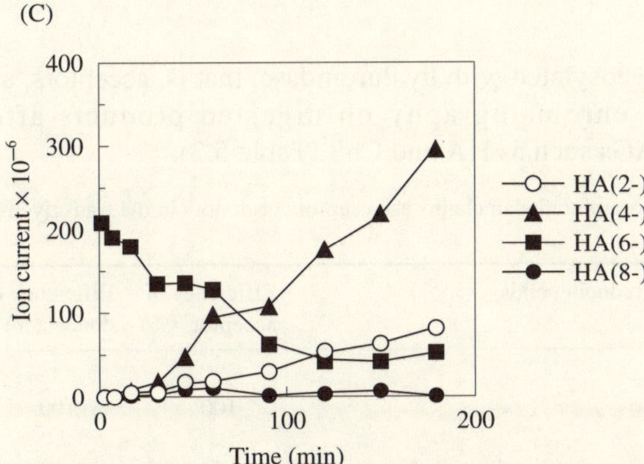

Fig. 5.3 Analysis of the reaction products by HPLC and mass spectrometry.
(A) HPLC of transglycosylation products released by hyaluronidase. PA- and non-PA-labeled hexasaccharides, which were used as the acceptor and donor, respectively, were incubated with hyaluronidase for 0 (a), 5 (b) and 30 min (c) in the same buffer system at 37°C and the reaction products were analyzed by HPLC on a Palpak type S column. Arrows indicate the elution position of various sizes of PA-HA oligosaccharides: 6, hexa-; 8, octa-; 10, deca-; 12, dodecasaccharide. (B) Ion-spray mass spectra of hyaluronidase digests. Hexasaccharide from HA was incubated with hyaluronidase for 0 min (a), 5 min (b), 30 min (c), 60 min (d), 90 min (e), 120 min (f), 150 min (g) and 180 min (h) in 0.2 M sodium acetate buffer, pH 5.0, at 37°C. An aliquot of the reaction mixture was then subjected to ion-spray mass spectrometry. Each oligosaccharide was quantified by integrating all ion intensities (C).
[Reprinted with permission from Takagaki, K. *et al.* (1994) *Biochemistry* **33**, 6504-6506, Copyright (1994) American Chemical Society]

Table 5.2 Major observed ion species of oligosaccharides from HA

	m/z			
	[M-H]$^-$	[M-2H]$^{2-}$	[M-3H]$^{3-}$	[M-4H]$^{4-}$
Oligosaccharide				
Disaccharide	396.3			
Tetrasaccharide	775.3	387.2		
Hexasaccharide	1155.0	577.1	384.3	
Octasaccharide		767.0	510.8	
Decasaccharide		956.0	637.2	
Dodecasaccharide		1144.8	763.2	572.2

[Reprinted with permission from Takagaki, K. *et al.* (1994) *Biochemistry* **33**, 6504, Copyright (1994) American Chemical Society]

B. Conditions for Transglycosylation Reaction

Transglycosylation reaction requires an acceptor sugar chain and a donor sugar chain. We describe here the conditions for transglycosylation reaction including acceptors and donors.

1) Acceptors

The sugar chains to be transglycosylated with hyaluronidase, that is, acceptors, are prepared by gel filtration chromatography on digested products after hyaluronidase treatment of GAGs such as HA and ChS (Table 5.3).

Table 5.3 Efficiency of various glycosaminoglycan chains as acceptors or donors in the transglycosylation by hyaluronidase

Structures of the sugar residues at the reducing ends of acceptors or donors		Efficiency of acceptor[a] (%)	Efficiency of donor[b] (%)
Hyaluronan			
GlcUAβ1-3GlcNAc-		100[c]	100
Chondroitin sulfate			
GlcUAβ1-3GalNAc-	(Chondroitin)	72.2	68.1
GlcUAβ1-3GalNAc (4-OSO_3^-)-	(Chondroitin 4-sulfate)	51.4	48.4
GlcUAβ1-3GalNAc (6-OSO_3^-)-	(Chondroitin 6-sulfate)	47.6	18.4
GlcUAβ1-3GalNAc (4-OSO_3^-, 6-OSO_3^-)-	(Chondroitin sulfate E)	34.1	–
Dermatan sulfate			
IduUAα1-3GalNAc (4-OSO_3^-)-	(Dermatan sulfate)	0	0
IduUAα1-3GalNAc-	(Desulfated dermatan sulfate)	–	10.9

[a]PA-HA (hexasaccharides) and [b]HA were used as acceptor and donor, respectively. [c]The activity shows the relative rate of the reaction product when the transglycosylated products using PA-HA (hexasaccharides) as the acceptor and hyaluronan as the donor is 100%.
[Reprinted with permission from Endo, M. *et al.* (2001) *Kikan Kagaku Sosetsu* **48**, 27]

Chain length of acceptors The minimum length of a GAG oligosaccharide as an acceptor is a tetrasaccharide. In the case of an acceptor labeled with PA, however, the minimum length is a hexasaccharide.[26] If the chain length of an acceptor is long, it is hydrolyzed during the transglycosylation reaction. The length of the acceptor must be shorter than that of the donor.

Sugar at the nonreducing end The nonreducing end of an acceptor must be a GlcUA if the transglycosylation is defined as the reverse reaction of the hydrolysis of hyaluronidase. In the case of DS as the substrate, disaccharide at the nonreducing end, IduUAα1-3GalNAc, was released if the disaccharide in tandem at the nonreducing end was desulfated completely. Therefore, it was shown that GAG chains with IduUA whose *N*-acetylgalactosamine does not have a sulfate group at the nonreducing end can be an acceptor[27)] (Table 5.3). This also demonstrated that a DS oligosaccharide also can be a substrate for an enzymatic reconstruction of GAG using hyaluronidase.

Sulfate group Since the position of the sulfate group in GAG has major implications for its biological activity,[1, 3)] the position of the sulfate group in an acceptor or donor is extremely important.[20)] Transglycosylation using nonsulfated oligosaccharides such as HA or Ch as both acceptor and donor is easy (Table 5.3). Transglycosylation reaction proceeds even if there are two sulfate groups in the *N*-acetylgalactosamine that is one of the components of the disaccharide unit.

Transglycosylation using GlcUA-GalNAc(6-OSO_3^-) is more difficult than using GlcUA-GalNAc(4-OSO_3^-). In DS, the transglycosylation reaction proceeds only after desulfation of a sulfate group at position 4 in *N*-acetylgalactosamine bound to IduUA. The efficiency of transglycosylation when various GAGs were used for the acceptors and the donors is shown in Table 5.3.

2) Donors

Disaccharide at the nonreducing end in a donor can be transferred to GlcUA at the nonreducing end in an acceptor immediately after the hexosaminide bond of hexuronic acid β1-3*N*-acetylhexosamines is hydrolyzed by hyaluronidase.[16)] Therefore, an *N*-acetylhexosaminide bond is an essential structure for hyaluronidase to act (Tables 5.1 and 5.3).

Chain length of donors Donors must be longer than acceptors, and the longer the better the efficiency of transglycosylation. The structure of the disaccharide at the nonreducing end in the donor should be determined.

Structure of a sugar chain as a donor It is easy to perform transglycosylation using HA and Ch that does not have any sulfate group in *N*-acetylhexosamine as the donor.[20)] Transglycosylation proceeds comparatively efficiently when Ch4S is used as the donor, and inefficiently with Ch6S. The efficiency of transglycosylation depends on the number and position of the sulfate groups in the disaccharide unit at the nonreducing end of the donor. Transglycosylation cannot proceed using disaccharide that has two sulfate groups as a donor.

Odd-numbered oligosaccharides as a tool for the termination reaction of transglycosylation The transglycosylation from oligosaccharides with GlcNAc at the nonreducing end was examined.[16)] Odd-numbered HA, heptasaccharides having a GlcNAc at the nonreducing end, were prepared by removing GlcUA with β-glucuronidase digestion of even-numbered HA. These heptasaccharides as donors and PA-hexasaccharides as acceptors were incubated with hyaluronidase, then

PA-nonasaccharides were obtained as transglycosylation products. When a nonasaccharide, which is a longer odd-numbered HA prepared in the same way, was used as the donor, PA-nonasaccharides were obtained as transglycosylation products (Fig. 5.4). These showed that when an oligosaccharide having a GlcNAc at the nonreducing end is used as a donor, the trisaccharide GlcNAc-GlcUA-GlcNAc is transferred to an acceptor. These results suggest that the sugar residue at the nonreducing end of the transglycosylation product becomes GlcNAc after the reaction by transferring the trisaccharides, ending the extension of the sugar chain. Therefore, we have since been using oligosaccharides having this structure to terminate the elongation of sugar chains.

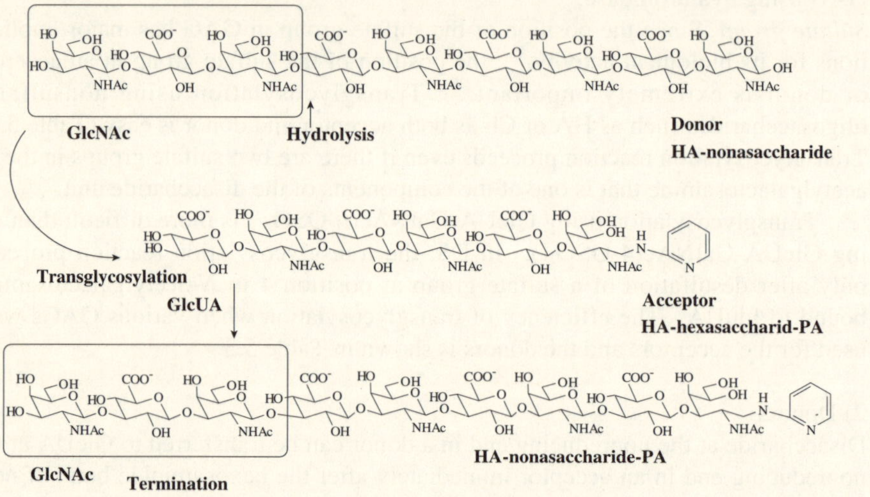

Fig. 5.4 Termination of transglycosylation using odd-numbered HA oligosaccharide as a donor. Transglycosylation by hyaluronidase was performed using nonasaccharide- and hexasaccharide-PA as donor and accepter, respectively. When an oligosaccharide having a GlcNAc at the nonreducing end is used as a donor, the trisaccharide, GlcNAc-GlcUA-GlcNAc, is transferred to an acceptor. The sugar residue at the nonreducing end of the transglycosylation product becomes GlcNAc, which is a structure to terminate the elongation of sugar chains.

3) pH

Figure 5.5 shows the effect of pH on the reactions of hydrolysis and transglycosylation of hyaluronidase. High molecular weight HA as the donor and PA-HA hexasaccharide as the acceptor was incubated with hyaluronidase at 37°C for 1 h at various pH values, and the reaction products were analyzed by HPLC. As a result, the optimal pH for sugar chain elongation was found to be pH 7.0, different from the optimal pH for hydrolysis pH 4.0. The maximum size of the transglycosylated product detected at this pH was docosasaccharide. The hydrolysis reaction of hyaluronidase requires NaCl, while the transglycosylation by hyaluronidase is inhibited in the presence of NaCl. The hydrolysis reaction progresses even under the optimal condition for this transglycosylation. Longer incubation does not always achieve longer GAG chains.[20] If conditions are controlled well, it is possible to increase the yield of transglycosylation reaction product.

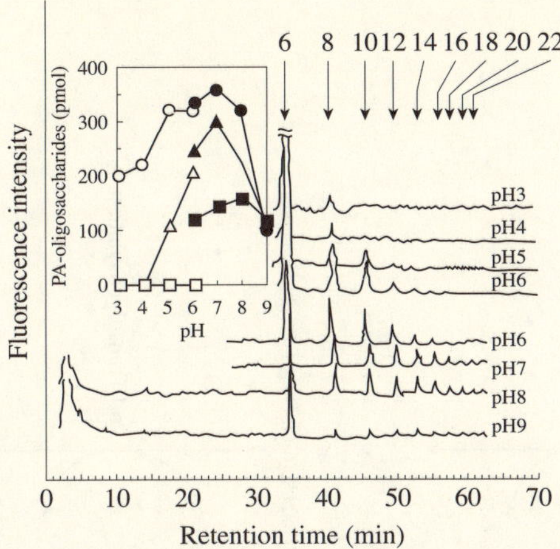

Fig. 5.5 PA-oligosaccharides produced by the transglycosylation reaction of testicular hyaluronidase at various pH values. PA-hexasaccharide as acceptor and HA as donor were incubated with hyaluronidase at 37°C for 1 h in sodium acetate buffer within the pH range of 3.0-6.0 and in Tris-HCl buffer, pH range of 6.0-9.0. The reaction mixtures were then subjected to HPLC on a PALPAK Type S column with fluorescence detection. The amount of each reaction procuct was calculated on the basis of fluorescence. Arrows indicate the elution positions of PA-HA oligosaccharide standards (6, hexa-; 8, octa-; 10, deca-; 12, dodeca-; 14, tetradeca-; 16, hexadeca-; 18, octadeca-; 20, eicosa-; 22, docosasaccharide). Inset, amounts of reaction products formed by the transglycosylation reaction in Fig. 5.4 were plotted. Open circle, octa-; open triangle, deca-; open square, hexadecasaccharide in sodium acetate buffer. Closed circle, octa-; closed triangle, deca-; closed square, hexadecasaccharide in 0.15 M Tris-HCl buffer. PA, 2-aminopyridine.
[Reprinted with permission from Saitoh, H. *et al.* (1995) *J. Biol. Chem.* **270**, 3743]

5.1.3 Reconstruction of GAG Chains Using Hyaluronidase

Hyaluronidase acts also on GAGs other than HA as substrates as mentioned above (section B1), suggesting GAGs other than the HA can be transferred. Various combinations of GAGs as donors and acceptors make it possible to synthesize various GAGs having *de novo* structure. Schematic synthesis of artificial proteoglycans using hyaluronidase is illustrated in Fig. 5.6. Ch4S disaccharide and Ch disaccharide was transferred to PA-Ch6S hexasaccharide sequentially (Fig. 5.6). The initial transglycosylation with hyaluronidase was performed using PA-Ch6S as the acceptor and Ch4S as the donor. The second transglycosylation was performed using this octasaccharide as the acceptor and Ch decasaccharide as the donor. As a result, an octasaccharide composed of GlcUAβ1-3GalNAc (Ch), GlcUAβ1-3GalNAc(4-OSO_3^-) (Ch4S), and GlcUAβ1-3GalNAc(6-OSO_3^-) (Ch6S) units was obtained. The reaction product in each step was collected by HPLC and its structure was analyzed by the exoglycosidase digestion method and mass spectrometry. Although the yield of each transglycosylation step was not so good, we succeeded in the enzymatic synthesis of GAGs as designed.

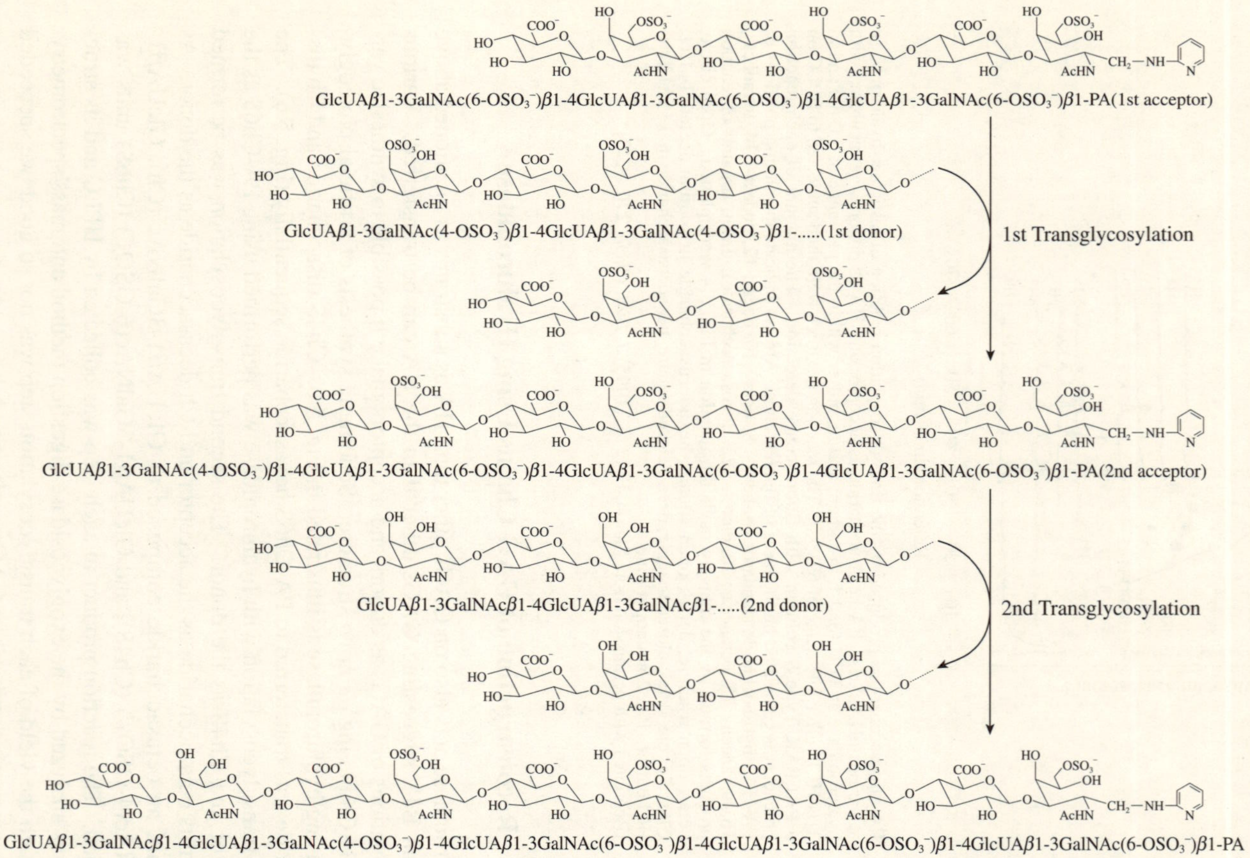

GlcUAβ1-3GalNAc(6-OSO_3^-)β1-4GlcUAβ1-3GalNAc(6-OSO_3^-)β1-4GlcUAβ1-3GalNAc(6-OSO_3^-)β1-PA(1st acceptor)

GlcUAβ1-3GalNAc(4-OSO_3^-)β1-4GlcUAβ1-3GalNAc(4-OSO_3^-)β1-.....(1st donor)

1st Transglycosylation

GlcUAβ1-3GalNAc(4-OSO_3^-)β1-4GlcUAβ1-3GalNAc(6-OSO_3^-)β1-4GlcUAβ1-3GalNAc(6-OSO_3^-)β1-4GlcUAβ1-3GalNAc(6-OSO_3^-)β1-PA(2nd acceptor)

GlcUAβ1-3GalNAcβ1-4GlcUAβ1-3GalNAcβ1-.....(2nd donor)

2nd Transglycosylation

GlcUAβ1-3GalNAcβ1-4GlcUAβ1-3GalNAc(4-OSO_3^-)β1-4GlcUAβ1-3GalNAc(6-OSO_3^-)β1-4GlcUAβ1-3GalNAc(6-OSO_3^-)β1-4GlcUAβ1-3GalNAc(6-OSO_3^-)β1-PA

Fig. 5.6 Scheme of the reconstruction method for the hybrid disaccharide having a GlcUA-GalNAc-GlcUA-GalNAc4S-GlcUA-GalNAc6S sequence positioned at the nonreducing end. GlcUA, glucuronic acid; GalNAc, *N*-acetylgalactosamine; PA, 2-aminopyridine, Ch, chondroitin; ChS, chondroitin sulfate. 0S, 4S and 6S indicate the disaccharide unit of chondroitin, chondroitin 4-sulfate and chondroitin 6-sulfate, respectively.

5.1.4 GAG Library

The GAG library consists of over 100 kinds of GAGs synthesized by enzymatic reconstruction of each disaccharide unit in HA, Ch, ChS and DS to date. Examples are shown in Fig. 5.7. Transglycosylation using endoglycosidase made it possible to reconstruct GAG chains by oligosaccharide units. A long GAG chain of at least eicosasaccharide length that has biological function, which is difficult to synthesize by current chemical synthesis, can be synthesized by this technique. GAG chains that are neither found in nature nor synthesized by chemical reaction have been synthesized by transglycosylation reaction with hyaluronidase. We have found that ChSE, GlcUAβ1-3GalNAc(4-OSO_3^-,6-OSO_3^-), has specific affinity to type V collagen.[28] In order to find the domain structure on ChSE required for this affinity, oligosaccharides obtained from ChSE by hyaluronidase digestion were purified, after which many kinds of oligosaccharides having certain sequences (octa- or hexasaccharides having different numbers and positions of sulfate groups) were enzymatically reconstructed. The affinity between these oligosaccharides and type V collagen was examined by surface plasmon resonance biosensor (BIA core), and the domain sequence on ChSE required for the affinity was determined.[29] This is one example of an application using the sugar chains reconstructed by hyaluronidase.

5.1.5 Construction of Neoglycan Attached to Core Protein by Transglycosylation with Hyaluronidase

Decorin is a small and simple proteoglycan that has only one DS chain linked to its core protein and is abundantly found in osseous tissue.[30, 31] This molecule has a high affinity to hydroxyapatite and collagen in osseous tissue and plays a key role in regulating calcification of bone. It is also widely distributed in other tissues and is suggested to be involved in many biological events, such as morphosis, tumor progression and wound healing, interacting with other molecules, *i.e.*, growth factors and adhesion molecules. Recently, we succeeded in rearranging nonreducing ends of ChS oligosaccharide chains linked to the core protein into HA or ChS with various patterns of sulfate groups by transglycosylation with hyaluronidase using ChS oligosaccharide chains with decorin core proteins as acceptors[32] (Fig. 5.8). It will be interesting to examine whether neoglycan attached to a core protein has new functions or specific affinity to interesting molecules, based on this new GAG structure.

It is also possible to introduce a sugar chain obtained by the transglycosylation into the protein by using endo-β-xylosidase,[33] as described below.

The techniques for reconstruction of GAGs using hyaluronidase and for introduction of sulfate groups into GAGs took us one step closer to our dream of giving proteoglycans a new biological function. The biological importance of such artificially synthesized proteoglycans will no doubt increase in the not-too-distant future.

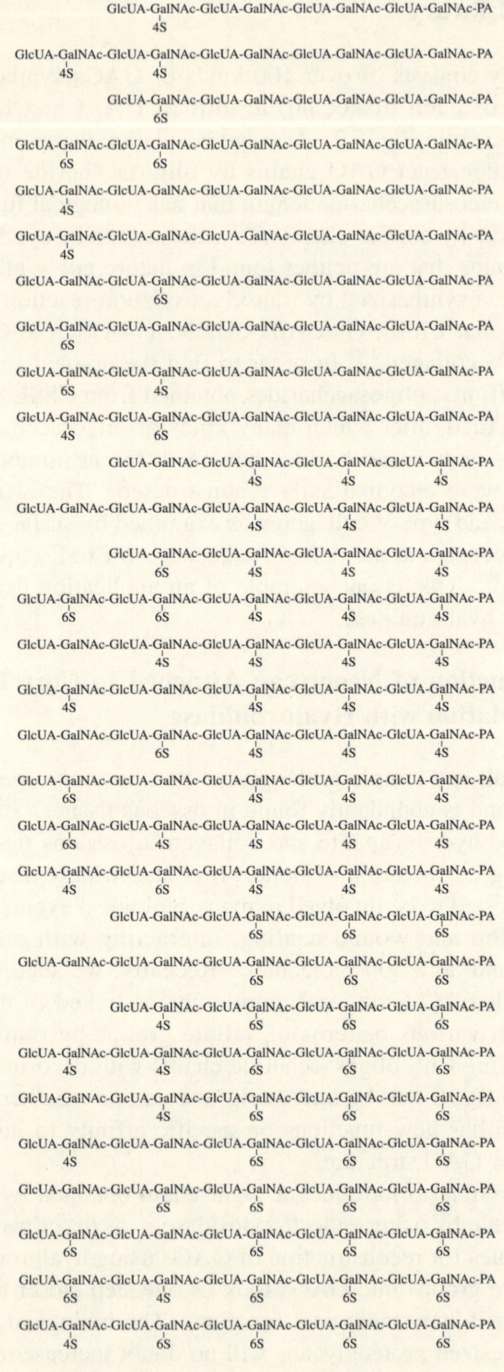

Fig. 5.7 Chondroitin sulfate oligosaccharide library.
(A) Nonsulfated and disulfated disaccharide units;

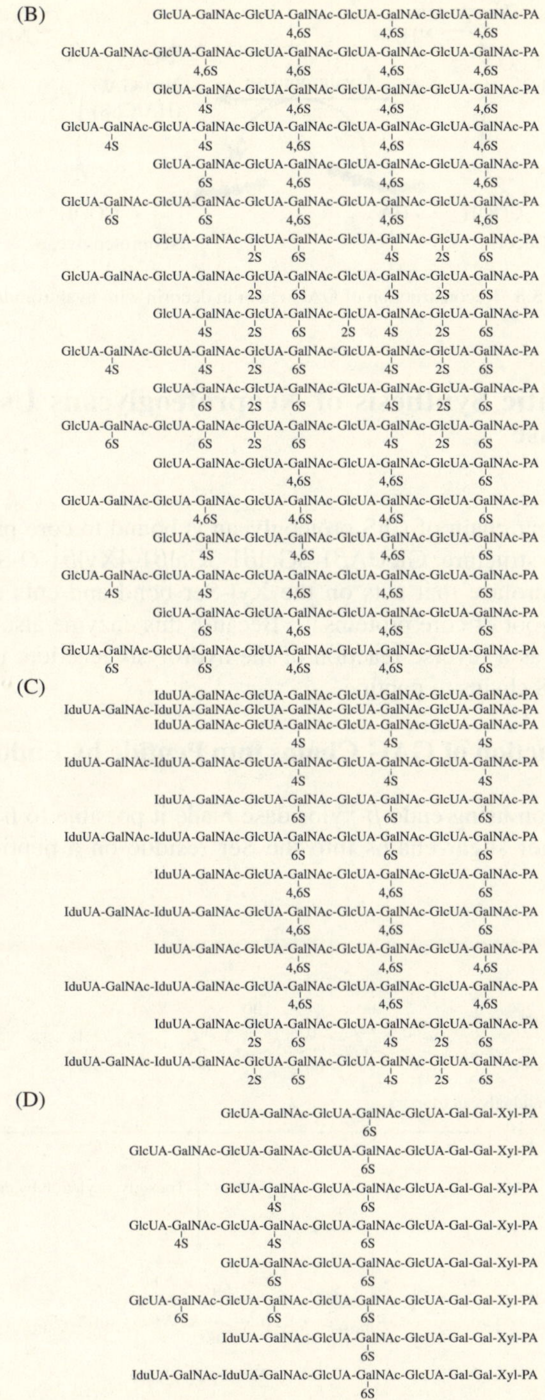

Fig. 5.7 Chondroitin sulfate oligosaccharide library. (B) polysulfated disaccharide units; (C) iduronic acid containing disaccharide units; (D) linkage region containing disaccharide units. [Reprinted with permission from Takagaki, K. *et al.* (2000) *Trends Glycosci. Glycotechnol.* **12**, 303]

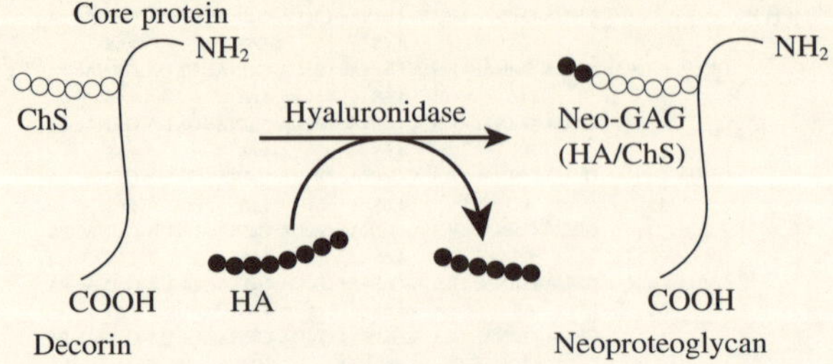

Fig. 5.8 Reconstruction of GAG chain in decorin with hyaluronidase.

5.2 Enzymatic Synthesis of Neoproteoglycans Using Endo-β-xylosidase

ChS that is a sugar chain of ChS proteoglycan is bound to core protein through a common linkage structure, GlcUAβ1-3Galβ1-3Galβ1-4Xylβ1-*O*-Ser. Endo-β-xylosidase is a hydrolase that acts on the Xyl-Ser bond and cuts out intact GAG chains from the root of core proteins.[14)] Because this enzyme also has transglycosylation activity as a reverse reaction of the hydrolysis reaction, it may be useful for attaching ChS chains to peptide.

5.2.1 Introduction of GAG Chains into Peptide by Endo-β-xylosidase

Transglycosylation using endo-β-xylosidase made it possible to transfer oligosaccharides or longer sugar chains into the Ser residue on a peptide[33)] (Fig. 5.9).

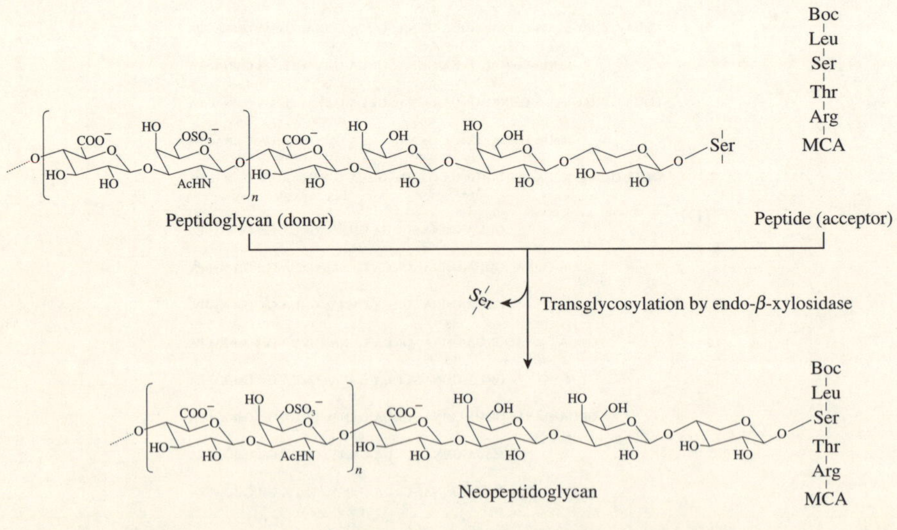

Peptidoglycan (donor)

Peptide (acceptor)

Transglycosylation by endo-β-xylosidase

Neopeptidoglycan

Fig. 5.9 Scheme of transglycosylation with endo-β-xylosidase.
Boc, *t*-butyloxycarbonyl; MCA, 4-methylcoumaryl-7-amide.

Either native ChS-GlcUAβ1-3Galβ1-3Galβ1-4Xylβ1-Ser structure or artificial ChS-GlcUAβ1-3Galβ1-3Galβ1-4Xylβ1-MU structure is acceptable for the introduction of sugar chains. Any ChS (Ch, Ch4S and Ch6S), DS, or heparan sulfate (HS), is acceptable as a donor (Table 5.4). The efficiency was ChS > DS > HS. HA, which does not have a protein moiety, did not transfer to the Ser residue on a peptide at all (Table 5.4). Although transglycosylation reaction products increased with time, the products were hydrolyzed and decreased gradually from 12 h after incubation. The optimum pH for the transglycosylation was 3.0, and it differed from the optimum pH 4.0 of the hydrolysis reaction of this enzyme. The donor should be a peptidoglycan attached to at least linkage hexasaccharides. The longer the donor GAG sugar chain is, the higher the efficiency of the transglycosylation (Table 5.5). When only sugar moiety, which is a long chain and has a linkage region, was used as the donor, the transglycosylation reaction did not proceed (Table 5.5), suggesting that the hydrolysis between a core protein and a GAG chain by the same enzyme is necessary for transglycosylation. This enzyme did not transfer an oligosaccharide to the Thr residue in a peptide, but transferred it to the Ser residue. Therefore, it was confirmed that long GAGs could be transferred to the peptide from GAGs that have a structure of Xyl-Ser or Xyl-MU.

Table 5.4 Effects on the GAG chain transfer reaction of various GAGs used as donors. Each peptidoglycan used as donor and [a]Boc-Leu-Ser-Thr-Arg-[b]MCA, a synthesized peptide, were incubated with endo-β-xylosidase under the typical conditions of transglycosylation. ChS, chondroitin sulfate; DS, dermatan sulfate; HS, heparan sulfate; HA, hyaluronan.

Donor	Relative activity[c] (%)
ChS (bovine tracheal cartilage, 30 kDa)	100
DS (pig skin, 32 kDa)	75
HS (bovine lung, 19 kDa)	43
HA (bacteria, 41 kDa)	0

[a] Boc, *t*-butyloxycarbonyl; [b] MCA, 4-methylcoumaryl-7-amide.
[c] The activity shows the relative rate of the reaction product when the products of transglycosylation products using intact peptide-ChS as the donor is 100%.
[Reprinted with permission from Ishido, K. *et al.* (2002) *J. Biol. Chem.* **277**, 11894]

Table 5.5 Effects of the length of GAG chain on transglycosylation. Each donor and Boc-Leu-Ser-Thr-Arg-MCA, a synthesized peptide, were incubated with endo-β-xylosidase under the typical conditions of transglycosylation. Boc, *t*-butyloxycarbonyl; MCA, 4-methylcoumaryl-7-amide; GlcUA, glucuronic acid; GalNAc, *N*-acetylgalactosamine; Gal, galactose; Xyl, xylose; Δ unsaturated.

Donor	Relative activity[a] (%)
(GlcUA-GalNAc)$_n$-GlcUA-GalNAc-GlcUA-Gal-Gal-Xyl-peptide	100
ΔGlcUA-GalNAc-GlcUA-Gal-Gal-Xyl-peptide	25
ΔGlcUA-Gal-Gal-Xyl-peptide	15
Xyl-peptide	0
(GlcUA-GalNAc)$_n$-GlcUA-GalNAc-GlcUA-Gal-Gal-Xyl	0

[a] The activity shows the relative rate of the reaction product when the product of transglycosylation using intact peptide-ChS as the donor is 100%.
[Reprinted with permission from Ishido, K. *et al.* (2002) *J. Biol. Chem.* **277**, 11894]

We experimentally examined the sensitivity of synthesized peptide to protease changes by introducing a GAG sugar chain as an application of this technique. Synthetic peptide with a GAG chain introduced by transglycosylation using endo-β-xylosidase was incubated with activated Protein C, a kind of protease. Interestingly, the sensitivity of this peptide to activated Protein C was low compared to the peptide without a GAG chain used as the control. This result suggests that sugar chains linked to proteoglycans function to protect the core protein. Using this technology in studies on proteoglycans, it may be possible to clarify the reason for GAG chains attaching to proteoglycans.

ChS chains including this linkage structure elongate from Xyl-MU as a primer when human skin fibroblasts are cultured with Xyl-MU.[34)] This product, ChS-MU, is collected and digested by hyaluronidase, then hexasaccharide-MU including the linkage region between the core protein and the ChS chain is obtained. GlcUA is located at the nonreducing end of the hexasaccharide-MU and this oligosaccharide can be an acceptor for transglycosylation (Fig. 5.2). The technique for transfer of Ch, Ch4S and Ch6S to hexasaccharide-MU as we designed has been established using transglycosylation of hyaluronidase. Therefore, it is possible to reconstruct an arbitrary oligosaccharide that has a certain specific activity on extension of the nonreducing end of the linkage structure with hyaluronidase and then to introduce it into a peptide by transglycosylation activity of endo-β-xylosidase. In consequence, the peptide may serve with a new activity as a proteoglycan whose oligosaccharide has been rearranged (Fig. 5.2).

5.2.2 Cellulase That Plays an Alternative Role as Endo-β-xylosidase

Most cellulases are endoglycosidases that act on the glucosyl β1-4glucose structure in the cellulose of a long chain.[35)] Recently it was found that a certain kind of cellulase can hydrolyze the Xyl-Ser structure in proteoglycan and liberate intact GAG chains.[36)] Therefore it was expected that the cellulase could transfer the long GAG chains to the Ser residue in peptidoglycan simultaneously with the hydrolysis like the endo-β-xylosidase from *Patinopecten yessoensis*.[14)] These findings led to the idea that certain kinds of cellulase and other glycosidases also have endo-β-xylosidase activity and can be applied to glycotechnology.

It was clarified that endo-β-xylosidase has transglycosylation activity which transfers an oligosaccharide to Ser residue in peptide. By using this enzyme, it became possible to introduce an oligosaccharide into the Ser residue in a genetically engineered protein whose sugar chains are deficient. How to selectively introduce a GAG chain into a given Ser residue remains to be solved.

References

1. Casu, B., Oreste, P., Torri, G., Zoppetti, G., Choay, J., Lormeau, J.C., Petitou, M., and Sinay, P. (1981) *Biochem. J.* **197**, 599-609
2. Thunberg, L., Bäckström, G., and Lindahl, U. (1982) *Carbohydr. Res.* **100**, 393-410
3. Maimone, M.M. and Tollefsen, D.M. (1990) *J. Biol. Chem.* **265**, 18263-18271
4. Alkhalil, A., Achur, R.N., Valiyaveettil, M., Ockenhouse, C.F., and Gowda, D.C. (2000) *J. Biol.*

Chem. **275**, 40357-40364
5. Petitou, M., Duchaussoy, P., Lederman, I., Choay, J., and Sinay, P. (1988) *Carbohydr. Res.* **179**, 163-172
6. Goto, F. and Ogawa, T. (1993) *Pure Appl. Chem.*, **65**, 793-801
7. Hirose, S., Ohsawa, T., Inagami, T., and Murakami, K. (1982) *J. Biol. Chem.* **257**, 6316-6321
8. Takegawa. K., Fujita, K., Fan, J.Q., Tabuchi, M., Tanaka, N., Kondo, A., Iwamoto, H., Kato, I., Lee, Y-C., and Iwahara, S. (1998) *Anal. Biochem.* **257**, 218-223
9. Ashida, H., Yamamoto, K., Murata, T., Usui, T., and Kumagai, H. (2000) *Arch. Biochem. Biophys.* **373**, 394-400
10. Murata, T., Morimoto, S., Zeng, X., Watanabe, S., and Usui, T. (1999) *Carbohydr. Res.* **320**, 192-199
11. Fujita, M., Shoda, S., and Kobayashi, S. (1998) *J. Am. Chem. Soc.* **120**, 6411-6412
12. Endo, M., Takagaki, K., and Nakamura, T. in *CRC Handbook of Endoglycosidases and Glycoamidases* (Takahashi, N. and Muramatsu, T., eds.), Chapter 5, pp. 105-132, CRC Press, Florida
13. Takagaki, K., Nakamura, T., Majima, M., and Endo, M. (1988) *J. Biol. Chem.* **263**, 7000-7006
14. Takagaki, K., Kon, A., Kawasaki, H., Nakamura, T., Tamura, S., and Endo, M. (1990) *J. Biol. Chem.* **265**, 854-860
15. Takagaki, K., Nakamura, T., Takeda, Y., Daidouji, K., and Endo, M. (1992) *J. Biol. Chem.* **267**, 18558-18563
16. Takagaki, K., Nakamura, T., Izumi, J., Saitoh, H., Endo, M., Kojima, K., Kato, I., and Majima, M. (1994) *Biochemistry* **33**, 6503-6507
17. Sudo, S., Takagaki, K., Munakata, H., Nakamura, W., Matsuya, H., and Endo, M. (1999) *J. Appl. Glycosci.* **46**, 299-302
18. Meyer, K. and Rapport, M.M. (1950) *Arch. Biochem. Biophys.* **27**, 287-293
19. Hoffman, P., Meyer, K., and Linker, A. (1956) *J. Biol. Chem.* **219**, 653-663
20. Saitoh, H., Takagaki, K., Majima, M., Nakamura, T., Matsuki, A., Kasai, M., Narita, H., and Endo, M. (1995) *J. Biol.Chem.* **270**, 3741-3747
21. Takagaki, K., Kojima, K., Majima, M., Nakamura, T., Kato, I., and Endo, M. (1992) *Glycoconj. J.* **9**, 174-179
22. Hase, S., Ikenaka, T., and Matsushima, Y. (1978) *Biochem. Biophys. Res. Commun.* **85**, 257-263
23. Kon, A., Takagaki, K., Kawasaki, H., Nakamura, T., and Endo, M. (1991) *J. Biochem.* **110**, 132-135
24. Takagaki, K., Nakamura, T., Kawasaki, H., Kon, A., Ohishi, S., and Endo, M. (1990) *J. Biochem. Biophys. Methods* **21**, 209-215
25. Takagaki, K., Munakata, H., Nakamura, W., Matsuya, H., Majima, M., and Endo, M. (1998) *Glycobiology* **8**, 719-724
26. Takagaki, K., Munakata, H., Kakizaki, I., Majima, M., and Endo, M. (2000) *Biochem. Biophys. Res. Commun.* **270**, 588-593
27. Takagaki, K., Munakata, H., Majima, M., and Endo, M. (1999) *Biochem. Biophys. Res. Commun.* **258**, 741-744
28. Munakata, H., Takagaki, K., Majima, M., and Endo, M. (1999) *Glycobiology* **9**, 1023-1027
29. Takagaki, K., Munakata, H., Kakizaki, I., Iwafune, M., Itabashi, T., and Endo, M. (2002) *J. Biol. Chem.* **277**, 8882-8889
30. Iwafune, M., Kakizaki, I., Yukawa, M., Kudo, D., Ota, S., Endo, M., and Takagaki, K. (2002) *Biochem. Biophys. Res. Commun.* **297**, 1167-1170
31. Hardingham, T.E. and Fosang, A. (1992) *FASEB J.* **6**, 861-870
32. Knudson, C.B. and Knudson, W. (2001) *Semin. Cell. Dev. Biol.* **12**, 69-78
33. Ishido, K., Takagaki, K., Iwafune, M., Yoshihara, S., Sasaki, M., and Endo, M. (2002) *J. Biol. Chem.* **277**, 11889-11895
34. Takagaki, K., Nakamura, T., Kon, A., Tamura, S., and Endo, M. (1991) *J. Biochem.* **109**, 514-519
35. Levy, I., Shani, Z., and Shoseyov, O. (2002) *Biomol. Eng.* **19**, 17-30
36. Takagaki, K., Iwafune, M., Kakizaki, I., Ishido, K., Kato, Y., and Endo, M. (2002) *J. Biol. Chem.* **277**, 18397-18403

6

Enzymatic Synthesis of Neoglycolipids

6.1 Enzymatic Synthesis of Neoglycolipids and Glycosphingolipids Derivatives Using Jellyfish Endoglycoceramidase

To analyze or modify the functions of glycosphingolipids, new methodology for generating various glycosphingolipid derivatives including fluorescence-labeled glycosphingolipids is required. Several chemical methods for synthesizing and modifying glycosphingolipids have been developed and improved in the past several decades.[1] However, chemical synthesis and modification of glycosphingolipids, especially those with long and branched sugar chain structures such as gangliosides, is still difficult. Alternatively, the method using enzymes for synthesis and modification of glycosphingolipids is time-saving and highly specific, giving rise to expectations of high yields and purity of products. In this context, the methods using transglycosylation and reverse hydrolysis reactions of glycosidases are widely studied and utilized for the synthesis of new glycoconjugates.[2] However, the application of glycosidases for synthesis of glycosphingolipids and their derivatives has been limited. This chapter describes the enzymatic synthesis of neoglycoconjugates (neoglycolipids) and the labeling of glycosphingolipids with fluorescence by the transglycosylation reaction of jellyfish endoglycoceramidase (EGCase).[3] The condensation of oligosaccharide and ceramide by the reverse hydrolysis reaction of the EGCase is also described. The transglycosylation reaction of EGCase (ceramide glycanase) was first reported using enzyme from leech[4] and coryneform bacteria[5].

6.1.1 Transfer of an Intact GM1a-oligosaccharide from GM1a to Various 1-Alkanols

As shown in Fig. 6.1, jellyfish EGCase hydrolyzed ganglioside GM1a to generate the oligosaccharide (lane shown as "water"), which is referred to as "GM1a-oligosaccharide" in this chapter. The structure of glycosphingolipids used is shown in Table 1.5 in Chapter 1 section 1.1.4. Interestingly, new spots were generated in addition to a GM1a-oligosaccharide when GM1a was incubated with the EGCase in the presence of 1-alkanols at a concentration of 20% (Fig. 6.1). These

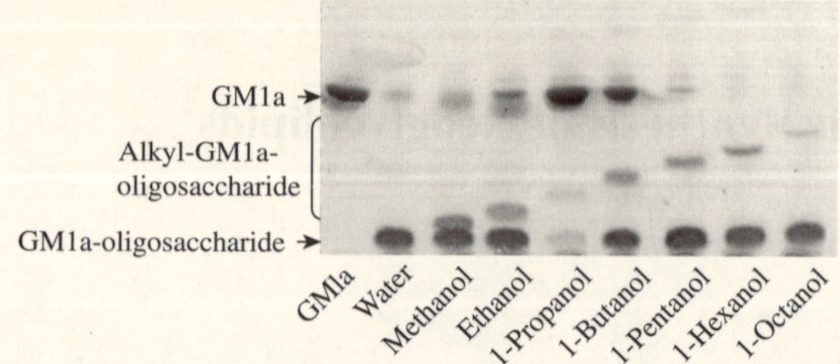

Fig. 6.1 Transfer of an intact sugar chain from GM1a to various 1-alkanols by the action of the jellyfish EGCase. The reaction mixtures contained 20 nmol of GM1, 0.5 milliunits of enzyme, and various 1-alkanols at a concentration of 20% in 40 μL of 25 mM sodium acetate solution, pH 3.0, containing 0.2% Triton X-100. After incubation at 37°C for 16 h, the reaction mixture was applied to a TLC plate, which was developed with chloroform/methanol/0.2% $CaCl_2$ (4/4/1, v/v/v). Carbohydrate-containing compounds were visualized by spraying the TLC plate with orcinol-H_2SO_4 reagent. GM1a, 20 nmol of standard GM1a. Water, hydrolysis of GM1a without 1-alkanols. Other lanes show the hydrolysis of GM1a in the presence of each 1-alkanol indicated. One milliunit of enzyme is defined as the enzyme amount capable of hydrolyzing 1 nmol GM1a per minute.

reaction products migrated faster than GM1a-oligosaccharide and their R_f became larger with an increase of carbon chain length of the 1-alkanol used, *i.e.*, methanol < ethanol < propanol < butanol *etc*. Thus, these products were likely to be an alkyl-GM1a-oligosaccharide in which the oligosaccharide was linked to each 1-alkanol used. To confirm this, the reaction product after reaction of the EGCase with GM1a in the presence of methanol was analyzed by MALDI-TOF MS using 2,5-dihydroxybenzoic acid as the matrix. The intense peaks $[M-H]^-$ were observed at *m/z* 997.1 and 1011.2, those attributed to GM1a-oligosaccharide and methyl-GM1a-oligosaccharide, respectively (Fig. 6.2). The former is the hydroly-

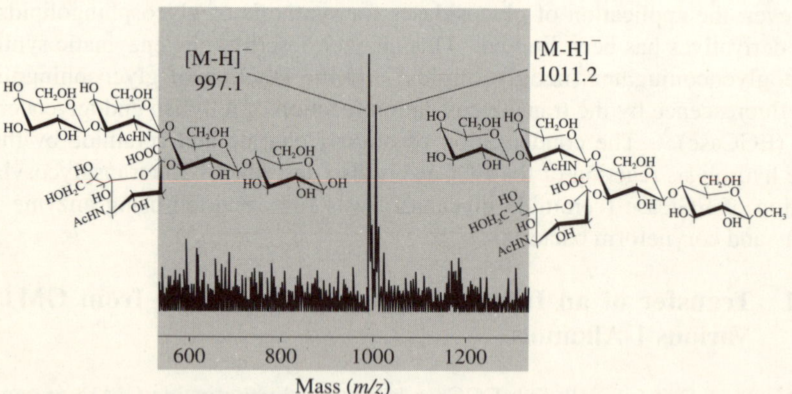

Fig. 6.2 MALDI-TOF-MS spectrum of GM1a-oligosaccharide and methyl-GM1a oligosaccharide. One hundred nmol of GM1a was incubated with 2.5 milliunits of the jellyfish EGCase in 200 μL of 25 mM sodium acetate solution, pH 3.0, containing 0.2% Triton X-100 in the presence of 25% methanol. After incubation at 37°C for 16 h, the reaction products were purified using a Sep-Pak Plus C18 cartridge then analyzed by MALDI-TOF-MS as described in Reference 3. The mass ions at *m/z* 997.1 and 1011.2 are consistent with the molecular mass of GM1a-oligosaccharide and methyl-GM1a oligosaccharide, respectively.

sis product of GM1a by the enzyme and the latter is consistent with its total mass for a GM1a-oligosaccharide with a methyl group, indicating that the enzyme transfers the oligosaccharide from GM1a to methanol. The acceptor specificity of the jellyfish enzyme is summarized in Table 6.1. Among various 1-alkanols tested, methanol was found to be the most preferential acceptor, followed by 1-hexanol, 1-pentanol and 1-octanol under a relatively short time incubation (1 h). However, under prolonged incubation (16 h) the apparent generation of these alkyl glycosides except methanol decreased somewhat. This suggests that hydrolysis of alkyl-glycosides could be faster than their synthesis if the alkyl-glycoside has a chain length of five or more carbons. In contrast, methyl-GM1a-oligosaccharide seems to be resistant to hydrolysis by the enzyme, thus gradually accumulating during incubation with the enzyme (Table 6.1). It should be noted that 1-propanol strongly inhibited both transglycosylation and hydrolysis reactions. The optimum concentration of organic solvents for the transglycosylation reactions was found to be 25% (v/v) for methanol, 12% (v/v) for ethanol and 10-70% (v/v) for 1-octanol. Various glycosphingolipids were tested for donor specificity (Table 6.2). As a result, GM1a was found to be the best donor, followed by GD1b and GT1b, when methanol was used as the acceptor. Neither globoside (Gb4Cer) nor glucosylceramide (GlcCer) is utilized by the enzyme as a donor substrate (Table 6.2). It is noteworthy that glucosylceramide is not hydrolyzed and goloboside is quite resistant to hydrolysis by the enzyme.[6)]

Table 6.1 Acceptor specificity for transglycosylation reaction of jellyfish EGCase

Acceptor	Alkyl-GM1a-oligosaccharide (%)	
	1 h	16 h
Control	–	–
Methanol	18.3	27.5
Ethanol	8.5	13.1
1-Propanol	0.6	2.1
1-Butanol	3.9	7.9
1-Pentanol	15.3	9.3
1-Hexanol	16.5	9.2
1-Octanol	14.1	3.0

GM1a was incubated at 37°C for 1 or 16 h with 0.5 milliunits of the enzyme in 40 μL of 25 mM sodium acetate solution, pH 3.0, containing 0.2% Triton X-100 in the presence of 20% 1-alkanol. One milliunit of the enzyme is defined as the enzyme amount which catalyzes the hydrolysis of 1 nmol of GM1a per min under the conditions described in Reference 6.

Table 6.2 Donor specificity for transglycosylation reaction of jellyfish EGCase

Substrate	Methyl-GSL Oligosaccharide (%)
GT1b	16.5
GD1b	22.9
GM1a	30.3
AsialoGM1	7.7
Globoside	0
LacCer	3.1
GlcCer	0

Various GSLs were incubated at 37˚C for 16 h with 0.5 milliunits of the enzyme in 20 μL of 25 mM sodium acetate solution, pH 3.0, containing 0.2% Triton X-100 in the presence of 25% methanol. One milliunit of the enzyme is defined as the enzyme amount which catalyzes the hydrolysis of 1nmol of GM1a per min under the conditions described in Reference 6.

6.1.2 Synthesis of Fluorescence-labeled Glycosphingolipids

Fluorescence-labeled glycosphingolipids are useful to analyze the metabolism and intracellular transport of glycosphingolipids as well as to determine the activities of various glycosphingolipid-hydrolyzing and -synthesizing enzymes. We found that the jellyfish EGCase transfers an oligosaccharide from various glycosphingolipids to the primary hydroxyl group of dodecanoylsphingosine, the dodecanoic acid of which at the ω position is coupled to NBD (4-nitrobenzo-2-oxa-1,3-diazole), a fluorescent ceramide (dodecanoyl-NBD-ceramide, C12-NBD-ceramide), producing various NBD-labeled glycosphingolipids (Fig. 6.3). The fluorescent glycosphingolipid generated from the incubation of GM1a with C12-NBD-ceramide in the presence of the enzyme was subjected to MALDI-TOF MS analysis. As a result, the pseudomolecular ion $[M–H]^-$ was found to be *m/z* 1640, which is consistent with the molecular mass of C12-NBD-GM1a. It is noted that the synthesized C12-NBD-GM1a was hydrolyzed by the *Rhodococcus* EGCase to produce C12-NBD-ceramide, suggesting that the C12-NBD-ceramide is coupled to an oligosaccharide *via* a β-glucosidic linkage, because the enzyme was demonstrated to specifically hydrolyze β-glucosidic linkage between the oligosaccharide and ceramide in glycosphingolipids.[7)] C6-NBD-ceramide (hexanoyl-NBD-ceramide) was also found to be utilized by the enzyme to produce C6-NBD-glycosphingolipids. The optimum concentration of NBD-ceramide was found to be 0.5 mM for both C12-NBD-ceramide and C6-NBD-ceramide as the acceptor. Interestingly, the transglycosylation activity of the enzyme using C12-NBD-ceramide as the acceptor was enhanced when acetone (15%) was added to the reaction mixture. On the other hand, the addition of acetone showed no significant increase of transglycosylation activity when C6-NBD-ceramide was used as the acceptor.

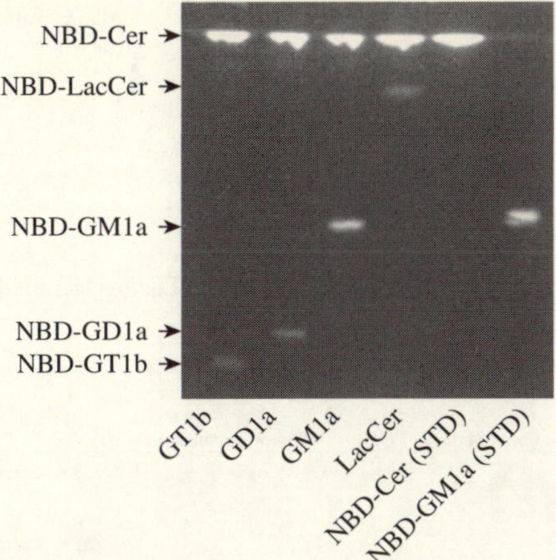

Fig. 6.3 Transfer of oligosaccharides from various glycosphingolipids to C12-NBD-ceramide by the action of the jellyfish EGCase. The reaction was carried out with 20 nmol of glycosphingolipid, 0.1 milliunits of enzyme and 200 pmol of C12-NBD-ceramide (NBD-C12 : 0, d18 : 1) in 20 μL of 25 mM sodium acetate solution, pH 3.0, containing 0.2% Triton X-100. Following incubation at 37°C for 1 h, the reaction mixture was evaporated to dryness and analyzed by TLC. Lane 1, GT1b + EGCase + C12-NBD-ceramide; lane 2, GD1b + EGCase + C12-NBD-ceramide; lane 3, GM1a + EGCase + C12-NBD-ceramide; lane 4, lactosylceramide + EGCase + C12-NBD-ceramide; lane 5, standard C12-NBD-ceramide; lane 6, standard C12-NBD-GM1.

6.1.3 Condensation Reaction

We investigated whether the jellyfish enzyme can catalyze not only transglycosylation but also reversed hydrolysis (condensation) reactions. The condensation reaction will be useful in synthesizing new glycosphingolipids using the appropriate oligosaccharides and ceramide. As shown in Fig. 6.4, jellyfish EGCase catalyzed the condensation of lactose and ceramide to produce lactosylceramide (LacCer). Pseudomolecular ion $[M+H]^+$ of the product was found to be *m/z* 890 by MALDI-TOF MS using the positive ion mode with 2,5-dihydroxybenzoic acid as the matrix. This was consistent with the molecular weight of lactosylceramide, in which ceramide is most likely octadecanoylsphingenine (C18:0, d18:1). The enzyme clearly catalyzed the reverse hydrolysis reaction in which lactose was condensed to ceramide generating lactosylceramide, although the reaction yield is quite low under the conditions used.

6.1.4 Comparison with EGCases (Ceramide Glycanases) from Other Sources

In addition to the jellyfish EGCase, transglycosylation activities of EGCase (ceramide glycanase) were found in enzymes from leeches[4] and *Corynebacterium* sp.[5] Compared with the transglycosylation reactions of other enzymes, jellyfish EGCase is unique in the following characteristics.[3] Methanol was found to be the

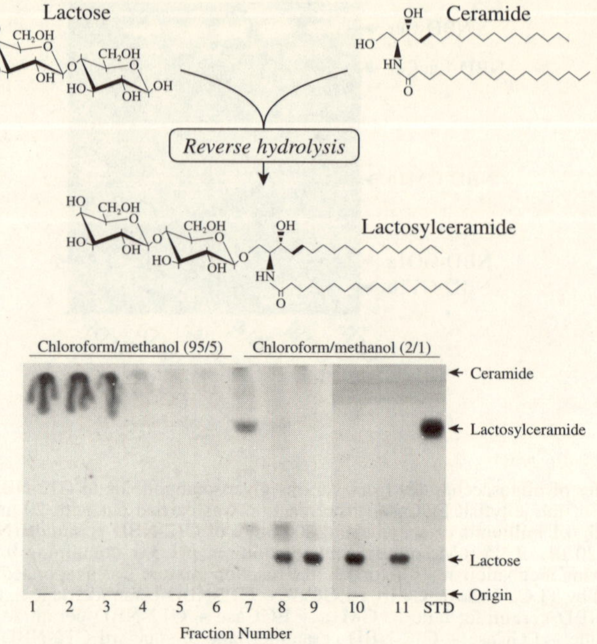

Fig. 6.4 Condensation of lactose with ceramide yielding lactosylceramide by the reverse hydrolysis reaction of jellyfish EGCase. The reaction mixture contained 5 mg of bovine brain ceramide, 10 mg of lactose and 2.5 milliunits of enzyme in 100 μL of sodium acetate solution, pH 3.0, containing 2% Triton X-100. Following incubation at 37°C for 40 h, the lactosylceramide generated was isolated using Sep-Pak Plus Silica column. The eluates from the column were checked by TLC. Ceramide was passed through the column using chloroform/methanol (95/5) (fractions 1 to 3). Lactosylceramide and unreacted lactose were eluted with chloroform/methanol (2/1) (fractions 4 to 10). Lane 11, standard lactose; STD, standard lactosylceramide. Details are described in Reference 3.

best acceptor for the jellyfish EGCase among the 1-alkanols tested (Fig. 6.1, Table 6.1). The leech enzyme was strongly inhibited by short chain 1-alkanols and no transfer product was observed when methanol was used as an acceptor.[4)] Ethanol was also utilized by the jellyfish EGCase as the acceptor (Fig. 6.1, Table 6.1), whereas EGCase from *Corynebacterium* sp. showed no transglycosylation activity with ethanol as the acceptor.[5)] These results indicate that jellyfish EGCase is much more suitable for synthesizing short-chain alkyl-glycosides than leech and bacterial EGCase (ceramide glycanase).

Leech enzyme was found to transfer the oligosaccharide from GM1a to $CF_3CO\text{-}NH(CH_2)_5CH_2OH$, $(CH_3)_3CO\text{-}CO\text{-}NH(CH_2)_5CH_2OH$, $CH_2{=}CH(CH_2)_7CH_2OH$, and $(HOCH_2)_3C\text{-}NH\text{-}CO\text{-}(CH_2)_4\text{-}COOMe$ as acceptors. Since these acceptors possess a primary hydroxyl group at one end and an additional functional group at the other end of the alkyl chain, they can be covalently linked to suitable proteins and matrices.[4)] *Corynebacterium* enzyme was reported to transfer the oligosaccharide from GM1a to *p*-nitrophenol.[5)]

We also found that the enzyme catalyzed the reverse hydrolysis reaction in which lactose was condensed to ceramide, yielding lactosylceramide. This reverse reaction has not yet been reported for leech and bacterial enzymes. Recently, glycosphingolipid-oligosaccharides can be synthesized on a large scale by combinations of various glycosyltransferases fixed on polymers[8] or by fermentation technology using microbes possessing specific glycosyltransferase genes.[9] Various sugar chains with glucose at the reducing end are expected to be condensed to ceramide by the reverse hydrolysis reaction of EGCase, yielding desired glycosphingolipids. It is, however, noted that globo-series glycosphingolipids are poor donor substrates for the transglycosylation reaction (Table 6.2) and this may also be the case for the condensation reaction because the enzyme hydrolyzes these glycosphingolipids only very slightly.[6]

The transglycosylation and condensation reactions of jellyfish EGCase are illustrated in Fig. 6.5. These unique reactions are useful for the preparation of gly-

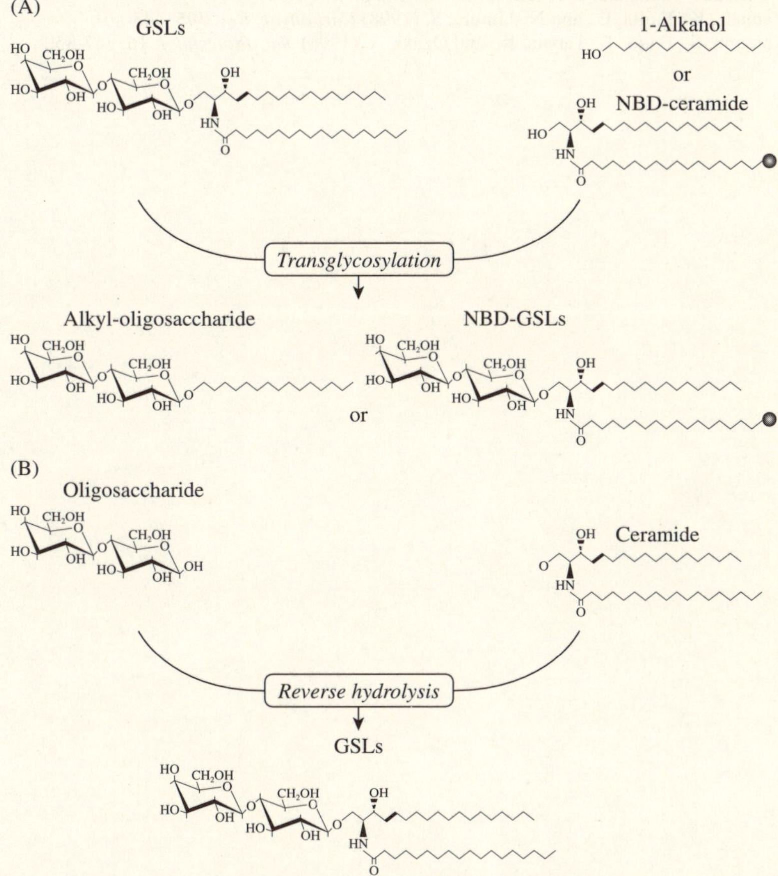

Fig. 6.5 Transglycosylation (A) and condensation (B) reactions catalyzed by jellyfish EGCase. (A), in the transglycosylation reaction, donor substrates, glycosphingolipids, are incubated with acceptor substrates, 1-alkanols or NBD-ceramide, in the presence of jellyfish EGCase, yielding alkyl-oligosaccharides or NBD-glycosphingolipids. (B), in reverse hydrolysis reaction, oligosaccharides are incubated with ceramides in the presence of jellyfish EGCase, yielding glycosphingolipids. GSLs; glycosphingolipids.

cosphingolipids and their derivatives including those that are fluorescence-labeled, and various neoglycoconjugtes possessing glycosphingolipid-sugar chains will thus facilitate the further development of glycosphingolipid research.

References

1. Ikami, T., Ishida, H., and Kiso, M. (2000) *Methods Enzymol.* **311**, 547-568
2. Cote, G.L. and Tao, B.Y. (1990) *Glycoconj. J.* **7**, 145-162
3. Horibata, Y., Higashi, H., and Ito, M. (2001) *J. Biochem.* **130**, 263-268
4. Li, Y.-T., Zhou, B., Rao, B.N.N., Schweinggruber, H., and Li, S.-C. (1991) *J. Biol. Chem.* **266**, 10723-10726
5. Ashida, H., Tsuji, Y., Yamamoto, K., Kumagai, H., and Tochikura, T. (1993) *Arch. Biochem. Biophys.* **305**, 559-562
6. Horibata, Y., Okino, N., Ichinose, S., Omori, A., and Ito, M. (2000) *J. Biol. Chem.* **275**, 31297-31304
7. Ito, M. and Yamagata, T. (1989) *J. Biol. Chem.* **264**, 9510-9519
8. Yamada, K., Fujita, E., and Nishimura, S. (1998) *Carbohydr. Res.* **305**, 443-461
9. Koizumi, S., Endo, T., Tabata, K., and Ozaki, A. (1998) *Nat. Biotechnol.* **16**, 847-850

7

Analysis of Sugar Structures Using Endoglycosidase

7.1 Analysis of Sugar Chain Structures Using Endo-β-*N*-acetyl-glucosaminidase

7.1.1 Analysis of Asparagine-linked Sugar Chains of Glycoproteins

Hydrazinolysis is an important technique for the nonselective release of unreduced asparagine-linked sugar chains in high yield from glycoproteins.[1] The method is required to minimize undesired side reactions.[1] Hydrazinolysis involves many steps for preparing released sugar chains for the following structural analysis. On the other hand, endo-β-*N*-acetylglucosaminidase (Endo-β-GlcNAc-ase) releases sugar chains without any side reactions, and the released sugar chains can be used for the structural analysis.

A. Specific Release of Asparagine-linked Sugar Chains Using Endo-β-GlcNAc-ase for the Structural Analysis of Sugar Chains

An easy way to determine structure is to compare the chromatography of released sugar chains with that of sugar chains of known structure as standards. The development of high-performance anion-exchange chromatography (HPAEC) has enabled quick and high-sensitive detection of sugar chains without labeling by radioisotope or fluorescent reagents. This method does not involve many steps and a small amount of starting material can be used for the structural analysis of sugar chains. Yeh *et al.* determined the structure of asparagine-linked sugar chains with 40 μg of hemolymph clottable protein of tiger shrimp.[2] Monosaccharide contents were determined with HPAEC after acid hydrolysis of the protein. Asparagine-linked sugar chains were released from the protein by Endo-H.[3] Ribonuclease B with sugar chains of known structure was hydrolyzed by Endo-H at the same time to give standards. Elution profiles of the sugar chains released from the clottable protein and the standards revealed the structure of the sugar chains of the clottable protein.

The release of sugar chains is usually performed with a soluble enzyme. On the other hand, immobilization makes it possible to stabilize an enzyme, perform a continuous enzyme reaction and recycle the enzyme. Kol *et al.* immobilized

Endo-β-GlcNAc-ase-B isolated from Basidiomycetes *Sporotrichum dimorphosporum* (Endo-B) on Sepharose 4B.[4] Immobilized Endo-B showed strong activity and good stability under conditions of heat denaturation and storage.

B. Release of Whole Asparagine-linked Sugar Chains with Combination of Endo-β-GlcNAc-ase and PNGase F for Structural Analysis of Sugar Chains

For the analysis of all types of asparagine-linked sugar chains, a combination of Endo-H and another enzyme is employed because Endo-H is specific for high-mannose and hybrid type sugar chains and cannot release complex type sugar chains. Complex type sugar chains of asparagine-linked form show structural diversity such as bi-, tri-, tetra- and pentaantennary and modifications of the side chains by NeuAc, Fuc or *N*-acetyllactosamine.[5] Endo-F_2 is specific for high-mannose and biantennary complex type sugar chains, while Endo-F_3 is specific for bi- and triantennary complex type sugar chains.[2] No Endo-β-GlcNAc-ase hydrolyzing all types of complex type sugar chains is currently available. Peptide-N^4-(*N*-acetyl-β-D-glucosaminyl)asparagine amidase F (PNGase F) is now usually used for releasing complex type sugar chains instead of Endo-β-GlcNAc-ase. This enzyme is an amidase that hydrolyzes the glycosylamine linkage of glycoproteins, releasing all types of asparagine-linked sugar chains from glycoproteins and glycopeptides.[3] PNGase F is commercially available. Endo-H-unsusceptible asparagine-linked sugar chains are susceptible to PNGase F and may be complex type sugar chains.

It is difficult to purify asparagine-linked sugar chains from a mixture of types of asparagine-linked sugar chains. Orberger *et al.* compared the structure of asparagine-linked sugar chains of the human transferrin receptor from placenta and the human hepatocarcinoma cell line.[6] Following digestion of the purified transferrin receptor with trypsin, glycopeptides were treated with Endo-H in order to release high-mannose and hybrid type sugar chains. The released sugar chains were thereafter separated based on differences of mannosylation. The remaining Endo-H-unsusceptible, complex type sugar chains were released from the glycopeptides with PNGase F. They were separated by ion-exchange chromatography and further by HPAEC after sialidase treatment. Purified sugar chains were further characterized by methylation analysis and by liquid secondary-ion MS.

Some glycoproteins have *O*-linked sugar chains along with asparagine-linked sugar chains. Nayak and Spiro attempted the complete characterization of type IV collagen.[7] They employed three methods to release sugar chains attached to the protein for structural analysis. The hydrazine/nitrous acid treatment is used for the release of hydroxylysine-linked sugar chains. Endo-H and PNGase F were used for releasing high-mannose type sugar chains and asparagine-linked sugar chains, respectively.

SDS-PAGE or two-dimensional electrophoresis gives high resolution in the separation of proteins. Transferred proteins on a membrane can be analyzed in terms of amino acid sequence and/or immunogenicity. Sugar chain analysis of transferred proteins has been performed "on membrane" using lectins[8, 9] of known specificity or the periodate method.[10] On the other hand, the protein on the membrane can be excised and dipped in glycosidase solution for "on membrane deglycosylation". Sugar chains are released into the solution and separated from the protein moiety that remains on the membrane. The structural information on released sugar chains can be obtained by microanalytical methods such as HPAEC or MALDI-TOF MS. Weitzhandler *et al.* reported a low pico mole analysis of sugar chains of glycoproteins which have been separated by SDS-PAGE and electrotransferred onto a PVDF membrane.[11] Endo-H digestion of bovine RNase B (30 μg) on membrane resulted in the release of high-mannose type sugar chains. The HPAEC profile of the released sugar chains was identical to that obtained from Endo-H digestion of RNase B in solution.

7.1.2 Typing of Asparagine-linked Sugar Chains of Glycoproteins

The susceptibility of glycoproteins to Endo-β-GlcNAc-ases of known specificity gives information about the sugar chain class (asparagine-linked or *O*-linked) and type (high-mannose, hybrid or complex). Removal of asparagine-linked sugar chains by Endo-β-GlcNAc-ases brings about a decrease in the molecular weight of glycoproteins. The deglycosylated protein of reduced molecular weight can be separated from intact glycoprotein by SDS-PAGE. The development of protein staining techniques such as silver staining and/or immunostaining after electroblotting make it possible to detect small amounts of proteins.

Takeda-Ezaki and Yamamoto determined the type of sugar chains of pro- and mature cathepsin E (CE) from human erythrocyte membrane and mature CE from rat erythrocyte membrane and rat spleen by examining the susceptibility to Endo-H followed by SDS-PAGE and immunoblotting.[12] The pro-CE and the mature CE from human erythrocyte membrane and the mature CE from rat erythrocyte membrane were revealed to be resistant to the action of Endo-H, indicating the presence of a complex type or a hybrid type sugar chain on the enzymes. In contrast, CE from rat spleen was susceptible to the Endo-H treatment, indicating the presence of high-mannose type sugar chains.

Some glycoproteins have more than two asparagine-linked sugar chains. SDS-PAGE of the products of the deglycosylation of these glycoproteins by enzyme treatment shows several intermediate deglycosylated protein bands of reduced molecular weight. The number of deglycosylated protein bands indicates the number of sugar chains in a glycoprotein. Dahms and Brzycki-Wessell expressed mannose 6-phosphate receptors in baculovirus-infected insect cells.[13] The recombinant protein was expressed as five species with molecular weights of 26000, 29000, 32000, 35000 and 39000. SDS-PAGE of the expressed protein after treatment with different amounts of Endo-H for various times showed that the number of species gradually decreased to one prominent band with a molecular weight of 26000, the size of that observed for the nonglycosylated form of the protein. These results indicated that the multiple forms in the protein were due to the difference in the number of attached high-mannose type sugar chains on the polypeptide of the protein.

7.1.3 Specific Release of Complex Type Sugar Chains from Glycoproteins by Endo-HS

At the end of this section, the usefulness of a novel enzyme, Endo-HS, for releasing asparagine-linked sugar chains of complex type from glycoproteins is introduced by showing the properties of Endo-HS, including the action of the enzyme on various native glycoproteins and the transglycosylation of complex type sugar chains from native glycoproteins.

A. Discovery of Endo-HS

The Endo-β-GlcNAc-ases so far presented are mainly specific for high-mannose and hybrid type sugar chains. Endo-H is a typical enzyme in terms of specificity. On the other hand, complex type sugar chains are typical of higher eukaryotic cells. They show high structural diversity caused by variation in numbers of side chains and modification of side chains. Endo-F_2 is known to be specific for high-mannose and biantennary complex types. Endo-F_3 is the only enzyme specific for bi- and triantennary complex type sugar chains.[3] No Endo-β-GlcNAc-ase specific for complex type sugar chains of bi-, tri- and tetraantennary was found. Furthermore, the protein moieties of native glycoproteins frequently hinder the action of Endo-β-GlcNAc-ases. Endo-β-GlcNAc-ases releasing sugar chains from native glycoproteins are very useful for analyzing both structure and function of sugar chains of native glycoproteins. So an Endo-β-GlcNAc-ase specifically releasing wide varieties of complex type sugar chain from native glycoproteins has been required. On the other hand, Endo-HS was found by the investigation of *in vivo* deglycosylation of human α-amylase as described below.

Human α-amylase is produced specifically in the salivary gland (salivary α-amylase) and the pancreas (pancreatic α-amylase), secreted into saliva or pancreatic juice. It is distributed in serum or urine. The activities of the two human α-amylases are correlated with some diseases, and they are diagnostic marker proteins of those diseases. Human pancreatic α-amylase has no sugar chain. On the other hand, human salivary α-amylase (HSA) is separated into multiple forms according to the difference of their isoelectric points.[14, 15] The multiple forms are divided into two groups of family A (HSA-A) and family B (HSA-B).[14, 15] HSA-A has an asparagine-linked biantennary complex type sugar chain containing the Lewis^X determinant and an α-Fuc on the proximal GlcNAc residue.[16] HSA-B does not have the sugar chain.[15, 16] The mechanism of the formation of the glycosylated and nonglycosylated HSA remains unknown. The deglycosylation of HSA-A to form HSA-B by the action of an endoglycosidase-like enzyme in human saliva has been found, as shown in Fig. 7.1[17] The activity disappears by ultracentrifugation, indicating that the enzyme exists in an insoluble form. The enzyme is assumed to release asparagine-linked sugar chains of complex type from native glycoproteins which are resistant to the action of known endoglycosidases such as Endo-H and Endo-F. The enzyme activity originates in the epithelial cells in human saliva peeled from the oral cavity epithelium.[18] The enzyme is solubilized by detergents and seems to be an intrinsic membrane protein.[17] Amino acid

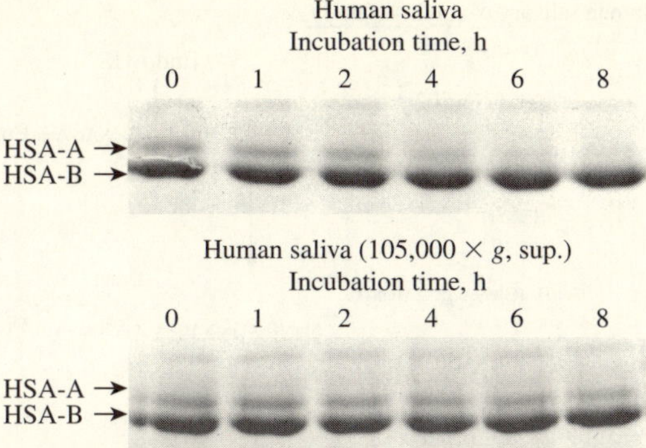

Fig. 7.1 Deglycosylation of human salivary α-amylase family A in human saliva. Human saliva was incubated at 37°C for the indicated times, then centrifuged at 105,000 × *g* for 60 min. The supernatant (10 μL) was subjected to SDS-PAGE with 7.5% gel. Supernatant of human saliva obtained by ultracentrifugation at 105,000 × *g* for 60 min was also incubated at 37°C for the indicated times, then the supernatant (10 μL) was subjected to SDS-PAGE. Human salivary α-amylases were stained by Coomassie Brilliant Blue. HSA-A, human salivary α-amylase family A; HSA-B, human salivary α-amylase family B.

and sugar composition of the enzyme reaction products of released sugar chain and protein moiety have been analyzed with HSA-A as the substrate. The amino acid composition of the reaction product of the protein moiety is the same as that of native HSA-A.[18)] On the other hand, the sugar compositions of released sugar chain and protein moiety indicate that this enzyme cleaves the sugar chains of HSA-A at the *N,N'*-diacetylchitobiose core to release residual sugar chains as shown in Fig. 7.2. This result has been confirmed by analysis of the reaction products of human transferrin tetraglycopeptide. This enzyme has been identified as an Endo-β-GlcNAc-ase and named Endo-HS.[18)]

B. Assay Method

Endo-HS can be assayed using native glycoproteins having a complex type sugar chain such as HSA-A as the substrate.[18)] HSA-A (10 μL; 0.2 mg/mL) in 0.2 M sodium-potassium phosphate buffer, pH 6.0, is mixed with 10 μL of the enzyme solution. The reaction mixture is incubated at 37°C. The enzyme reaction is stopped by the addition of 20 μL of a sample buffer of SDS-PAGE following incubation at 95°C for 3 min. Aliquots of 20 μL are subjected to SDS-PAGE with 7.5% polyacrylamide gel. The amount of the deglycosylated protein of deduced molecular weight is determined with a densitometer after staining with Coomassie Brilliant Blue (Fig. 7.3). One unit of the enzyme activity is defined as the enzyme amount that deglycosylates 1 μg of HSA-A per min. Human transferrin having two biantennary complex type sugar chains can also be used as the substrate instead of HSA-A.

C. Purification Procedure[18)]

Solubilization Saliva (2750 mL) collected in the presence of NaN_3 and stored at –20°C is centrifuged at 105,000 × *g* for 60 min. The precipitate is washed with

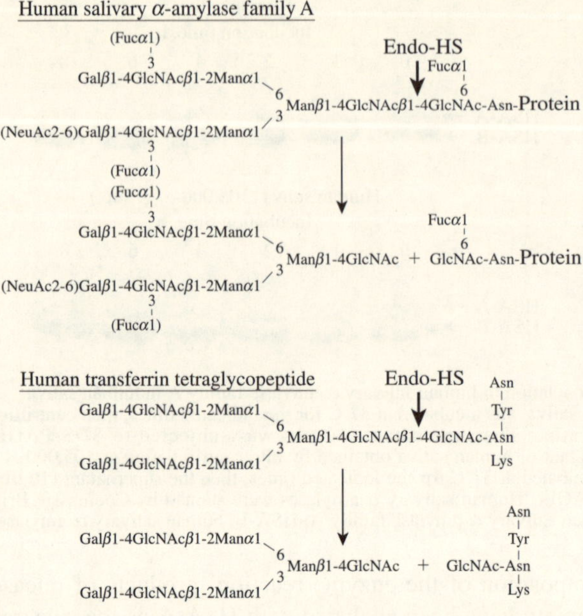

Fig. 7.2 Action of Endo-HS on asparagine-linked sugar chains.

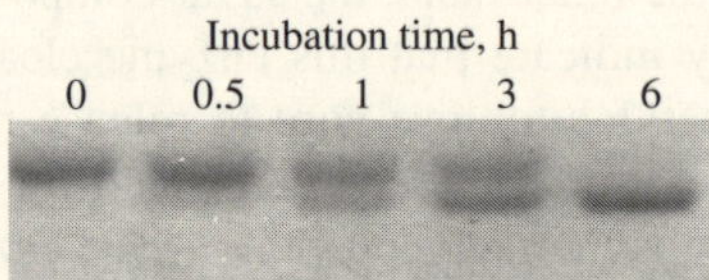

Fig. 7.3 Successive reaction of Endo-HS with human salivary α-amylase family A. Human salivary α-amylase family A (0.1 mg/mL) was incubated with Endo-HS at 37˚C for the indicated times. Aliquot of the reaction mixture was subjected to SDS-PAGE with 7.5% gel. Proteins were stained by Coomassie Brilliant Blue.

10 mM Tris-HCl buffer, pH 7.2. It is suspended in 100 mL of 1% 3-[(3-cholamidopropyl)dimethylammonio]-1-propane sulfonate (CHAPS) in the same buffer solution. The suspension is mixed for 60 min at 4˚C and centrifuged at 105,000 × *g* for 60 min. The supernatant is collected. The precipitate is resuspended in 100 mL of the CHAPS solution and centrifuged at 105,000 × *g* for 60 min. The supernatant is collected. The procedure is performed until about 80% of the Endo-HS activity is solubilized.

Sephacryl S-300 HR The solubilized Endo-HS solution is concentrated by ultrafiltration. The concentrate is applied on a column of Sephacryl S-300 HR(ϕ2.2 × 95.0 cm) equilibrated with 1% CHAPS in 10 mM Tris-HCl buffer, pH 7.2, 50 mM NaCl. The Endo-HS-active fractions are pooled.

Hydroxyapatite Endo-HS preparation obtained by Sephacryl S–200 column chromatography is applied on a column of hydroxyapatite (ϕ 1.2 × 21.5 cm) equilibrated with 1% CHAPS in 10 mM sodium-potassium phosphate buffer, pH 6.0.

The adsorbed Endo-HS is eluted with a linear gradient of 10 mM to 0.2 M sodium-potassium phosphate buffer, pH 6.1 (100 mL – 100 mL). The enzyme-active fractions are pooled. The Endo-HS preparation obtained is free from proteases and exoglycosidases.

D. Characteristics of Endo-HS

Endo-HS shows an optimum pH of around 6.0 and a molecular weight of 270,000 and is stable up to 40°C.

The specificity of Endo-HS for the structure of the sugar chains has been investigated using HSA-A with trimmed-sugar chain as substrate. The sugar chain of HSA-A has been sequentially trimmed from the nonreducing end by sequential exoglycosidase digestion, and the trimming has been confirmed by reduction of the molecular weight of HSA-A and reactivity of the treated HSA-As with various lectines.[18] Activities of Endo-HS for the sugar chain-trimmed HSAs are listed (Table 7.1).[18] Endo-HS acts on desialylated and defucosylated HSA-A (sugar chain ii) with the same velocity as that for native HSA-A(sugar chain i). On the other hand, Endo-HS is less active for HSA-A having an agalactobiantennary complex type sugar chain (sugar chain iii). Furthermore, Endo-HS activity for HSA-A having the "core" structure that is common in all asparagine-linked sugar chains (sugar chain iv) is reduced to 5% of the reaction rate for native HSA-A. Endo-HS shows high specificity for the mature asparagine-linked sugar chains

Table 7.1 Action of Endo-HS on sugar chain-trimmed human salivary α-amylase family A

	Sugar chain	Relative activity (%)
i	(Fucα1) 3 Galβ1-4GlcNAcβ1-2Manα1 ↘6 Fucα1 6 ↓ Manβ1-4GlcNAcβ1-4GlcNAc (NeuAc2-6) Galβ1-4GlcNAcβ1-2Manα1 ↗3 3 (Fucα1)	100
ii	Galβ1-4GlcNAcβ1-2Manα1 ↘6 Fucα1 6 ↓ Manβ1-4GlcNAcβ1-4GlcNAc Galβ1-4GlcNAcβ1-2Manα1 ↗3	99.0
iii	GlcNAcβ1-2Manα1 ↘6 Fucα1 6 ↓ Manβ1-4GlcNAcβ1-4GlcNAc GlcNAcβ1-2Manα1 ↗3	13.4
iv	Manα1 ↘6 Fucα1 6 ↓ Manβ1-4GlcNAcβ1-4GlcNAc Manα1 ↗3	5.0

(complex type) but not for immature (high-mannose type and hybrid type).

The specificity of Endo-HS for the sugar chain structure is distinct from those of the microbial enzyme such as Endo-D,[19] Endo-H,[20] Endo-CI, Endo-CII[21] and Endo-M[22] and the animal enzyme from hen oviduct,[23] rat liver[24] and human fibroblast.[25] Endo-F preparation[26] is separated into three distinct activities of Endo-F_1, Endo-F_2 and Endo-F_3.[27] Endo-F_1 and Endo-F_2 act on high-mannose type sugar chains, while only Endo-F_3 can act preferentially on both bi- and triantennary complex type sugar chains.[27, 28] Endo-F_3 is similar to Endo-HS in specificity for the sugar chain structure.

The origin of Endo-HS activity is thought to be epithelial cells peeling from the epithelium into saliva, and it is solubilized with detergents but not with chelating reagents or salts.[17] It is likely that Endo-HS is an intrinsic membrane protein and integrated on the surface of the epithelial cells. The microbial enzymes are secreted into culture medium.[20, 29] Animal enzymes in the cytosol[30] or the lysosome[31] are thought to be soluble enzymes. Endo-HS is distinct from those enzymes in the *in vivo* condition.

E. Deglycosylation of Native Glycoproteins by Endo-HS

The action of Endo-HS on various native glycoproteins has been investigated by detecting the deglycosylated proteins by SDS-PAGE. Table 7.2 summarizes the action of Endo-HS on various native glycoproteins.[18] Endo-HS releases bi- and triantennary complex type sugar chains from human transferrin, calf fetuin and human RNase UL. Endo-HS can also release sugar chains from human lactoferrin and human salivary α-amylase family A. The sugar chains of human lactoferrin and human salivary α-amylase family A have a Fuc residue attached to the proximal GlcNAc residue,[16] indicating that Endo-HS is not hindered from releasing sugar chains by the Fuc residue. It can release complex type sugar chains from native glycoproteins even in the presence of Fuc at the proximal GlcNAc residue although Fuc is generally thought to be a hindrance to the action of Endo-β-GlcNAc-ase. Endo-HS can also act on sheep IgG, indicating the action of complex type sugar chains having bisecting GlcNAc residue.

An intermediate with one mole of sugar chain formed by Endo-HS is observed with human transferrin and human lactoferrin because they have two moles of biantennary complex type sugar chains. Two intermediates are detected in the course of a reaction with calf fetuin having three moles of triantennary complex type. Human RNase UL also shows multiple intermediates in the course of deglycosylation by Endo-HS. Human RNase UL has three moles of sugar chains and they are bi-, tri-, and tetraantennary complex types. Endo-HS can completely remove sugar chains from the glycoproteins as mentioned above. On the other hand, Endo-HS cannot act on ovalbumin, Taka-amylase and RNase B which have high-mannose or hybrid type sugar chains.

F. Deglycosylation of Highly Glycosylated Native Glycoprotein by Endo-HS

The action of Endo-HS on highly glycosylated glycoproteins has been also inves-

Table 7.2 Action of Endo-HS on various native glycoproteins

Glycoproteins	Action
Human salivary α-amylase family A (2-C, +Fuc)	+
Human lactoferrin (2-C, +NeuAc, +Fuc)	+
Human transferrin (2-C, +NeuAc)	+
Calf fetuin (3-C, +NeuAc)	+
Human urinary RNase UL (2-C, +NeuAc; 3-C, +NeuAc; 4-C, +NeuAc)	+
Sheep IgG (2-C, +bisecting GlcNAc)	+
Ovalbumin (M, H)	–
Bovine pancreatic RNase B (M)	–
Taka-amylase (M)	–

+, hydrolysis; –, no hydrolysis; 2-C, biantennary complex type sugar chain; 3-C, triantennary complex type sugar chain; 4-C, tetraantennary complex type sugar chain; +NeuAc, NeuAc binding to nonreducing end of Gal residue; +Fuc, fucose binding to the proximal GlcNAc residue; M, high-mannose type sugar chain; H, hybrid type sugar chain.

tigated. Human α_1-acid glycoprotein has five moles of sugar chains.[32] The structure of the sugar chain is bi-, tri- or tetraantennary complex type. Each sugar chain has Fuc attached to the proximal GlcNAc residue. Fig. 7.4 shows the successive reaction products formed by the action of Endo-HS. Endo-HS can act on native human α_1-acid glycoprotein stepwise. Five deglycosylated protein bands appeared during the reaction, indicating the number of attached asparagine-linked sugar chains on human α_1-acid glycoprotein.

Native glycoproteins show resistance to the complete removal of sugar chains by sugar chain-cleaving enzymes. Denaturation of glycoproteins and/or a larger amount of enzymes are required for complete removal of sugar chains from glycoproteins. It is remarkable that Endo-HS can readily remove all complex type sugar chains from native glycoproteins without denaturing the glycoproteins. Endo-HS is very useful for the specific release of complex type sugar chains from native glycoproteins. Furthermore, determination of the number of deglycosylated inter-

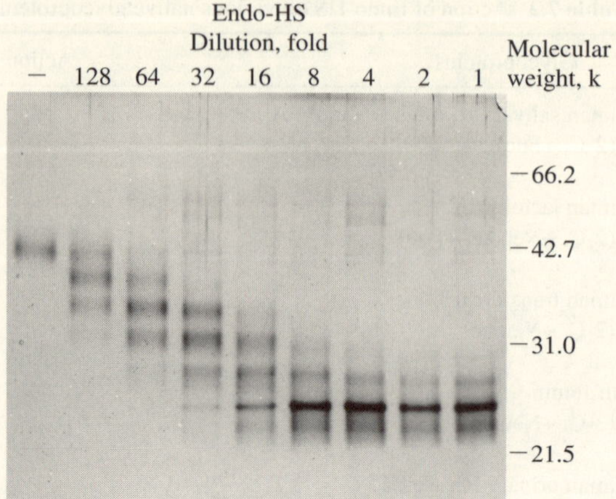

Fig. 7.4 Action of Endo-HS on human α_1-acid glycoprotein.
Human α_1-acid glycoprotein (0.1 mg/mL) was incubated at various concentrations of Endo-HS at 37°C for 60 min. Aliquot of the reaction mixture was subjected to SDS-PAGE with 12.5% gel. Proteins were stained by Coomassie Brilliant Blue.

mediates by Endo-HS shows the number of the complex type sugar chains attached to glycoproteins. Combined use of Endo-HS, Endo-H and PNGase F is a powerful tool for the analysis of the structure of asparagine-linked sugar chains.

G. Transglycosylation Activity of Endo-HS

Endo-β-GlcNAc-ase can cleave the *N,N'*-diacetylchitobiose linkage with hydroxyl groups of some suitable agents to transfer released sugar chains to the agents. Transglycosylation of asparagine-linked sugar chains was first reported with Endo-F.[33] It transfers the sugar chain from asparagine-linked sugar chain prepared from ovalbumin to glycerol. Furthermore, it has been revealed that other microbial Endo-β-GlcNAc-ases such as Endo-A[34] or Endo-M[35] can transfer asparagine-linked sugar chains to other molecules having hydroxyl groups. The transglycosylation by Endo-β-GlcNAc-ase makes it possible to synthesize molecules having asparagine-linked sugar chains. The transferred sugar chains seem to be ruled by the specificity of hydrolysis activity of each enzyme. As described above, Endo-HS is specific for complex type sugar chains and can release bi-, tri- and tetraantennary complex type sugar chains from glycopeptides, Asn-oligosaccharides and even native glycoproteins.[18] Endo-HS[36] is expected to transfer various complex type sugar chains to some other molecules having hydroxyl groups. Some properties of the transglycosylation activity of Endo-HS using native glycoproteins as oligosaccharide donors and *p*-nitrophenyl (*p*NP) monosaccharides as oligosaccharide acceptors are described below.

Transglycosylation reaction is performed in 200 μL of reaction mixture containing sugar chain donor (0–200 μM), sugar chain acceptor (0–100 mM), Endo-HS (3.5 units/mL) and 0.1 M sodium-potassium phosphate buffer, pH 6.0, at 37°C for 48 h. HPLC of the aliquots of the reaction mixture is performed with a column of Superdex Peptide 10/30 (Amarsham Pharmacia Biotech, UK) at a flow rate of

0.4 mL/min. The transglycosylation products are determined by measuring the absorbance at 300 nm.

Transglycosylation activity has been determined with *p*NP-β-Glc as a sugar chain acceptor and human transferrin as a sugar chain donor. The reaction product detected by HPLC appeared as the concentration of *p*NP-β-Glc increased. The product is confirmed to be the transglycosylation product of the asialobiantennary complex type sugar chain of human transferrin to *p*NP-β-Glc by successive exoglycosidase digestion. The same result is obtained using *p*NP-β-Gal as an acceptor. Endo-HS is confirmed to show transglycosylation reaction.

Specificity of transglycosylation of Endo-HS for sugar chain acceptor has been determined with 100 mM of various *p*NP-monosaccharides as the acceptor and 250 μM of human transferrin as the sugar chain donor. Endo-HS can transfer the sugar chain of human transferrin to *p*NP of α-Glc, α-Gal, β-Gal, β-Man, β-Xyl, and β-GlcNAc. Transglycosylation product is also detected with *p*NP-glycerol as a sugar chain acceptor. Significant difference in the amount of transglycosylation product is not observed. The K_m and V_{max} values of the transglycosylation for the sugar chain donor (human transferrin) has been measured with different sugar chain acceptors, *p*NP-β-Glc and *p*NP-glycerol. No significant difference in K_m and V_{max} values is observed with the two different sugar chain acceptors.

Transglycosylation by Endo-HS has been investigated with other native glycoproteins as sugar chain donors and *p*NP-β-Glc as sugar chain acceptor. The transglycosylation product is detected with calf fetuin having triantennary complex type oligosaccharides. The elution position of the product on HPLC is faster than that obtained with human transferrin as sugar chain donor. Successive exoglycosidase digestion of the product indicates that the asialotriantennary complex type sugar chain is transferred to *p*NP-β-Glc. Furthermore, three transglycosylation products are formed with human α_1-acid glycoprotein as the sugar chain donor. Human α_1-acid glycoprotein has bi-, tri- and tetraantennary complex type sugar chains.[32] Successive exoglycosidase digestion of the products indicates that asialobi-,asialotri- and asialotetraantennary complex type sugar chains are transferred to *p*NP-β-Glc from human α_1-acid glycoprotein. The transglycosylation from glycopeptide has been also determined. Its peptide moiety is Lys-Val-Ala-Asn-Lys-Thr, and a sialylated biantennary complex type sugar chain is attached at the Asn.[37] The transglycosylation product is eluted at the same position as that of the transglycosylation product obtained with human transferrin as the donor. The pattern of exoglycosidase digestion of the transglycosylation product is the same as that of the transglycosylation product using human transferrin as sugar chain donor. The structure of the oligosaccharide transferred from the glycopeptide is the asialobiantennary complex type.

The results described above demonstrate that Endo-HS shows low specificity for sugar chain donors and can transfer bi-, tri- and tetraantennary complex type sugar chains directly from native glycoproteins and glycopeptides to various monosaccharides. Endo-HS is expected to be an important reagent for enzymatic synthesis of neoglycoconjugates having complex type sugar chains.

7.2 Analysis of Sugar Chain Structures Using Endo-α-*N*-acetylgalactosaminidase

7.2.1 Release of *O*-Linked Sugar Chain from Glycoprotein

The structure of *O*-linked sugar chains in glycoproteins has conventionally been analyzed by releasing, using alkaline sodium borohydride treatment, the sugar chain component as sugar alcohol, and analyzing that sugar alcohol.[38, 39] However, derivatization of the releasing sugar alcohol is difficult, and analytical methods are limited. Applying hydrazinolysis, which has come to be used frequently in the structural analysis of *N*-linked sugar chains, to *O*-linked sugar chains[40,41] is desirable because nonmodified free sugar chains with a reducing end can be obtained. However, considerable broken down products result, so purification is extremely difficult. In any case, alkalinolysis and hydrazinolysis, which are chemical breakdown methods, cause damage to the protein component and a part of the carbohydrate moiety of glycoproteins, so they cannot be used for analyzing the structure and function of the protein moiety.

Endo-α-*N*-acetylgalactosaminidase (Endo-α-GalNAc-ase) is an endo-type glycosidase that releases sugar chains by acting on sites of conjugation between *O*-linked sugar chains that bind to serine or threonine residue in protein, and the protein moiety. Several types of Endo-α-GalNAc-ase have been reported.[42-47] These Endo-α-GalNAc-ases are useful in the research of the structure and function of *O*-linked sugar chains on the cell surface because they can release *O*-linked sugar chains from cell surface glycoproteins without damaging the cell, but the substrate specificities and aglycon specificities of these enzymes are different.[48-51] Endo-α-GalNAc-ase from *Clostridium perfringens*, *Streptococcus pneumoniae*, *Alcaligenes* sp., and *Bacillus* sp. reportedly releases only the disaccharide Galβ1-3GalNAc that binds to serine or threonine in the protein moiety.[42–45, 47] Endo-α-GalNAc-ases from *Streptococcus pneumoniae*[44, 45] and *Alcaligenes* sp.[43] are marketed and often used for analyzing the structure and function of the sugar chain and the protein moiety in *O*-linked glycoproteins. On the other hand, Endo-α-GalNAc-ase from *Streptomyces* sp. (Endo-α-GalNAc-ase-S) releases not only the disaccharide Galβ1-3GalNAc, but also sugar chains with an even greater molecular weight.[46, 51-53] Endo-α-GalNAc-ase-S has yet to be marketed because it is difficult to optimally control conditions when culturing the *Streptomyces* sp., and the assay method for the Endo-α-GalNAc-ase-S is extremely complicated, so it has not been fully purified.

Observation of the structure of *O*-glycan reveals that the structure of the *O*-linked sugar chain is not as irregular as the *N*-linked sugar chain. Several core (Core 1 to Core 8) structures were detected as the main core without a common overall structure being detected. Fetuin, a glycoprotein having Core 1 and Core 2 structures, was selected as the substrate, and Endo-α-GalNAc-ase was reacted with it.

7.2.2 Release of *O*-Linked Sugar Chain from Asialofetuin Using Endo-α-GalNAc-ase

A. Specific Release of *O*-Linked Sugar Chain Using Endo-α-GalNAc-ase

Endo-α-GalNAc-ase from *Streptococcus pneumoniae* was incubated at 37°C for 2 h with 2 mg asialofetuin and an appropriate amount of buffer for a total volume of 0.5 mL. After incubation, the reaction mixture was boiled at 100°C for 10 min. Following centrifugation at 6,000 × *g* for 5 min, the supernatant was applied to a Sephadex G-15 column, eluted with 40 mM sodium phosphate buffer, pH 7.0, and 0.5 mL fractions were collected. A portion of the sugar chain was desalted by gel filtration on a column and lyophilized before Fast Atom Bombardment Mass spectrometry (FAB-MS) analysis. FAB-MS was performed in the positive-ion mode on an Analytica of JEOL JMS-HX 110 A mass spectrometer. The charged mass ion $[M+H]^+$ was at *m/z* 384 (Fig. 7.5). The molecular mass of 383 Da calculated from this value coincided with the theoretical value of Galβ1-3GalNAc.

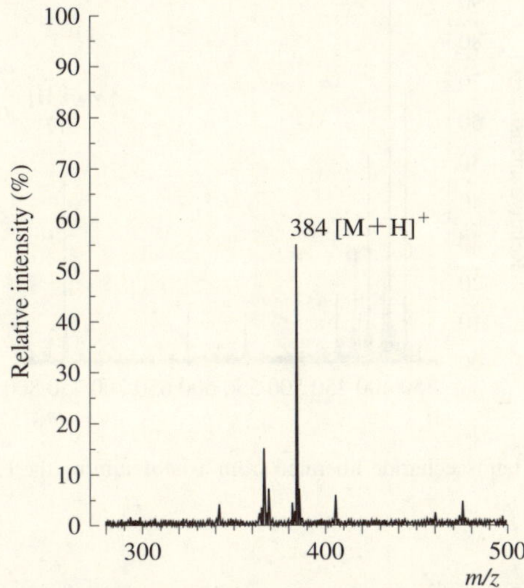

Fig. 7.5 FAB-MS of the disaccharide liberated from asialofetuin by the Endo-α-GalNAc-ase from *Streptococcus pneumoniae*.

Endo-α-GalNAc-ases from *Clostridium perfringens*, *Alcaligenes* sp. and *Bacillus* sp. were reacted with asialofetuin glycoprotein and glycopeptide, resulting in the release of a Galβ1-3GalNAc residue only as per the above. These Endo-α-GalNAc-ases were reacted with protein having uncertain sugar chain binding to reveal low-molecularization of the protein, indicating it highly likely that the Galβ1-3GalNAc disaccharide had bound. However, low-molecularization of protein is also sometimes seen due to contamination of these Endo-α-GalNAc-ases with protease, so it is important to check for the presence of released sugar chains.

B. Release of *O*-Linked Sugar Chain Using Endo-α-GalNAc-ase-S

Next, Endo-α-GalNAc-ase-S was reacted with asialofetuin. Two milligrams of asialofetuin, 0.03 mL of enzyme solution and an appropriate amount of buffer were incubated in a total volume of 0.2 mL. After incubation at 37°C for 6 h, the reaction mixture was boiled at 100°C for 10 min and centrifuged at 10,000 × *g* for 10 min. Following centrifugation, the supernatant was applied to a Sephadex G-15 column above and eluted under the same conditions as above. A portion of the sugar chain was desalted on a gel filtration column and lyophilized before FAB-MS analysis. FAB-MS was performed in the positive-ion mode on MS spectrometer. The charged mass ion $[M+H]^+$ was at *m/z* 384 and *m/z* 749 (Fig. 7.6). The molecular masses of 383 and 748 Da calculated from these values coincided with the theoretical values of Galβ1-3GalNAc and Galβ1-3(Galβ1-4GlcNAcβ1-6)GalNAc, respectively.

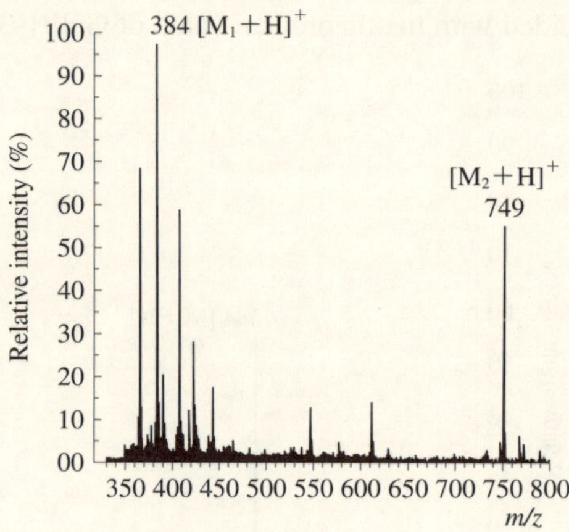

Fig. 7.6 FAB-MS of the tetrasaccharide liberated from asialofetuin by the Endo-α-GalNAc-ase from *Streptomyces* sp.

7.2.3 Release of *O*-Linked Sugar Chain from Fetuin Using Endo-α-GalNAc-ase

A. Release of *O*-Linked Sugar Chain with Combination of Endo-α-GalNAc-ase and Several Exoglycosidases

Endo-α-GalNAc-ases from *Clostridium perfringens*, *Alcaligenes* sp., *Streptococcus pneumoniae* and *Bacillus* sp. do not liberate *O*-linked sugar chains having sialic acid from fetuin glycoprotein having sialic acid. It is possible to liberate *O*-linked sugar chains from fetuin glycoprotein if several types of exo-glycosidases are incubated with fetuin having sialic acid before Endo-α-GalNAc-ase from *Clostridium perfringens*, or *Alcaligenes* sp., or *Streptococcus pneumoniae* or *Bacillus* sp. was incubated with fetuin having sialic acid. Sialic acid, galactose, and *N*-acetylgalactosamine were liberated from fetuin by incubating fetuin glycoprotein having sialic acid with the sialidase from *Arthrobacter ureafaciens*, β-galactosidase from *Streptococcus pneumonia* and β-*N*-acetyl hexosaminidase from *Streptococcus pneumonia*. Next, Galβ1-3GalNAc disaccharide was liberated when Endo-α-GalNAc-ase from *Clostridium perfringens*, *Alcaligenes* sp., *Streptococcus pneumoniae* and *Bacillus* sp. was reacted with residual fetuin glycoprotein. The fetuin *O*-linked sugar chains should in theory all be liberated without damage to the fetuin protein component. It goes without saying that the interaction of enzymes and substrate, such as the tendency for enzymes to approach the glycoprotein as the substrate, largely affect the types and amounts of sugar chains that do become liberated.

B. Release of *O*-Linked Sugar Chain Using Endo-α-GalNAc-ase-S

Endo-α-GalNAc-ase-S that has broad substrate specificity compared to marketed Endo-α-GalNAc-ase is introduced, showing the effect of this enzyme on sialoglycoprotein and the properties of the enzyme including transglycosylation activity.

1) Discovery

The major role of the sugar chain in mucin is to protect mucus glycoproteins from degradation. The portion of molecules densely present in such a sugar chain which covers the sugar chain even if it is thoroughly digested by protease remains as macromolecules without enzymatic effect. In order to develop low-molecularizing enzymes by efficiently degrading mucin having such a property, *Streptomyces* sp. OH-11242 was isolated from soil at Jogashima. The *Streptomyces* sp. OH-11242 was grown using porcine gastric mucin (PGM) as the sole carbon source, and multiple types of sugar chains in the culture solution were liberated. They were purified by Sephadex G-25 column chromatography and applied on to a silica gel 60 plate (Fig. 7.7). The reducing end of the liberated sugar chain was fluorescently labeled and hydrolyzed, and the reducing end sugar was analyzed. The findings indicated that multiple types of endo-type enzymes were present. One of these was Endo-α-GalNAc-ase, which became the source of focus.

2) Assay method

As noted above, multiple endo-type and exo-type enzymes are present in the culture solution of the *Streptomyces* sp. OH-11242. The activity of Endo-α-GalNAc-ase, which is the target, must be assayed from among these. In establishing an assay method for useful Endo-α-GalNAc-ase activity it is important to develop a method that enables the simple detection in small amounts of sugar chains that fulfill the following two criteria: i) the molecular weight of the liberated sugar chains should be greater than Galβ1-3GalNAc disaccharide and ii) the reducing end of the liberating sugar chain must be *N*-acetylgalactosamine. The assay method for an Endo-α-GalNAc-ase that selectively liberates mucin-type sugar chains from glycoprotein was conducted as follows. Twenty micrograms of PGM were incubated with 5 μL of an enzyme solution and an appropriate amount of buffer, in a total volume of 20 μL. After incubation at 37°C for 4 h, the reaction mixture was boiled at 100°C for 5 min and pyridylaminated (PA), as previously described.[54)]

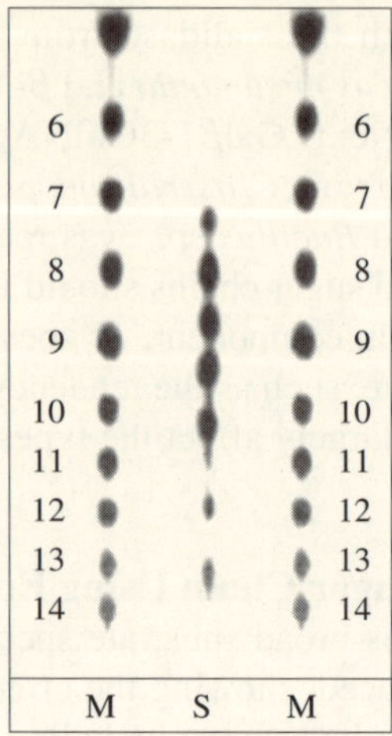

Fig. 7.7 Thin-layer chromatogram of the sugar chains in the culture solution of *Streptomyces* sp. The silica gel 60 plate was developed at room temperature for 20 h using the following solvent system: *n*-propanol/ acetic acid/ water (3/3/2, v/v/v). Sugar chains were visualized with phenol-H_2SO_4 reagent. M, mixture of standard glucose oligomers. Vertical numbers indicate glucose units. S, sugar chains prepared from the culture solution of *Streptomyces* sp. OH-11242.

To determine the reducing end sugars of the sugar chains liberated from PGM through enzyme reactions, the sugar chain-PA was hydrolyzed with 4 M TFA-HCl at 100°C for 4 h in a sealed tube, and the produced PA-monosaccharides were chromatographed on a reverse-phase column following reacetylation of the amino groups. The above method is summarized in Fig. 7.8.

3) Partial purification
The culture fluid of *Streptomyces* sp. OH-11242 was dialyzed against distilled water. The dialysate was brought to 80% (w/v) saturation by adding solid ammonium sulfate with stirring. The precipitate that formed was collected by centrifugation at 9,500 × *g* for 20 min, dissolved in a minimum amount of water and dialyzed against distilled water. The crude enzyme was prepared by 80% (v/v) ammonium sulfate precipitation, gel chromatofocusing, DEAE-Toyopearl and *N*-(*p*-aminophenyl)-oxamic acid-agarose chromatography as previously described.[52, 55] Some of the pooled solution was concentrated by dialysis against polyethylene glycol 20000.

4) Deglycosylation of sialoglycoprotein
Fetuin (50 mg) in the presence of 500 μM 2,3-dehydro-2-deoxy-*N*-acetylneuraminic acid was incubated at 37°C for 6 h with enzyme solution (0.1 mL) of Endo-α-GalNAc-ase-S and an appropriate amount of buffer in a total volume of 5 mL. After incubation, the reaction mixture was boiled at 100°C for 10 min. Following centrifugation at 6,000 × *g* for 10 min, the supernatant was applied to a Bio-Gel P-4 column eluted with 0.1 M pyridine acetate buffer, pH 5.2, and 2 mL

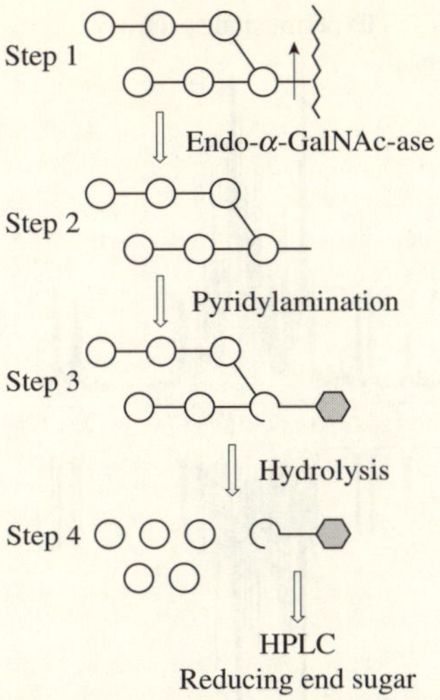

Step 1
Endo-α-GalNAc-ase
Step 2
Pyridylamination
Step 3
Hydrolysis
Step 4
HPLC
Reducing end sugar

Fig. 7.8 Flow chart of the procedure for the assay of the Endo-α-GalNAc-ase activity. Sugar chains liberated from glycoproteins and glycopeptides were separated according to size using TLC and/or HPLC in Step 2 or Step 3 in the figure. Of the sugar chains liberated from glycoproteins and glycopeptides, the reducing end sugar of those having a greater molecular weight than Galβ1-3GalNAc was analyzed. open circles, sugar; shaded hexagons, 2-aminopyridine.

fractions were collected. A portion of the sugar chains (I, II and III) was pyridylaminated[54] (I-PA, II-PA and III-PA) after being concentrated and then further analyzed by HPLC.[52] In order to reach a decisive conclusion about the structure of the sugar chains (I-PA, II-PA and III-PA), ^{1}H-NMR spectral data were collected and analyzed.[52, 56, 57] The ^{1}H-NMR spectrum observed for II-PA is given in Fig. 7.9 (experimental spectrum) as a typical example, together with the computer-calculated spectrum (Fig. 7.9, simulated spectrum).[58, 59] Assignment for the all protons of the sugar chains I-PA and III-PA were performed by the above method. From the assignment for the all protons of sugar chains I-PA, II-PA and III-PA, the structures of the PA-sugar chains I-PA, II-PA and III-PA were determined as the PA-derivatives of NeuAcα2-3Galβ1-3NeuAcα2-3Galβ1-4GlcNAcβ1-6) GalNAc, NeuAcα2-3Galβ1-3(NeuAcα2-6)GalNAc and NeuAcα2-3Galβ1-3GalNAc, respectively. In ^{1}H-NMR spectra, the fact that all protons can be assigned does not simply mean that only the primary structure of sugar chains can be reliably determined.[60] Structural information is increased by assignment of all the protons in sugar chain,[61] so the various possibilities related to "functional analysis" expand.

5) Transglycosylation activity

Endo-α-GalNAc-ase-S was able to prepare either naturally or non-naturally present glycopeptides by transglycosylation activity, which is the activity that transfers sugar chains to peptides.[62] The next issue will be the development of technologies that selectively introduce the sugar chain at specific positions in the sugar

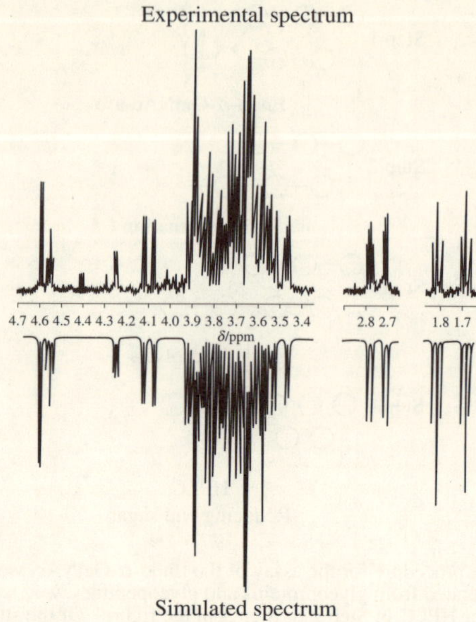

Fig. 7.9 Comparison of ^{1}H NMR spectrum (experimental spectrum) of II-PA and simulated ^{1}H NMR spectrum of NeuAcα2-3Galβ1-3(NeuAcα2-6) GalNAc-PA.

residue, and selectively introduce the sugar chain to specific amino acids of a peptide. As a result, it will no doubt become possible to freely design and synthesize mucin type sugar chains, glycopeptides and glycoproteins having effective functions.

Analysis of sugar chain structures that combines Endo-α-GalNAc-ase and the nondestructive NMR analysis method allows for a broad range of applications. It is therefore anticipated that applying this method to natural glycoproteins and sugar chains will lead to the discovery and development of more useful endo-type enzymes that act on mucus glycoprotein.

7.3 Analysis of Sugar Chain Structures Using Endo-β-xylosidase Activity of Cellulase

Structural analysis of sugar chains is absolutely indispensable to investigate the function of oligosaccharides, to assess the quality of genetically engineered glycoprotein, and for the application to diagnosis of disease caused by structural abnormalities in sugar chains. It is not difficult to determine the sequence of proteins and nucleic acids using a commercial automatic sequencer. Structural analysis is, however, very difficult and complex, so there is no other way but for researchers to carry it out by themselves. A glycomic approach, which involves exhaustive research analyzing glycoproteins including carbohydrate chains, is especially important in this post genomic era. Separation or structural determination of sugar chains of glycoprotein by two-dimensional mapping techniques has been established since the 1980s[63-66] and more than 160 kinds of oligosaccharide libraries have already been reported. This method is in practical use for structure estimation of oligosaccharides with unknown structures. However, the two-dimensional mapping for proteoglycans has never been tried, largely because of the difficulty of purification of endoglycosidases which release intact glycosaminoglycan (GAG) chains from core proteins.

Cellulase (most coded for endo-1,4-β-glucanase, EC 3.2.1.4) is an endoglycosidase that is widely distributed in plants and microorganisms such as fungi, and acts on the glucosylβ1-4glucose structure in cellulose of a long chain.[67] The glucosylβ1-4glucose configuration is similar to the Xyl-Ser linkage between the GAG chain and the core protein in proteoglycan. It is known that some glycosidases recognize both xyloside bonds and glucoside bonds.[68, 69] It has also been suggested that β-glucosidase from *Aspergillus sojae* shows β-xylosidase activity.[70] Recently, the author and his research team demonstrated that cellulase from *Aspergillus niger* can hydrolyze the Xyl-Ser linkage in proteoglycan and liberate intact GAG chains from glycosaminoglycan peptide (GAG-peptide) by its endo-β-xylosidase activity.[71] The cellulase acted on the Xyl-Ser linkage between the core protein and GAG chain regardless of the sugar chain type; it hydrolyzed the Xyl-Ser linkage of chondroitin sulfate peptidoglycan (ChS-peptide), dermatan sulfate peptidoglycan (DS-peptide), and heparan sulfate peptidoglycan (HS-peptide). This enzyme is easily obtained from commercial sources. We succeeded in isolating GAG chains from core protein using this enzyme, and in making two-dimensional oligosaccharide maps of the chains.[72]

Here, we introduce an experimental example in which we investigated the relation between the molecular weight and the number of sulfate groups per disaccharide unit of GAGs of unknown structure.

In order to isolate intact GAGs from the target proteoglycan, proteoglycans from various sources were incubated with cellulase after digestion with protease. The GAG chains were labeled with 2-aminopyridine (PA),[53, 73] then fractionated by HPLC, anion-exchange chromatography, on a TSKgel SAX column. Three kinds of GAG-PA, HS-PA, ChS/DS-PA and Hep-PA were identified and the molar ratio of each GAG was determined (Fig. 7.10, Table 7.3).

Each GAG chain fraction was then analyzed by two kinds of HPLC with different principles of separation to obtain information on molecular weights and the number of sulfate groups per disaccharide unit. The molecular weight of each GAG was estimated by HPLC on a Shodex OHpak SB-803 gel filtration column, and the sulfate content of each fraction was determined by HPLC on a TSKgel DEAE 5-PW anion-exchange column after chondroitinase ABC digestion.[74] The number of sulfate groups per disaccharide was calculated from the results of HPLC. These two relations can be shown as a two-dimensional map (Fig. 7.11). Various PA-GAGs, PA-Ch6S, PA-Ch4S, PA-HS and PA-Hep, whose molecular weights are known, were used as standards.

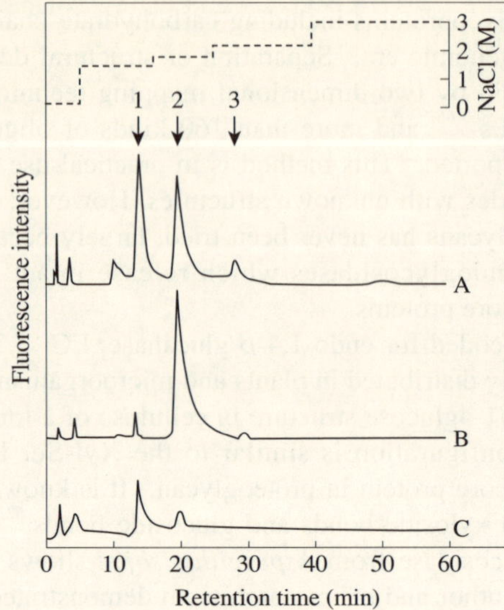

Fig. 7.10 HPLC of GAG chains obtained from bovine lung (A), tracheal cartilage (B) and cerebrum (C). GAGs from various tissues were analyzed by HPLC on a TSKgel SAX column with fluorescence detection. The arrows indicate the elution position of PA-HS (1), PA-ChS/DS (2) and PA-Hep (3). PA, 2-aminopyridine; GAG, glycosaminoglycan; HS, heparan sulfate; ChS, chondroitin sulfate; DS, dermatan sulfate; Hep, heparin.
[Reprinted from *Anal. Biochem.* **325**, Iwafune, M. *et al.*, 37, Copyright (2004), with permission from Elsevier]

Table 7.3 GAG composition of bovine tissues

Tissue	HS	ChS/DS	Heparin
	nmol/g wet tissue		
Lung	8.3	44.0	23.2
	[a](11.0)	(58.3)	(30.7)
Tracheal cartilage	4.1	128.5	5.5
	(3.0)	(93.0)	(4.0)
Cerebrum	7.4	5.2	–
	(58.7)	(41.3)	–

[a] Numbers in parentheses represent the percentage of the total.
[Reprinted from *Anal. Biochem.* **325**, Iwafune, M. *et al.*, 37, Copyright (2004), with permission from Elsevier]

The combination of endo-β-xylosidase activity of cellulase and two-dimensional mapping has made it possible to analyze various GAG chains encyclopedically.

This new approach will greatly contribute to not only biological but also clinical research based on the analysis of GAG chains; we expect breakthroughs in unknown functions of the GAG portions of proteoglycan and comparison of the structures of GAGs from proteoglycans in sick and normal tissues, and we hope that it will help to clarify details of the mechanisms of GAG synthesis, including

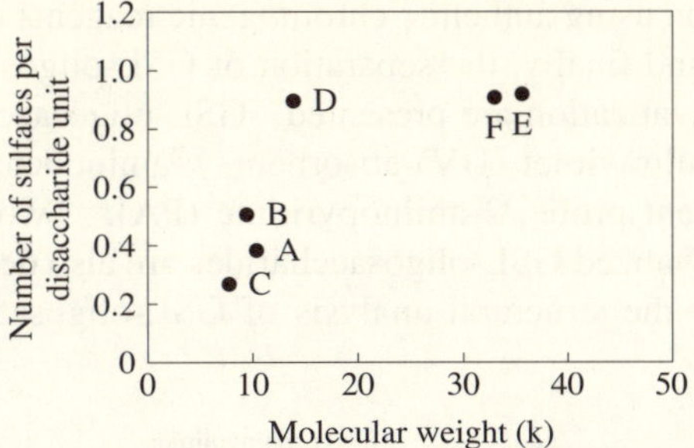

Fig. 7.11 Two-dimensional polysaccharide chain map for PA-GAGs obtained from each tissue. Symbols indicate the retention times of fractionated HS (A-C) and ChS/DS (D-F) from lung (A and D), tracheal cartilage (B and E) and cerebrum (C and F). PA, 2-amino pyridine; GAG, glycosaminoglycan, HS, heparan sulfate, ChS, chondroitin sulfate, DS, dermatan sulfate. [Reprinted from *Anal. Biochem.* **325**, Iwafune, M. *et al.*, 39, Copyright (2004), with permission from Elsevier]

phosphorylation, sulfation and other modifications.

7.4 Analysis of Sugar Chain Structures Using an Endoglycoceramidase

Glycosphingolipids (GSLs) consist of oligosaccharides and heterogeneous ceramides. Thus chromatographic analysis of natural GSLs often gives complex results. Analysis of oligosaccharides cleaved from GSLs is expected to provide simpler chromatographic separation depending solely on the oligosaccharide structure. Although ozonolysis[75, 76] and periodate oxidation[77] have been conducted to obtain intact oligosaccharides from GSLs, these methods require an alkaline condition and are not suitable for preparing alkaline-labile oligosaccharides. Nor are they suitable for preparing oligosaccharides from small sample quantities. They are also time consuming. This section presents a simple and reproducible method for obtaining intact oligosaccharides from GSLs with high recovery using endoglycoceramidase (EGCase). The specificity of EGCase is quite wide, so this method can be applied to all neutral and acidic GSLs in which an oligosaccharide is linked to ceramide *via* β-glucosidic linkage.[78] To analyze the oligosaccharide structure of GSLs, the GSL fraction should be extracted from natural sources and separated from the other amphiphilic compounds. The classic method of GSL isolation includes alkaline treatment of the lipid extracts, which means that alkaline-labile *O*-acylated GSLs can be degraded, *e.g.*, *O*-acetylated derivatives of the sialic acid moiety are often found in natural GSLs. In this section, an improved method for the isolation of GSLs using phenyl boronate-conjugate column is first introduced, followed by a description of the enzymatic release of intact oligosaccharides from GSLs by the action of EGCase and purification of released oligosaccharides. Several chromatographic protocols are available for the separation of GSL-derived oligosaccharides. Separation of GSL-oligosaccharides by

TLC and visualization using authentic chromogenic reagents and specific antibody are then described, and finally, the separation of GSL-oligosaccharides by HPLC before and after derivatization are presented. GSL-oligosaccharides are quantitatively labeled with ultraviolet (UV)-absorbent, *p*-aminobenzoic acid ethyl ester (ABEE) or fluorescent probe, 2-aminopyridine (PA). Two-dimensional HPLC mappings of the derivatized GSL-oligosaccharides are also described.

A flow chart of the structural analysis of GSL-oligosaccharide is shown in Fig. 7.12.

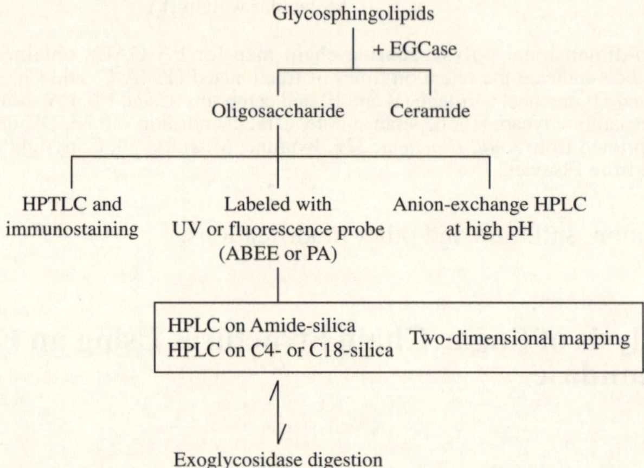

Fig. 7.12 Structure analysis of oligosaccharides from GSLs. Flow chart of the structure analysis is shown.

7.4.1 Isolation of GSLs from Biological Samples Using a Phenyl Boronate-conjugated Matrix Column

GSLs are usually isolated from crude extracts after treatment with alkaline by which abundant glycerophospholipids are eliminated. However, recent careful analysis of GSLs revealed that alkaline-labile GSLs are present in natural sources, although they are usually minor constituents. For example, *O*-acetylated sialyl residues, such as 4-*O*-acetyl Neu5Gc[79-81] and 9-*O*-acetyl Neu5Ac,[82-84] are often found in alkaline-labile GSLs. Some of them are tumor-associated or developmentally regulated antigens. To isolate such alkaline-labile GSLs, the authors introduced phenyl boronate agarose (PBA) gels[85, 86] according to the method of Krohn *et al.*[87] using phenyl boronate polystyrene. The method is based on the strong affinity of *cis*-diol groups of carbohydrates to boronic acid (Fig. 7.13). Now, however, phenyl boronate conjugated polyacrylamide beads, Immobilized Boronic acid Gel (Pierce), and phenyl boronate-conjugated glass beads, ProSep-PB Media (Millipore), are commercially available and either phenyl boronate matrix (PBM) can be used as an alternative for PBA.

Oligosaccharide moieties of most natural GSLs contain *cis*-diol groups, which

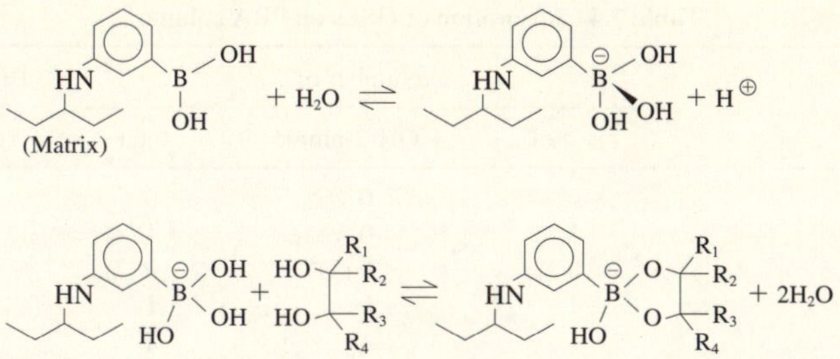

Fig. 7.13 Reaction of matrix-immobilized boronic acid and *cis*-diol group.

have high affinity to boronic acid of PBM. Anhydrous solvent is preferable for the binding and the bound GSLs are readily released from the matrix into hydrous solvent. For the binding of GSLs to PBM, a chloroform/methanol mixture 9/1 to 8/1 (v/v) in volume is effective. To recover the bound GSLs, a chloroform/methanol/water mixture of 5/1/1 (v/v/v) is used. The procedure is shown in Fig. 7.14A. GSLs of 20 mg, which contain 16 μmol hexose, are purified from porcine erythrocyte lipid extract by 1 mL of PBA60 using chloroform/methanol (9/1, v/v) as the loading solvent. Gangliosides of 42 mg, which contain 8.4 μmol of sialic acid, are purified from the crude amphiphilic lipid fraction of bovine brain (Folch's upper phase) by 1 mL of the matrix using chloroform/methanol (8/2, v/v) as the loading solvent. Since glucose possesses no *cis*-diol group, GlcCer does not bind to PBM. However, even though sulfatide does not contain a *cis*-diol group, it binds to PBM as well as GalCer. 1-OH-2-Sulfate group may interact to boronic acid (Table 7.5). The 1-OH-2-amide group of *N*-acetylhexosamines or sialic acids may also interact with boronic acid (Table 7.4). In addi-

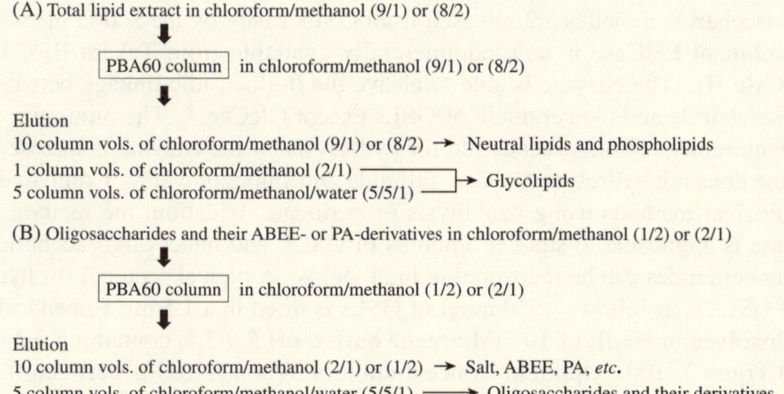

Fig. 7.14 Purification of GSLs (A), oligosaccharides or their derivatives (B) by phenyl boronate matrix chromatography. PBA60 contains 60-100 μmol boron/mL gel. The gel should be dehydrated by methanol washing then suspended in the indicated solvent. Chloroform/methanol/water mixture of ratio in volume. Samples containing gangliosides or relatively hydrophilic GSLs are dissolved in chloroform/methanol(2/1) or (1/1) then chloroform is added to a ratio of 8/1. Oligosaccharide samples are dissolved in methanol then chloroform is added to a ratio of 1/2.

Table 7.4 Adsorption of GSLs on PBA column

Material	Number of			Binding capacity
	cis-diol	1-OH-2-amide	total	(μmol/mL gel)
GlcCer	0	0	0	0
GalCer	1	0	1	3.0
Sulfatide	0	0 (2)[a]	0 (2)	3.4
LacCer	1	0	1	8.3
Gb_4Cer	1	1	2	16.5
NeuGc-LacCer	2	2	4	11.9
GM1	2	2	4	18.8
GD1a	4	3	7	12.6
GT1b	4	4	8	5.0
L-α-Phosphatidyl-inositol	1	0	1	2.2

PBA60 contains 60-100 μmol boron/mL gel.
Chloroform/methanol in a ratio of 8/2 (v/v) was used for the loading of glycolipids while the ratio was 1/2 (v/v) for lactose.
[a]1-OH-2-sulfate.

tion to GSLs, PBM adsorbs phosphatidylinositol, which has a *cis*-diol group. Moreover, using PBM, Nagatsuka *et al.* isolated a novel microdomain-signaling lipid, phosphatidylglucose,[88, 89] although the binding mechanism remains unknown. PBM can also be used for the purification of GSL-oligosaccharides as described below.

7.4.2 Enzymatic Release of GSL-oligosaccharides by EGCase and Their Purification

Oligosaccharide moieties are released from GSLs easily by the action of EGCase. Recombinant EGCase is now commercially available from Takara Bio, Japan. (rEGCase II). The enzyme is able to cleave the β-glucosidic linkage between the oligosaccharide and the ceramide of GSLs except GlcCer.[78] The minimum structure requirement of oligosaccharide for hydrolysis by the enzyme is lactose. The enzyme does not hydrolyze GalCer, sulfatide and sphingomyelin. Compared with the classical methods using ozonolysis or periodate oxidation, the method using EGCase is applicable to smaller amounts of GSLs, and intact oligosaccharides as well as ceramides can be recovered in high yields. A typical protocol for hydrolysis of GSLs is as follows[78]: 50 nmol of GSLs is dried in a 1.5 mL Eppendorf tube and dissolved in 40 μL of 10 mM acetate buffer, pH 5.0-5.5, containing 0.2-0.4 % (w/v) Triton X-100. Optimum concentration of Triton X-100 is between 0.2 and 0.4 % of reaction mixture. The solution was subjected to sonic treatment for 1-2 min. If difficult to dissolve, the sample should be heated briefly. Recombinant EGCase II solution of 0.75-3.75 μL (1.5-7.5 milliunits) is added to GSL solution and the reaction mixture is incubated at 37°C for 16 h. Since globo-series GSLs are somewhat resistant to hydrolysis by EGCase,[78] use of 7.5 milliunits enzyme is recommended for 50 nmol of GSLs.

Released oligosaccharides are recovered in the upper layer of Folch's partition by adding 5 volumes of mixture of chloroform/methanol (2/1, v/v). Ceramide and uncleaved GSLs are recovered in the lower layer. Alternatively, the oligosaccharides and the hydrophobic compounds can be separated by Sep-Pak C_{18} cartridge.[90] Contamination with salt may disrupt the results, *e.g.*, MS analysis and ion-exchange chromatography. To remove salt and nonsaccharide compounds from the oligosaccharide fraction, a PBM column is useful (Fig. 7.14B, Table 7.5) as an alternative for Sephadex G-15 column chromatography.[91] Glucose is not adsorbed to PBM, so it can be separated from the other saccharides (Table 7.5).

Table 7.5 Adsorption of saccharides on PBA column

Saccharide	*cis*-diol	1-OH-2-amide	total	(*μ*mol/mL gel)
Glucose	0	0	0	1.9
Galactose	1	0	1	25
D-Glucuronic acid	0	0	0	5.5
D-Galacturonic acid	1	0	1	7.4
N-Acetyl-D-glucosamine	0	1	1	4.4
N-Acetyl-D-galactosamine	1	1	2	27.8
Lactose	1	0	1	11.6
Chondro *Δ*Di 0S	0	1	1	1.23
Chondro *Δ*Di 4S	0	1	1	2.3
Chondro *Δ*Di 6S	0	1	1	2.6
Maltotriose	0	0	0	0.6
GM1 oligosaccharide	2	2	4	>2.5
GD1a oligosaccharide	4	3	7	5.4
GT1b oligosaccharide	4	4	8	7.7
NeuGc-nLc4	2	2	4	2.2

Chloroform/methanol (1/2, v/v) was used for the loading of oligosaccharides.

7.4.3 Thin-layer Chromatography (TLC) of GSL-oligosaccharides and Immunostaining

Oligosaccharides released from GSLs by EGCase can be analyzed by TLC.[92] The freeze-dried reaction mixture is dissolved in 10 μL of 50% aqueous methanol and applied to silica gel 60 HPTLC or TLC plate (Merck) and developed with 1-butanol/acetic acid/water (2/2/1, v/v/v). All the oligosaccharides are detected by orcinol-H_2SO_4 or sialylated saccharides by the resorcinol-HCl method. They can be quantitated by densitometry with a detection range of 0.5-5 nmol.

Oligosaccharides separated on TLC plate can be detected by antibodies or lectins similar to the case of native GSLs.[91] For this purpose, the NH_2-HPTLC plate (Merck) is used. Amino groups on the plate and the reducing end of the oligosaccharides form Schiff's base and the saccharides are immobilized on the plate. Schiff's base is reduced to stabilize the coupling.[91] Oligosaccharides in methanolic or aqueous solutions are applied to the NH_2-HPLC plate and developed with acetonitrile/15 mM potassium phosphate buffer, pH 5.6 (1/1, v/v). The plate is dried and the reducing reaction is carried out by incubation with 0.2 M sodium cyanoborohydride in 1,2-dichloroethane/methanol (5/1, v/v) containing a 1/12 volume of methanolic 0.1 M HCl for 16 h at room temperature. After the incubation, the plate is washed with methanol and dried. Enzyme-immunostaining of the immobilized oligosaccharides on NH_2-HPLC is carried out by a method similar to that for native GSLs.[93, 94] The plate is soaked for 1 h at room temperature in PBS containing 1% ovalbumin, 1% polyvinylpyrrolidone, 0.02% NaN_3 (solution A) to block nonspecific antibody binding. Ovalbumin may be replaced by bovine serum albumin according to the specificity of the antibodies or lectins. Antibody or lectin is diluted in solution A and applied to the TLC plate in a plastic bag and incubated for 1-2 h at 23-37˚C. The plate is washed with PBS containing 0.05% Tween 20 (solution B) and incubated with horseradish peroxidase labeled-second antibody in solution B for 1-2 h at 23-37˚C. The plate is washed with solution B and a color developing substrate solution is added to the plate. The substrate solution is prepared as follows: 3 mg of 4-chloro-1-naphthol dissolved in 1 mL methanol is added to 5 mL of 100 mM Tris-HCl buffer, pH 7.2, then H_2O_2 is added to a final concentration of 0.01%. The sensitivity of the detection is about 1-10 pmol. Chemiluminescent substrate for the enzyme may be used to increase the sensitivity. An avidin-biotin system or alkaline phosphatase-labeled antibody may also be used.

7.4.4 Anion-exchange HPLC of GSL-oligosaccharides with a Strong Alkaline Solvent System

The technique of high-pH anion-exchange chromatography with pulsed amperometric detection is a powerful method for resolving closely related oligosaccharides.[95] Beside acidic oligosaccharides, neutral oligosaccharides can be analyzed by the system without a derivatization (Fig. 7.15). Oligosaccharides with Neu5Ac or Neu5Gc, are clearly separated (Fig. 7.15C). Due to the use of a strong alkaline developing solvent for the chromatography, *O*-acyl groups of oligosaccharides are cleaved off during analysis and *O*-acylated oligosaccharides are undetectable. Pulsed amperometric detection has a sensitivity range of 10-100 pmol. It produces not equal but similar responses for a wide range of saccharide structures (Fig. 7.15).[95] This method can also detect sugar alcohols, amino acids and peptides. These contaminants sometimes disrupt the results. Desalting of the sample is recommended.

7.4.5 HPLC Separation of GSL-oligosaccharides Labeled with UV-absorptive or Fluorescent Probes and Two-dimensional HPLC Mapping

Since oligosaccharide itself has neither specific UV absorption nor fluorescence, labeling with UV absorbent or fluorescent probes is necessary for sensitive detection. UV-absorbent ABEE[96, 97] and 2-aminopyridine, a fluorescent probe,[53, 98]

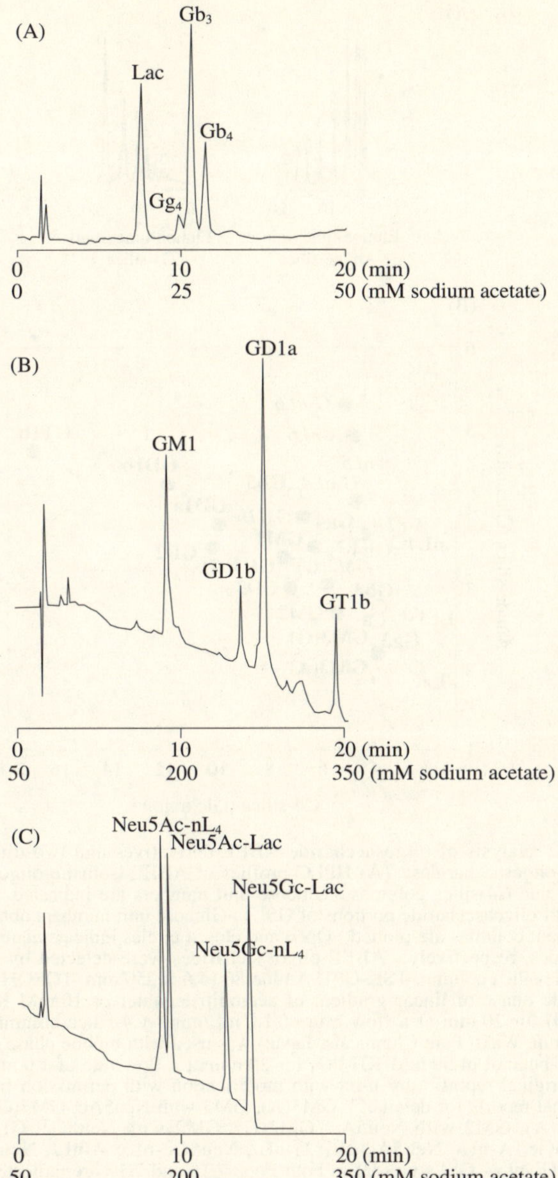

Fig. 7.15 Separation of oligosaccharides by anion-exchange HPLC using strong alkaline mobile phase. Bio LC system (Dionex, USA) with Carbopac PA-1 column (4.6 × 250 mm) was used. Mobile phase of 100 mM NaOH containing the indicated concentrations of sodium acetate in linear gradient was used at a flow rate of 1 mL/min. (A) Oligosaccharide mixture containing Lac (2.5 nmol), Gb_3 (2.5 nmol), Gb_4 (1.5 nmol) and Gg_4 (0.4 nmol) was analyzed. (B) Oligosaccharide mixture from bovine brain ganglioside was analyzed. (C) Oligosaccharide mixture containing 1 nmol each of Neu5Ac-Lac, Neu5Gc-Lac, Neu5Ac-nL_4 and Neu5Gc-nL_4 was analyzed.

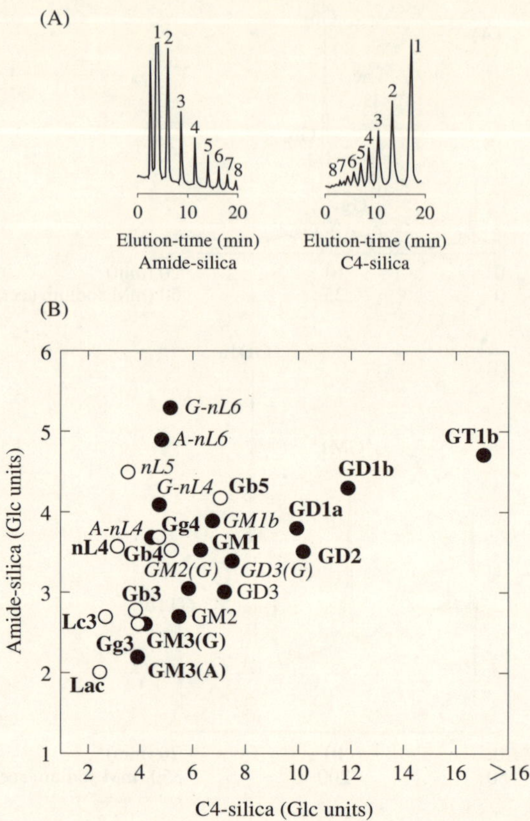

Fig. 7.16 HPLC analysis of oligosaccharide ABEE-derivatives and two-dimensional mapping of GSL-oligosaccharides. (A) HPLC profiles of ABEE-isomalto-oligosaccharides on amide-silica and C4-silica columns. Glucose unit numbers are indicated. (B) Two-dimensional map of oligosaccharide portions of GSLs. Glucose unit numbers obtained by HPLC on two different columns are plotted. Open and closed circles indicate neutral and acidic oligosaccharides, respectively. ABEE-oligosaccharides were detected by absorption at 304 nm. Amide-silica column, TSK-GEL Amide 80 (4.6 × 250 mm, TOSOH, Japan) was used with mobile phase of linear gradient of acetonitrile/water or 10 mM KH_2PO_4 (75/25, v/v to 50/50) for 20 min at a flow rate of 1.0 mL/min. C4-silica column, Wakosil 5C$_4$ (4.6 × 250 mm, Wako Pure Chemicals, Japan) was used with mobile phase of linear gradient of 4-8% 1-butanol in 50 mM KH_2PO_4 for 20 min at a flow rate of 1.0 mL/min. The figures of the original report[96)] are used with modification with permission from Elsevier. See the original reports for details.[96)] GM3(A), GM3 with Neu5Ac; GM3(G), GM3 with Neu5Gc; GM2(A), GM2 with Neu5Ac; GM2(G), GM2 with Neu5Gc; GD3(G), GD3 with two Neu5Gc; A-nL$_4$, Neu5Ac-nL$_4$; G-nL$_4$, Neu5Gc-nL$_4$; A-nL$_6$, Neu5Ac-nL$_6$; and G-nL$_6$, Neu5Gc-nL$_6$. GSLs present in both Figs. 7.16 and 7.17 are indicated in bold letters while those present only in Fig. 7.16 are indicated in italics.

have been successfully used for the HPLC of GSL-oligosaccharides. Amounts as low as 4-5 pmol of ABEE-oligosaccharides and 50 fmol of pyridylamino (PA) oligosaccharides are detectable. ABEE-oligosaccharides possess constant molar absorption coefficients when a hexose is present in the reducing end of the oligosaccharide.[99)] Fluorescence efficiency of PA-oligosaccharides is almost the same. In both cases, the reducing ends of oligosaccharides form Schiff's base

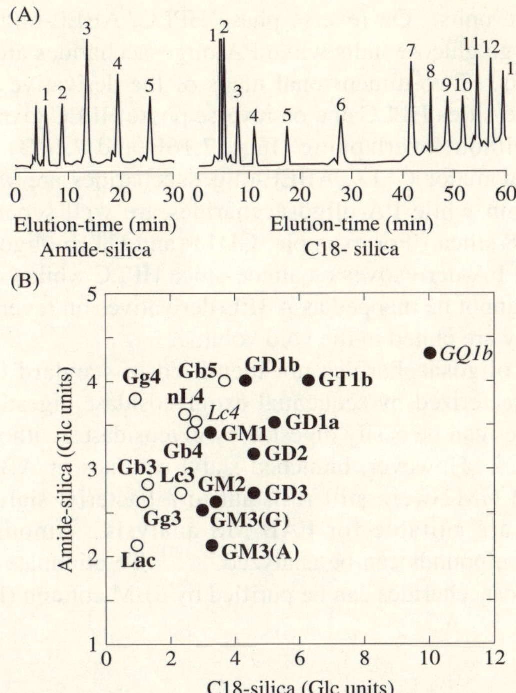

Fig. 7.17 HPLC analysis of oligosaccharide PA-derivatives and two-dimensional mapping of GSL-oligosaccharides. (A) HPLC profiles of PA-isomalto-oligosaccharides on amide-silica and C18-silica columns. Glucose unit numbers are indicated. (B) Two-dimensional map of oligosaccharide portions of GSLs. Glucose unit numbers obtained by HPLC on two different columns are plotted. Open and closed circles indicate neutral and acidic oligosaccharides, respectively. PA-oligosaccharides were detected by fluorescence at 380-400 nm by excitation with 310-320 nm wave length. Amide-silica column, Palpak Type S (4.6 × 250 mm, Takara Bio, Japan) was used with mobile phase of linear gradient of acetonitrile/200 mM acetic acid-triethylamine buffer, pH 7.3 (75/25, v/v to 62.5/37.5) for 25 min at a flow rate of 1.0 mL/min. C18-silica column, Palpak Type R (4.6 × 250 mm, Takara Bio) was used with mobile phase of linear gradient of 0-0.25% 1-butanol in 50 mM acetic acid-triethylamine buffer, pH 5.0, for 50 min at a flow rate of 1.0 mL/min. The figures of the original report[98] are used with modification with permission from Elsevier. See the original reports for details.[98] Abbreviations for the GSLs are shown in the legend to Fig. 7.16. GSLs present in both Figs. 7.16 and 7.17 are indicated in bold letters while those present only in Fig. 7.17 are indicated in italics.

with the amino group of the probes and the base is reduced to induce covalent coupling. Usually, the reductive amination reaction is preferable in acidic pH, although sialyl groups of oligosaccharides are labile at pH~6. Thus, the reaction condition is optimized to not degrade sialyl compounds.[96, 100] PA derivatization system is commercially available from Takara Bio, Japan.

Acidic ABEE-oligosaccharides are completely separated from neutral ones on amide-silica column using mobile phase without salt.[96] Using salt-containing mobile phase, acidic ABEE-oligosaccharides are well separated on the same column. Separation of standard isomalto-oligosaccharides are shown in Figs. 7.16A and 7.17A. On amide-silica column, both ABEE- and PA-oligosaccharides are eluted from small glucose units. On reverse phase HPLC, ABEE-oligosaccharides are eluted out from large glucose units while PA-oligosaccharides are eluted out from small glucose units. Two-dimensional maps of the derivative blotting glucose units on both amide-silica HPLC and on reverse phase HPLC give good separation of major GSLs without overlapping (Figs. 7.16B and 7.17B). Generally, for oligosaccharides of major GSLs, ABEE-oligosaccharides are well separated on amide-silica column while PA-oligosaccharides are well separated on reverse phase column, C18-silica. For example, GD1a- and GT1b-oligosaccharides cannot be separated as PA-derivatives on amide-silica HPLC while GT1b- and GQ1b-oligosaccharides cannot be mapped as ABEE-derivatives on reverse phase C4-silica HPLC since they are eluted in the void volume.

Structures of oligosaccharides not identified in standard two-dimensional maps may be characterized by sequential exoglycosidase digestion. ABEE- and PA-oligosaccharides can be easily digested by glycosidases without detergent, unlike to native GSLs. However, branched sialyl residues of ABEE-oligosaccharides of GM1 and GM2 were still resistant to a bacterial sialidase.[96)] ABEE-oligosaccharides are suitable for FAB-MS analysis. Amounts of less than 100 pmol of the compounds can be analyzed.[99, 101)] To eliminate noise on HPLC, ABEE- or PA-oligosaccharides can be purified by PBM column (Fig. 7.14B).

References

1. Patel, T.P. and Parekh, R.B. (1994) *Methods Enzymol.* **230**, 57-66
2. Yeh, M.-S., Huang, C.-J., Leu, J.-H., Lee, Y.C., and Tsai, I.-H. (1999) *Eur. J. Biochem.* **266**, 624-633
3. Tarentino, A.L. and Plummer, T.H., Jr. (1994) *Methods Enzymol.* **230**, 44-57
4. Kol, O., Brasart, C., Spik, G., Montreuil, J., and Bouqelet, S. (1989) *Glycoconj. J.* **6**, 333-348
5. Kornfeld, R. and Kornfeld, S. (1985) *Annu. Rev. Biochem.* **54**, 631-664
6. Orberger, G., Geyer, R., Stirm, S., and Tauber, R. (1992) *Eur. J. Biochem.* **205**, 257-267
7. Nayak, B.R. and Spiro, R.G. (1991) *J. Biol. Chem.* **266**, 13978-13987
8. Rohringer, R. and Holden, D.W. (1985) *Anal. Biochem.* **144**, 118-127
9. Haselbeck, A., Schickaneder, E., von der Eltz, H., and Hosel, W. (1990) *Anal. Biochem.* **191**, 25-30
10. Segrest, J.P. and Jackson, R.L. (1972) *Methods Enzymol.* **28**, 54-62
11. Weitzhandler, M., Kadlecek, D., Avdalovic, N., Forte, J.G., Chow, D., and Townsend, R.R. (1993) *J. Biol. Chem.* **268**, 5121-5130
12. Takeda-Ezaki, M. and Yamamoto, K. (1993) *Arch. Biochem. Biophys.* **304**, 352-358
13. Dahms, N.M. and Brzycki-Wessell, M.A. (1995) *Arch. Biochem. Biophys.* **317**, 497-503
14. Kauffman, D.L., Zager, N.I., Cohen, E., and Keller, P.J. (1970) *Arch. Biochem. Biophys.* **137**, 325-339
15. Keller, P.J., Kauffman, D.L., Allan, B.J., and Williams, B.L. (1971) *Biochemistry* **10**, 4867-4874
16. Yamashita, K., Tachibana, Y., Nakayama, T., Kitamura, M., Endo, Y., and Kobata, A. (1980) *J. Biol. Chem.* **255**, 5635-5642
17. Ito, K., Li, L.-F., Nishiwaki, M., Okada, Y., and Minamiura, N. (1992) *J. Biochem.* **112**, 88-92
18. Ito, K., Okada, Y., Ishida, K., and Minamiura, N. (1993) *J. Biol. Chem.* **268**, 1674-1681
19. Koide, N. and Muramatsu, T. (1974) *J. Biol. Chem.* **249**, 4897-4904
20. Tarentino, A.L., Plummer, T.H., Jr., and Maley, F. (1972) *J. Biol. Chem.* **247**, 2629-2631
21. Tai, T., Yamashita, K., and Kobata, A. (1977) *Biochem. Biophys. Res. Commun.* **78**, 434-441
22. Kadowaki, S., Yamamoto, K., Fujisaki, M., and Tochikura, T. (1991) *J. Biochem.* **110**, 17-21
23. Tarentino, A.L. and Maley, F. (1976) *J. Biol. Chem.* **251**, 6537-6543
24. Tachibana, Y., Yamashita, K., and Kobata, A.(1982) *Arch. Biochem. Biophys.* **214**, 199-210
25. Tachibana, Y., Yamashita, K., Kawaguchi, M., Arashima, S., and Kobata, A. (1981) *J. Biochem.* **90**, 1291-1296

26. Tarentino, A.L., Gomez, C.M., and Plummer, T.H., Jr. (1985) *Biochemistry* **24**, 4665-4671
27. Plummer, T.H., Jr. and Tarentino, A.L. (1991) *Glycobiology* **1**, 257-263
28. Trimble, R.B. and Tarentino, A.L. (1991) *J. Biol. Chem.* **266**, 1646-1651
29. Muramatsu, T. (1971) *J. Biol. Chem.* **246**, 5535-5537
30. Pierce, R.J., Spik, G., and Montreuil, J. (1980) *Biochem. J.* **185**, 261-264
31. Baussant, T., Strecker, G., Wieruszeski, J.M., Montreuil, J., and Michalski, J.C. (1986) *Eur. J. Biochem.* **159**, 381-385
32. Yoshima, H., Matsumoto, A., Mizuochi, T., Kawasaki, T., and Kobata, A. (1981) *J. Biol. Chem.* **256**, 8476-8484
33. Trimble, R.B., Atkinson, P.H., Tarentino, A.L., Plummer, T.H., Jr., Maley, F., and Tomer, K.B. (1986) *J. Biol. Chem.* **261**, 12000-12005
34. Takegawa, K., Yamaguchi, S., Kondo, A., Iwamoto, H., Nakoshi, M., Kato, I., and Iwahara, S. (1991) *Biochem. Int.* **24**, 849-855
35. Yamamoto, K., Kadowaki, S., Watanabe, J., and Kumagai, H. (1994) *Biochem. Biophys. Res. Commun.* **203**, 244-252
36. Ito, K. and Minamiura, N. (1999) *Glycoconj. J.* **16**, 53
37. Seko, A., Koketsu, M., Nishizono, M., Enoki, Y., Ibrahim, H.R., Juneja, L.R., Kim, M., and Yamamoto, T. (1997) *Biochim. Biophys. Acta* **1335**, 23-32
38. Endo, Y., Yamashita, K., Han, Y.N., Iwanaga, S., and Kobata, A. (1977) *J. Biochem.* **82**, 545-550
39. Carlson, D.M. (1968) *J. Biol. Chem.* **243**, 616-626
40. Kuraya, N. and Hase, S. (1992) *J. Biochem.* **112**, 122-126
41. Patel, T., Bruce, J., Merry, A., Bigge, C., Wormald, M., Jaques, A., and Parekh, R. (1993) *Biochemistry* **32**, 679-693
42. Huang, C.C. and Aminoff, D. (1972) *J. Biol. Chem.* **247**, 6737-6742
43. Yamamoto, K., Fan, J.-Q., Kadowaki, S., Kumagai, H., and Tochikura, T. (1987) *Agric. Biol. Chem.* **51**, 3169-3171
44. Endo, Y. and Kobata, A. (1976) *J. Biochem.* **80**, 1-8
45. Bhavanandan, V.P., Umemoto, J., and Davidson, E.A. (1976) *Biochem. Biophys. Res. Commun.* **70**, 738-745
46. Iwase, H., Ishii, I., Ishihara, K., Tanaka, Y., Omura, S., and Hotta, K. (1988) *Biochem. Biophys. Res. Commun.* **151**, 422-428
47. Ashida, H., Yamamoto, K., Murata, T., Usui, T., and Kumagai, H. (2000) *Arch. Biochem. Biophys.* **373**, 394-400
48. Fun, J.-Q., Kadowaki, S., Yamamoto, K., Kumagai, H., and Tochikura, T. (1988) *Agric. Biol. Chem.* **52**, 1715-1723
49. Umemoto, J., Bhavanandan, V.P., and Davidson, E.A. (1977) *J. Biol. Chem.* **252**, 8609-8614
50. Glasgow, L.R., Paulson, J.C., and Hill, R.L. (1977) *J. Biol. Chem.* **252**, 8615-8623
51. Ishii-Karakasa, I., Iwase, H., Hotta, K., Tanaka, Y., and Omura, S. (1992) *Biochem. J.* **288**, 475-482
52. Ishii-Karakasa, I., Iwase, H., and Hotta, K. (1997) *Eur. J. Biochem.* **247**, 709-715
53. Tanaka, Y., Takahashi, Y., Shinose, M., Omura, S., Ishii-Karakasa, I., Iwase, H., and Hotta, K. (1998) *J. Ferment. Bioeng.* **85**, 381-386
54. Hase, S., Ikenaka, T., and Matsushima, Y. (1978) *Biochem. Biophys. Res. Commun.* **85**, 257-263
55. Cuatrecasas, P. and Illiano, G. (1971) *Biochem. Biophys. Res. Commun.* **44**, 178-184
56. Kamerling, J.P. and Vliegenthart, J.F.G.(1992) *Biol. Magn. Reson.* **10**, 1-287
57. Ishii-Karakasa, I. (2001), in *New Developments in Glycomedicine* (Endo, M., Harata, S., Saito, Y., Munakata, A., Sasaki, M., and Tsuchida, S., eds.) pp. 151-159, Elsevier Science B.V., Amsterdam
58. Van Halbeek, H., Dorland, L., Veldink, G.A., and Vliegenthart, J.F.G. (1982) *Eur. J. Biochem.* **127**, 1-6
59. De Tar, D.F. (1968) in *Computer Programs for Chemistry* vol.1 (De Tar, D.F., ed.) pp.10-53, Benjamin, W.A., New York
60. Rao, V.S.R., Qasba, P.K., Balaji, P.V., and Chandrasekaran, R. (1998) *Conformation of Carbohydrates*, Harwood Academic Publishers
61. Ishii-Karakasa, I. (2003) *Anal. Sci.* **19**, 93-97
62. Ajisaka, K., Miyasato, M., and Ishii-Karakasa, I. (2001) *Biosci. Biotechnol. Biochem.* **65**, 1240-1243
63. Hase, S., Sugimoto, T., Takemoto, H., Ikenaka, T., and Schmid, K. (1986) *J. Biochem.* **99**, 1725-1733
64. Hase, S., Natsuka, S., Oku, H., and Ikenaka, T. (1987) *Anal. Biochem.* **167**, 321-326

65. Hase, S., Ikenaka, K., Mikoshiba, K., and Ikenaka, T. (1988) *J. Chromatogr.* **434**, 51-60
66. Tomiya, N., Awaya, J., Kurono, M., Endo, S., Arata, Y., and Takahashi, N. (1988) *Anal. Biochem.* **171**, 73-90
67. Levy, I., Shani, Z., and Shoseyov, O. (2002) *Biomol. Eng.* **19**, 17-30
68. Notenboom, V., Birsan, C., Warren, R.A., Withers, S.G., and Rose, D.R. (1998) *Biochemistry* **37**, 4751-4758
69. Robinson, D. and Abrahams, H.E. (1967) *Biochim. Biophys. Acta* **132**, 212-214
70. Kimura, I., Yoshida, N., and Tajima, S. (1999) *J. Biosci. Bioeng.* **87**, 538-541
71. Takagaki, K., Iwafune, M., Kakizaki, I., Ishido, K., Kato, Y., and Endo, M. (2002) *J. Biol. Chem.* **277**, 18397-18403
72. Iwafune, M., Kakizaki, I., Nakazawa, H., Nukatsuka, I., Endo, M., and Takagaki, K. (2004) *Anal. Biochem.* **325**, 35-40
73. Kon, A., Takagaki, K., Kawasaki, H., Nakamura, T., and Endo, M. (1991) *J. Biochem.* **110**, 132-135
74. Suzuki, S., Saito, H., Yamagata, T., Anno, K., Seno, N., Kawai, Y., and Furuhashi, T. (1968) *J. Biol. Chem.* **243**, 1543-1550
75. Wiegandt, H. and Baschang, G.Z. (1965) *Z. Naturforsch. [B]*. **20b**, 164-166
76. Wiegandt, H. and Bucking, H.W. (1970) *Eur. J. Biochem.* **15**, 287-292
77. Hakomori, S.I. (1966) *J. Lipid Res.* **7**, 789-792
78. Ito, M. and Yamagata, T. (1989) *J. Biol. Chem.* **264**, 9510-9519
79. Gasa, S., Makita, A., and Kinoshita, Y. (1983) *J. Biol. Chem.* **258**, 876-881
80. Miyoshi, I., Higashi, H., Hirabayashi, Y., Kato, S., and Naiki, M. (1986) *Mol. Immunol.* **23**, 631-638
81. Yachida, Y., Tsuchihashi, K., and Gasa, S. (1997) *Carbohydr. Res.* **298**, 201-212
82. Hirabayashi, Y., Li, Y.-T., and Li, S.-C. (1983) *FEBS Lett.* **161**, 127-130
83. Hirabayashi, Y., Hirota, M., Suzuki, Y., Matsumoto, M., Obata, K., and Ando, S. (1989) *Neurosci. Lett.* **106**, 193-198
84. Zhang, G., Ji, L., Kurono, S., Fujita, S.C., Furuya, S., Hirabayashi, Y., Hirota, M., Suzuki, Y., Matsumoto, M., Obata, K., Ando, S., Li, Y.-T., and Li, S.-C. (1997) *Glycoconj. J.* **14**, 847-857
85. Higashi, H., Ikuta, K., Ueda, S., Kato, S., Hirabayashi, Y., Matsumoto, M., and Naiki, M. (1984) *J. Biochem.* **95**, 785-794
86. Higashi, H., Hirabayashi, Y., Fukui, Y., Naiki, M., Matsumoto, M., Ueda, S., and Kato, S. (1985) *Cancer Res.* **45**, 3796-3802
87. Krohn, K., Eberlein, K., and Gercken, G. (1978) *J. Chromatogr. A* **153**, 550-552
88. Nagatsuka, Y., Kasama, T., Ohashi, Y., Uzawa, J., Ono, Y., Shimizu, K., and Hirabayashi, Y. (2001) *FEBS Lett.* **497**, 141-147
89. Nagatsuka, Y., Hara-Yokoyama, M., Kasama, T., Takekoshi, M., Maeda, F., Ihara, S., Fujisawa, S., Oshima, E., Ishii, K., Kobayashi, T., Shizimu, K., and Hirabayashi, Y. (2003) *Proc. Natl. Acad. Sci. USA* **100**, 7454-7459
90. Kubo, H. and Hoshi, M. (1985) *J. Lipid Res.* **26**, 638-641
91. Higashi, H., Hirabayashi, Y., Ito, M., Yamagata, T., Matsumoto, M., Ueda, S., and Kato, S. (1987) *J. Biochem.* **102**, 291-296
92. Ito, M. and Yamagata, T. (1989) *Methods Enzymol.* **179**, 488-496
93. Higashi, H., Fukui, Y., Ueda, S., Kato, S., Hirabayashi, Y., Matsumoto, M., and Naiki, M. (1984) *J. Biochem.* **95**, 1517-1520
94. Higashi, H., Sugii, T., and Kato, S. (1988) *Biochim. Biophys. Acta* **963**, 333-339
95. Hardy, M.R. and Townsend, R.R. (1988) *Proc. Natl. Acad. Sci. USA* **85**, 3289-3293
96. Higashi, H., Ito, M., Fukaya, N., Yamagata, S., and Yamagata, T. (1990) *Anal. Biochem.* **186**, 355-362
97. Higashi, H. (1990) *Trends Glycosci. Glycotechnol.* **2**, 514-519
98. Ohara, K., Sano, M., Kondo, A., and Kato, I. (1991) *J. Chromatogr.* **586**, 35-41
99. Matsuura, F. and Imaoka, A. (1988) *Glycoconj. J.* **5**, 13-26
100. Kondo, A., Suzuki, J., Kuraya, N., Hase, S., Kato, I., and Ikenaka, T. (1990) *Agric. Biol. Chem.* **54**, 2169-2170
101. Wang, W.T., LeDonne, N.C., Jr., Ackerman, B., and Sweeley, C.C. (1984) *Anal. Biochem.* **141**, 366-381

Appendix

Specificities of endoglycosidases and list of commercially available endoglycosidases

Endo-α-*N*-acetylgalactosaminidase

Enzyme (Origin)	Substrate specificity	Maker	References
Endo-α-*N*-acetylgalactosaminidase (*Clostridium perfringens*)	Galβ1-3GalNAc-Ser (or Thr) ↑ in glycoproteins		1
Endo-α-*N*-acetylgalactosaminidase (*Alcaligenes* sp.)	Galβ1-3GalNAc-Ser (or Thr) ↑ in glycopeptides and glycoproteins	Seikagaku Kogyo	2, 3
Endo-α-*N*-acetylgalactosaminidase (*Streptococcus pneumoniae*)	Galβ1-3GalNAc-Ser (or Thr) ↑ in glycopeptides and glycoproteins	Calbiochem Sigma	4, 5, 6
Endo-α-*N*-acetylgalactosaminidase (*Streptomyces* sp.)	Galβ1-3GalNAc-Ser (or Thr) and ↑ Galβ1-3(Galβ1-4GlcNAcβ1-6) GalNAc-Ser (or Thr) ↑ in glycoproteins		7, 8
Endo-α-*N*-acetylgalactosaminidase (*Bacillus* sp.)	Galβ1-3GalNAc-Ser (or Thr) ↑ in glycopeptides and glycoproteins		9

Endo-β-*N*-acetylglucosaminidase

Enzyme (Origin)	Substrate specificity	Maker	References
Endo-β-*N*-acetylglucosaminidase A (*Arthrobacter protophormiae*)	High-mannose type sugar chains in glycopeptides and glycoproteins		10
Endo-β-*N*-acetylglucosaminidase (*Aspergillus oryzae*)	High-mannose and hybrid type sugar chains in glycopeptides and glycoproteins		11

Enzyme (Origin)	Substrate specificity	Maker	References
Endo-β-*N*-acetylglucosaminidase (*Bacillus alvei*)	High-mannose type sugar chains in glycopeptides and glycoproteins		12
Endo-β-*N*-acetylglucosaminidase (Keratanase II) (*Bacillus circulans*)	Releases Galβ1-4GlcNAc and Galβ1-4GlcNAcβ1-3Galβ1-4 GlcNAc from keratan sulfate		13
Endo-β-*N*-acetylglucosaminidase (Keratanase II) (*Bacillus* sp. Ks36)	Releases Galβ1-4GlcNAc and Galβ1-4GlcNAcβ1-3Galβ1-4 GlcNAc from keratan sulfate	Seikagaku Kogyo	14, 15
Endo-β-*N*-acetylglucosaminidase (*Bacillus circulans*)	High-mannose type sugar chains in glycopeptides and glycoproteins		16
Endo-β-*N*-acetylglucosaminidase CI (*Clostridium perfringens*)	Core structure of complex-type and M5 high-mannose type sugar chains of glycopeptides		17
Endo-β-*N*-acetylglucosaminidase CII (*Clostridium perfringens*)	High-mannose type sugar chains of glycopeptides		17
Endo-β-*N*-acetylglucosaminidase D (*Streptococcus pneumoniae*)	Core structure of complex type and M5 high-mannose type sugar chains of glycopeptides	Seikagaku Kogyo ICN Calbiochem	18, 19, 20
Endo-β-*N*-acetylglucosaminidase (*Flavobacterium* sp.)	High-mannose and hybrid type sugar chains of glycopeptides and glycoproteins	Seikagaku Kogyo	21, 22
Endo-β-*N*-acetylglucosaminidase F (*Flavobacterium meningosepticum*)	(Endo-β-*N*-acetylglucosaminidase F1 contaminating endo-β-*N*-acetylglucosaminidase F2) High-mannose, hybrid without GlcNAcβ1-4Man, and bi-antennary complex type sugar chains of glycopeptides	ICN Roche	23, 24
Endo-β-*N*-acetylglucosaminidase F_1 (*Flavobacterium meningosepticum*)	High-mannose and hybrid type sugar chains of glycopeptides Core fucose reduces enzyme activity by 50-fold	Calbiochem Sigma	25, 26
Endo-β-*N*-acetylglucosaminidase F_2 (*Flavobacterium meningosepticum*)	High-mannose and biantennary complex type sugar chains of glycopeptides	Calbiochem Sigma	25, 26
Endo-β-*N*-acetylglucosaminidase F_3 (*Flavobacterium meningosepticum*)	Biantennary and triantennary complex type, and trimannosyl core structure sugar chains of glycopeptides	Calbiochem Sigma	25, 26

Enzyme (Origin)	Substrate specificity	Maker	References
Endo-β-*N*-acetylglucosaminidase F-I (Fig latex)	Core trimannosyl structure and high-mannose (M5 and M6) type sugar chains of glycopeptides		27
Endo-β-*N*-acetylglucosaminidase F-II (Fig latex)	High-mannose (M5 and M6) type sugar chains of glycopeptides		27
Endo-β-*N*-acetylglucosaminidase H (*Streptomyces plicatus*)	High-mannose and hybrid type sugar chains of glycopeptides and glycoproteins	Calbiochem ICN Roche Seikagaku Kogyo Sigma	28, 29, 30, 31
Endo-β-*N*-acetylglucosaminidase HO (Hen oviduct)	High-mannose type sugar chains of glycopeptides and glycoproteins		32, 33
Endo-β-*N*-acetylglucosaminidase HS (Human saliva)	Complex type sugar chains of glycoproteins		34
Endo-β-*N*-acetylglucosaminidase M (*Mucor hiemalis*)	High-mannose, hybrid and complex type sugar chains of glycopeptides and glycoproteins	Tokyo Kasei Kogyo	35, 36, 37

Endo-β-galactosidase

Enzyme (Origin)	Substrate specificity	Maker	References
Endo-β-galactosidase (*Bacteroides fragilis*)	R-GlcNAcβ1-3Galβ1-4GlcNAc (or Glc) ↑ -GlcNAcβ1-6Galβ1-4GlcNAc β1-6Galβ1-4GlcNAcβ1- ↑ ↑ in polylactosaminoglycan		38, 39
Endo-β-galactosidase (*Clostridium perfringens*)	Releases GlcNAcα1-4Gal from porcine gastric mucin		40, 41
Endo-β-galactosidase (*Clostridium perfringens*)	Releases GalNAcα1-3 (Fucα1-2)Gal and Galα1-3 (Fucα1-2)Gal from blood group A and B oligosaccharides, respectively		42
Endo-β-galactosidase C (*Clostridium perfringens*)	Releases Galα1-3Gal from the xenoantigen		43

Enzyme (Origin)	Substrate specificity	Maker	References
Endo-β-galactosidase (*Escherichia freundii*)	R-GlcNAcβ1-3Galβ1-4GlcNAc ↑ (or Glc) in keratan sulfates, glycoproteins, milk oligosaccharides and glycosphingolipids	Seikagaku Kogyo ICN Calbiochem	44, 45, 46, 47, 48
Endo-β-galactosidase (*Flavobacterium keratolyticus*)	R-GlcNAcβ1-6Galβ1-4GlcNAcβ1- ↑ R- (± Fuc(α1-4)GlcNAcβ1-3Gal β1-4GlcNAc (or Glc) ↑ in keratan sulfates, glycoproteins, milk oligosaccharides, and glycosphingolipids		49, 50, 51
Endo-β-galactosidase (Human urine)	R-GlcNAcβ1-3Galβ1-4Glc (or GlcNAc) ↑ in keratan sulfates, glycoproteins		52
Endo-β-galactosidase (*Patinopecten*)	Hydrolyzes Galβ1-3Gal in the linkage region of peptidoglycan		53, 54
Endo-β-galactosidase (Keratanase) (*Pseudomonas* sp.)	R-($6SO_3^-$)-GlcNAcβ1-3Galβ 1-4Glc in keratan sulfates ↑	Seikagaku Kogyo Sigma	55
Endo-β-galactosidase (*Streptococcus pneumoniae*)	Releases GalNAcα1-3 (Fucα1-2) Gal and Galα1-3 (Fucα1-2) Gal from blood group A and B oligosaccharides, respectively		56

Endo-β-glucuronidase

Enzyme (Origin)	Substrate specificity	Maker	References
Endo-β-glucuronidase (Rabbit liver)	-4GlcUAβ1-3GalNAc ($6SO_3^-$) β1-4GlcUAβ1-3Galβ1-3Galβ1-4Xyl ↑ in peptidoglycan chondroitin sulfate		57, 58

Endoglycoceramidase (Ceramide glycanase)

Enzyme (Origin)	Substrate specificity	Maker	References
Endoglycoceramidase II (*Rhodococcus* sp.)	R-Galβ1-4Glcβ1-Cer ↑ Cleaves the linkage between the oligosaccharide and ceramide of various acidic and neutral glycosphingolipids	Calbiochem Sigma Takara	59, 60
Endoglycoceramidase II with activator (*Rhodococcus* sp.)	R-Galβ1-4Glcβ1-Cer ↑ Hydrolyzes the glycosphingolipids on the surfaces of intact cells	Takara	61
Ceramide glycanase (*Macrobdella decora*)	R-Galβ1-4Glcβ1-Cer ↑ Hydrolyzes intact glycan chains from various glycosphingolipids in which the glycan chain is linked to the ceramide through β-glucosyl linkage	Calbiochem	62, 63

Endo-α-mannosidase

Enzyme (Origin)	Substrate specificity	Maker	References
Endo-α-mannosidase (Rat liver)	Releases Glcα1-3Man from $Glc_1Man_9GlcNAc$		64

Endo-β-mannosidase

Enzyme (Origin)	Substrate specificity	Maker	References
Endo-β-mannosidase (*Lilium longiflorum*)	Man_n-Manα1-6Manβ-$GlcNAc_2$- ↑ (PA, Peptides, OH)		65, 66, 67

Endo-β-xylosidase

Enzyme (Origin)	Substrate specificity	Maker	References
Endo-β-xylosidase (*Patinopecten*)	Hydrolyzes Xylβ1-*O*-Ser in the linkage region of peptidoglycans		68

References

1. Huang, C.C. and Aminoff, D. (1972) *J. Biol. Chem.* **247**, 6737-6742

2. Yamamoto, K., Fan, J.Q., Kadowaki, S., Kumagai, H., and Tochikura, H. (1987) *Agric. Biol. Chem.* **51**, 3169-3171
3. Fan, J.Q., Kadowaki, S., Yamamoto, K., Kumagai, H., and Tochikura, T. (1988) *Agric. Biol. Chem.* **52**, 1715-1723
4. Endo, Y. and Kobata, A. (1976) *J. Biochem.* **80**, 1-8
5. Bhavanandan, V.P., Umemoto, J., and Davidson, E.A. (1976) *Biochem. Biophys. Res. Commun.* **70**, 738-745
6. Glasgow, L.R., Paulson, J.C., and Hill, R.L. (1977) *J. Biol. Chem.* **252**, 8615-8623
7. Iwase, H., Ishii, I., Ishihara, K., Tanaka, Y., Omura, S., and Hotta, K. (1988) *Biochem. Biophys. Res. Commun.* **151**, 422-428
8. Ishii-Karaskasa, I., Iwase, H., Hotta, K., Tanaka, Y., and Omura, S. (1992) *Biochem. J.* **288**, 475-482
9. Ashida, H., Yamamoto, K., Murata, T., Usui, T., and Kumagai, H. (2000) *Arch. Biochem. Biophys.* **373**, 394-400
10. Takegawa, K., Nakoshi, M., Iwahara, S., Yamamoto, K., and Tochikura, T. (1989) *Appl. Environ. Microbiol.* **55**, 3107-3112
11. Hitomi, J., Murakami, Y., Saitoh, F., Shigemitsu, N., and Yamaguchi, H. (1985) *J. Biochem.* **98**, 527-533
12. Morinaga, T., Kitamikado, M., Iwase, H., Li, S.-C., and Li, Y.-T. (1983) *Biochim. Biophys. Acta* **749**, 211-213
13. Yamagishi, K., Suzuki, K., Imai, K., Mochizuki, H., Morikawa, K., Kyogashima, M., Kimata, K., and Watanabe, H. (2003) *J. Biol. Chem.* **278**, 25766-25772
14. Hashimoto, N., Morikawa, K., Kikuchi, H., Yoshida, K., and Tokuyasu, K. (1988) *Seikagaku* **60**, 935 (in Japanese)
15. Nakazawa, K., Ito, M., Yamagata, T., and Suzuki, S. (1989) in *Keratan Sulfate* (Eds., Greiling, H. and Scott, J.E.) pp. 99-110, The Biochemical Society, London
16. Cohen, R.E., Zhang, W.J., and Ballou, C.E. (1982) *J. Biol. Chem.* **257**, 5730-5737
17. Ito, S., Muramatsu, T., and Kobata, A. (1975) *Arch. Biochem. Biophys.* **171**, 78-86
18. Muramatsu, T. (1971) *J. Biol. Chem.* **246**, 5535-5537
19. Koide, N. and Muramatsu, T. (1974) *J. Biol. Chem.* **249**, 4897-4904
20. Tai, T., Yamashita, K., Ogata-Arakawa, M., Koide, N., Muramatsu, T., Iwashita, S., Inoue, Y., and Kobata, A. (1975) *J. Biol. Chem.* **250**, 8569-8575
21. Yamamoto, K., Kadowaki, S., Takegawa, K., Kumagai, H., and Tochikura, T. (1986) *Agric. Biol. Chem.* **50**, 421-429
22. Yamamoto, K., Takegawa, K., Kumagai, H., and Tochikura, T. (1986) *Agric. Biol. Chem.* **50**, 2167-2169
23. Elder, J.H. and Alexander, S. (1982) *Proc. Natl. Acad. Sci. USA* **79**, 4540-4544
24. Tarentino, A.L., Gómez, C.M., and Plummer, T.H., Jr. (1985) *Biochemistry* **24**, 4665-4671
25. Trimble, R.B. and Tarentino, A.L. (1991) *J. Biol. Chem.* **266**, 1646-1651
26. Plummer, T.H., Jr. and Tarentino, A.L. (1991) *Glycobiology* **1**, 257-263
27. Li, S.-C., Asakawa, M., Hirabayashi, Y., and Li, Y.-T. (1981) *Biochim Biophys Acta* **660**, 278-283
28. Tarentino, A.L., Plummer, T.H., Jr., and Maley, F. (1972) *J. Biol. Chem.* **247**, 2629-2631
29. Tarentino, A.L. and Maley, F. (1974) *J. Biol. Chem.* **249**, 811-817
30. Tarentino, A.L., Plummer, T.H., Jr., and Maley, F. (1974) *J. Biol. Chem.* **249**, 818-824
31. Trumbly, R.J., Robbins, P.W., Belfort, M., Ziegler, F.D., Maley, F., and Trimble, R.B. (1985) *J. Biol. Chem.* **260**, 5683-5690
32. Tarentino, A.L. and Maley, F. (1976) *J. Biol. Chem.* **251**, 6537-6543
33. Kato, T., Hatanaka, K., Mega, T., and Hase, S. (1997) *J. Biochem.* **122**, 1167-1173
34. Ito, K., Okada, Y., Ishida, K., and Minamiura, N. (1993) *J. Biol. Chem.* **268**, 16074-16081
35. Kadowaki, S., Yamamoto, K., Fujisaki, M., Izumi, K., Tochikura, T., and Yokoyama, T. (1990) *Agric. Biol. Chem.* **54**, 97-106
36. Kadowaki, S., Yamamoto, K., Fujisaki, M., and Tochikura, T. (1991) *J. Biochem.* **110**, 17-21
37. Yamamoto, K., Kadowaki, S., Fujisaki, M., Kumagai, H., and Tochikura, T. (1994) *Biosci. Biotech. Biochem.* **58**, 72-77
38. Scudder, P., Uemura, K., Dolby, J., Fukuda, M.N., and Feizi, T. (1983) *Biochem. J.* **213**, 485-494
39. Hanisch, F.G., Uhlenbruck, G., Peter-Katalinic, J., Egge, H., Dabrowski, J., and Dabrowski, U. (1989) *J. Biol. Chem.* **264**, 872-833
40. Ashida, H., Anderson, K., Nakayama, J., Maskos, K., Chou, C.W., Cole, R.B., Li, S.-C., and Li, Y.-T. (2001) *J. Biol. Chem.* **276**, 28226-28232

41. Ashida, H., Maskos, K., Li, S.-C., and Li, Y.-T. (2002) *Biochemistry* **41**, 2388-2395
42. Anderson, K.M., Ashida, K., Maskos, K., Dell, A., Li, S.-C., and Li, Y.-T. (2005) *J. Biol. Chem.* **280**, 7720-7728
43. Ogawa, H., Muramatsu, H., Kobayashi, T., Morozumi, K., Yokoyama, I., Kurosawa, N., Nakao, A., and Muramatsu, T. (2000) *J. Biol. Chem.* **275**, 19368-19374
44. Kitamikado, M., Ueno, R., and Nakamura, T. (1970) *Bull. Jpn. Soc. Sci. Fish.* **36**, 592-596
45. Kitamikado, M. and Ueno, R. (1970) *Bull. Jpn. Soc. Sci. Fish.* **36**, 1175-1180
46. Fukuda, M.N. and Matsumura, G. (1975) *Biochem. Biophys. Res. Comm.* **64**, 465-471
47. Fukuda, M.N. and Matsumura, G. (1976) *J. Biol. Chem.* **251**, 6218-6255
48. Fukuda, M.N., Watanabe, K., and Hakomori, S. (1978) *J. Biol. Chem.* **253**, 6814-6819
49. Kitamikado, M., Ito, M., and Li, Y.-T. (1981) *J. Biol. Chem.* **256**, 3906-3909
50. Fukuda, M.N. (1981) *J. Biol. Chem.* **256**, 3900-3905
51. Amano, J., Straehl, P., Berger, E.G., Kochibe, N., and Kobata, A. (1991) *J. Biol. Chem.* **266**, 11461-11477
52. DeGasperi, R., Li, Y.-T., and Li, S.-C. (1986) *J. Biol. Chem.* **261**, 5696-5698
53. Takagaki, K., Kon, A., Kawasaki, H., Nakamura, T., Tamura, S., and Endo, M. (1990) *Biochem. Biophys. Res. Commun.* **169**, 15-21
54. Takagaki, K., Nakamura, T., Takeda, Y., Daidouji, K., and Endo, M. (1992) *J. Biol. Chem.* **267**, 18558-18563
55. Nakazawa, K. and Suzuki, S. (1975) *J. Biol. Chem.* **250**, 912-917
56. Takasaki, S. and Kobata, A. (1976) *J. Biol. Chem.* **251**, 3603-3609
57. Takagaki, K., Nakamura, T., Majima, M., and Endo, M. (1985) *FEBS Lett.* **181**, 271-274
58. Takagaki, K., Nakamura, T., Majima, M., and Endo, M. (1988) *J. Biol. Chem.* **263**, 7000-7006
59. Ito, M. and Yamagata, T. (1986) *J. Biol. Chem.* **261**, 14278-14282
60. Ito, M. and Yamagata, T. (1989) *J. Biol. Chem.* **264**, 9510-9519
61. Ito, M., Ikegami, Y., and Yamagata, T. (1991) *J. Biol. Chem.* **266**, 7919-7926
62. Li, S.-C., DeGasper, R., Muldrey, J.E., and Li, Y.-T. (1986) *Biochem. Biophys. Res. Commun.* **141**, 346-352
63. Zhou, B., Li, S.-C., Laine, R.A., Huang, R.T.C., and Li, Y.-T. (1989) *J. Biol. Chem.* **264**, 12272-12277
64. Lubas, W. and Spiro, R. (1987) *J. Biol. Chem.* **262**, 3775-3781
65. Sasaki, A., Yamagishi, M., Mega, T., Norioka, S., Natsuka, S., and Hase, S. (1999) *J. Biochem.* **125**, 363-367
66. Ishimizu, T., Sasaki, A., Okutani, S., Maeda, M., Yamagishi, M., and Hase, S. (2004) *J. Biol. Chem.* **279**, 38555-38562
67. Sasaki, A., Ishimizu, T., and Hase, S. (2005) *J. Biochem.* **137**, 87-93
68. Takagaki, K., Kon, A., Kawasaki, H., Nakamura, T., Tamura, S., and Endo, M. (1990) *J. Biol. Chem.* **265**, 854-860

List of Abbreviations

Following abbreviations are used in this book.

ABEE	*p*-aminobenzoic acid ethyl ester
Asn	L-asparagine
Boc	*tert*-butyloxycarbonyl
CD	circular dichroism
Cer	ceramide
Ch	chondroitin
Ch4S	chondroitin 4-sulfate
Ch6S	chondroitin 6-sulfate
ChS	chondroitin sulfate
DMSO	dimethyl sulfoxide
DNS	dansyl
DS	dermatan sulfate
EDTA	ethylenediaminetetraacetic acid
EGCase	endoglycoceramidase
Endo-A, -C, -CI, -CII, -D, -F, -H, -M, -HS	endo-*β*-*N*-acetylglucosaminidase A, C, CI, CII, D, F, H, M, HS
Endo-*α*-GalNAc-ase	endo-*α*-*N*-acetylgalactosaminidase
Endo-*β*-Gal-ase	endo-*β*-galactosidase
Endo-*β*-GlcNAc-ase	endo-*β*-*N*-acetylglucosaminidase
FAB-MS	fast atom bombardment mass spectrometry
Fmoc	9-fluorenylmethyloxycarbonyl
Fuc	L-fucose
GAG	glycosaminoglycan
Gal	D-galactose
GalNAc	2-acetamido-2-deoxy-D-galactose, *N*-acetyl-D-galactosamine
Gb4Cer	globotetraosylceramide
Glc	D-glucose
GlcNAc	2-acetamido-2-deoxy-D-glucose, *N*-acetyl-D-glucosamine
GlcUA	D-glucuronic acid
Gln	L-glutamine
GM	ganglioside
GM1a	monosialosyl gangliotetraosylceramide

GT1b	disialosyl gangliotetraosylceramide
HA	hyaluronan
Hep	heparin
HPLC	high performance liquid chromatography
HS	heparan sulfate
IduUA	L-iduronic acid
Lac	lactosyl
Lc4Cer	lactotetraosylceramide
MALDI-TOF	matrix-assisted laser-desorption ionization-time-of-flight
Man	D-mannose
MS	mass spectrometry
MU	4-methylumbelliferone
NBD	4-nitrobenzo-2-oxa-1,3-diazole
NeuAc, Neu5Ac	*N*-acetylneuraminic acid
NeuGc, Neu5Gc	*N*-glycolylneuraminic acid
NMR	nuclear magnetic resonance
PA	2-aminopyridine, pyridylamino
PAD	pulsed amperometric detector
PAGE	polyacrylamide gel electrophoresis
PCR	polymerase chain reaction
PG	proteoglycan
PNGase	peptide-N^4-(*N*-acetyl-β-D-glucosaminyl)asparagine amidase
*p*NP	*p*-nitrophenyl
RNase	ribonuclease
SDS	sodium dodecyl sulfate
Ser	L-serine
SGP	sialyl glycopeptide
Sia	sialic acid
Thr	L-threonine
TLC	thin-layer chromatography
Xyl	D-xylose

Index

F

G

U, V

X